METHODS IN MOLECULAR BIOLOGY

Series Editor
John M. Walker
School of Life and Medical Sciences
University of Hertfordshire
Hatfield, Hertfordshire, UK

For further volumes:
http://www.springer.com/series/7651

For over 35 years, biological scientists have come to rely on the research protocols and methodologies in the critically acclaimed *Methods in Molecular Biology* series. The series was the first to introduce the step-by-step protocols approach that has become the standard in all biomedical protocol publishing. Each protocol is provided in readily-reproducible step-by-step fashion, opening with an introductory overview, a list of the materials and reagents needed to complete the experiment, and followed by a detailed procedure that is supported with a helpful notes section offering tips and tricks of the trade as well as troubleshooting advice. These hallmark features were introduced by series editor Dr. John Walker and constitute the key ingredient in each and every volume of the *Methods in Molecular Biology* series. Tested and trusted, comprehensive and reliable, all protocols from the series are indexed in PubMed.

Phage Engineering and Analysis

Methods and Protocols

Edited by

Huan Peng

Cellular Signaling Laboratory International Research, Center for Sensory Biology and Technology of MOST, Key Laboratory of Molecular Biophysics of MOE, College of Life Science and Technology, Huazhong University of Science and Technology (HUST), Wuhan, Hubei, China

Jianfeng Liu

College of Life Science and Technology, Huazhong University of Science and Technology (HUST), Wuhan, Hubei, China

Irene A. Chen

Department of Chemical and Biomolecular Engineering, Department of Chemistry and Biochemistry, University of California, Los Angeles, CA, USA

Humana Press

Editors
Huan Peng
Cellular Signaling Laboratory
International Research, Center for
Sensory Biology and Technology of
MOST, Key Laboratory of Molecular
Biophysics of MOE, College of Life
Science and Technology
Huazhong University of Science
and Technology (HUST)
Wuhan, Hubei, China

Irene A. Chen
Department of Chemical and
Biomolecular Engineering, Department
of Chemistry and Biochemistry
University of California
Los Angeles, CA, USA

Jianfeng Liu
College of Life Science and Technology
Huazhong University of Science
and Technology (HUST)
Wuhan, Hubei, China

ISSN 1064-3745 ISSN 1940-6029 (electronic)
Methods in Molecular Biology
ISBN 978-1-0716-3797-5 ISBN 978-1-0716-3798-2 (eBook)
https://doi.org/10.1007/978-1-0716-3798-2

This Humana imprint is published by the registered company Springer Science+Business Media, LLC, part of Springer Nature.
The registered company address is: 1 New York Plaza, New York, NY 10004, U.S.A.

Paper in this product is recyclable.

Preface

Phages are making a comeback. While phages (bacteriophages) were a hotbed for fundamental research in the early days of molecular biology, giving us the genetic switch of lambda lysogeny and the all-important restriction enzymes by way of their bacterial hosts, by 1970 the excitement over phages had petered out as attention focused on more complex biological organisms. Interest in phage research remained essentially flat for decades (Fig. 1), for what else could be gained? Plenty, as it turns out. The humble phage has become an indispensable member of today's synthetic biology toolkit. Phages were recognized as one of the simplest ways to physically couple a gene to the protein it encodes, enabling the artificial selection of active variants out of large, diverse libraries. The technique of phage display took off in the mid-1990s and is now a popular method for the development of peptide and antibody affinity reagents. Fundamental interest in phages and their "arms race" with bacteria got a major boost with the momentous discovery of CRISPR biology and its ramifications for genome editing in the 2010s. These discoveries did not escape the attention of the Nobel committees in 2018 and 2020, and the scientific community continues to plumb the depths of phage biology in search of new mechanisms arising from this coevolution. Even phage therapy, the potential use of phages to treat bacterial infections, long neglected in favor of antibiotics, has garnered attention (and respectability as a scientific field) as antibiotic-resistant pathogens inexorably gain ground in our communities and hospitals.

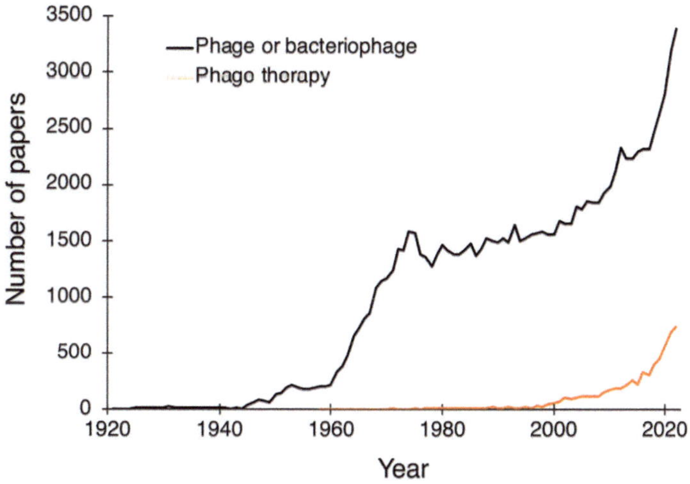

Fig. 1 Number of papers in the PubMed database matching the query [phage OR bacteriophage] (black) or [phage therapy] (orange)

The number of researchers working on phages is not large, but it is growing quickly. Practitioners both new and old benefit greatly from sharing protocols and technical insights as the field brings together scientists from many backgrounds, from microbiologists to biophysicists to chemical engineers. The first half of this volume collects protocols on phage engineering, beginning with phage display, selection and evolution (Chaps. 1, 2, 3, 4, and 5), and related applications of genetic modification of phages (Chaps. 6, 7, and 8). Two chapters describe expansions of phage engineering beyond the protein space to display new chemical moieties (Chaps. 9 and 10). The second half of this volume describes methods to study phage biology, analyzing structures by electron microscopy (Chaps. 11, 12, and 13), characterizing phage transcripts and proteins (Chaps. 14 and 15), and finally, probing the biology of whole phages (Chaps. 16 and 17). We are profoundly grateful for the authors who contributed the knowledge and hard-won expertise found in these chapters. The coming years will be an exciting time for phage science and engineering.

The editors would like to thank Mr. Zhijun Chen for his assistance in assembling the book. Dr. Huan Peng is grateful for support from the National Natural Science Foundation of China (Grant No. 32201100).

Wuhan, Hubei, China — *Huan Peng*
Wuhan, Hubei, China — *Jianfeng Liu*
Los Angeles, CA, USA — *Irene A. Chen*

Contents

Contributors

TAREQ AHMED • *Department of Chemistry, University of Alberta, Edmonton, AB, Canada*

RANEEM AKEL • *Department of Chemistry, University of Toronto, Mississauga, ON, Canada*

LILIAN COSTA ALECRIM • *Biochemistry Department, Institute of Chemistry, University of Sao Paulo, Sao Paulo, Brazil*

MARYAM ALI • *Department of Chemistry, University of Toronto, Mississauga, ON, Canada*

FABIO C. L. ALMEIDA • *Institute of Medical Biochemistry, Federal University of Rio de Janeiro, Rio de Janeiro, Brazil*

DAMAYANTI BAGCHI • *Department of Chemical Physiology and Biochemistry, Oregon Health and Science University, Portland, OR, USA; Department of Chemical and Biomolecular Engineering, University of California, Los Angeles, CA, USA*

MAARTEN BOON • *Laboratory of Gene Technology, Department of Biosystems, KU Leuven, Leuven, Belgium*

MARIA J. BOTERO • *Department of Chemistry, University of Toronto, Mississauga, ON, Canada*

ERIC J. CARPENTER • *Department of Chemistry, University of Alberta, Edmonton, AB, Canada*

IRENE A. CHEN • *Department of Chemical and Biomolecular Engineering, University of California, Los Angeles, CA, USA; Department of Chemistry and Biochemistry, University of California, Los Angeles, CA, USA*

SHUQING CHEN • *School of Pharmaceutical Sciences, Zhejiang University, Hangzhou, China*

JEFFREY K. CORNUAULT • *Meinig School of Biomedical Engineering, Cornell University, Ithaca, NY, USA*

RATMIR DERDA • *Department of Chemistry, University of Alberta, Edmonton, AB, Canada*

AMANDA C. FREISE • *Department of Microbiology, Immunology & Molecular Genetics, University of California, Los Angeles, CA, USA*

RICARDO JOSE GIORDANO • *Biochemistry Department, Institute of Chemistry, University of Sao Paulo, Sao Paulo, Brazil*

HANNE HENDRIX • *Laboratory of Gene Technology, Department of Biosystems, KU Leuven, Leuven, Belgium*

TZVI HOLTZMAN • *Department of Clinical Microbiology and Immunology, School of Medicine, Tel Aviv University, Tel Aviv, Israel; Department of Biotechnology, Israel Institute for Biological Research, Ness Ziona, Israel*

DANISH INTIZAR • *Laboratory of Gene Technology, Department of Biosystems, KU Leuven, Leuven, Belgium*

DEBORAH JACOBS-SERA • *Department of Biological Sciences, University of Pittsburgh, Pittsburgh, PA, USA*

ZHENWEI JIANG • *State Key Laboratory of Tribology in Advanced Equipment, Tsinghua University, Beijing, China*

JING YI LAI • *Institute for Research in Molecular Medicine, Universiti Sains Malaysia, Penang, Malaysia*

ROB LAVIGNE • *Laboratory of Gene Technology, Department of Biosystems, KU Leuven, Leuven, Belgium*

TAORAN LI • *Department of Chemistry, University of Alberta, Edmonton, AB, Canada*

THEAM SOON LIM • *Institute for Research in Molecular Medicine, Universiti Sains Malaysia, Penang, Malaysia; Analytical Biochemistry Research Centre, Universiti Sains Malaysia, Penang, Malaysia*
CHIH-LAN LIN • *Department of Chemistry, University of Alberta, Edmonton, AB, Canada*
YUAN MA • *State Key Laboratory of Tribology in Advanced Equipment, Tsinghua University, Beijing, China*
ANGY LISETH DAVALOS MACIAS • *Biochemistry Department, Institute of Chemistry, University of Sao Paulo, Sao Paulo, Brazil*
RAM NECHOOSHTAN • *Department of Biotechnology, Israel Institute for Biological Research, Ness Ziona, Israel*
LIQIANG PAN • *School of Pharmaceutical Sciences, Zhejiang University, Hangzhou, China*
HUAN PENG • *Cellular Signaling Laboratory, International Research Center for Sensory Biology and Technology of MOST, Key Laboratory of Molecular Biophysics of MOE, College of Life Science and Technology, Huazhong University of Science and Technology, Wuhan, Hubei, China*
LEENA PUTZEYS • *Laboratory of Gene Technology, Department of Biosystems, KU Leuven, Leuven, Belgium*
UDI QIMRON • *Department of Clinical Microbiology and Immunology, School of Medicine, Tel Aviv University, Tel Aviv, Israel*
KRISANAVANE REDDI • *Department of Microbiology, Immunology & Molecular Genetics, University of California, Los Angeles, CA, USA*
KAAT SCHROVEN • *Laboratory of Gene Technology, Department of Biosystems, KU Leuven, Leuven, Belgium*
SEA-PHAGES • *Science Education Alliance Phage Hunters Advancing Genomics and Evolutionary Science Program (SEA-PHAGES), Howard Hughes Medical Institute, Chevy Chase, MD, USA*JUMI A. SHIN • *Department of Chemistry, University of Toronto, Mississauga, ON, Canada*
MIKAEL SKURNIK • *Department of Bacteriology and Immunology, Human Microbiome Research Program, Faculty of Medicine, University of Helsinki, Helsinki, Finland*
BRENDA PEI CHUI SONG • *Institute for Research in Molecular Medicine, Universiti Sains Malaysia, Penang, Malaysia*
JIRAPAT THONGCHOL • *Center for Phage Technology, Department of Biochemistry and Biophysics, Texas A&M University, College Station, TX, USA*
MARLEEN VOET • *Laboratory of Gene Technology, Department of Biosystems, KU Leuven, Leuven, Belgium*
XING WAN • *Department of Bacteriology and Immunology, Human Microbiome Research Program, Faculty of Medicine, University of Helsinki, Helsinki, Finland; Department of Microbiology, Faculty of Agriculture and Forestry, University of Helsinki, Helsinki, Finland*
YUJIAO WANG • *State Key Laboratory of Tribology in Advanced Equipment, Tsinghua University, Beijing, China*
YANXI YANG • *Department of Chemical and Biomolecular Engineering, University of California, Los Angeles, CA, USA*
IDO YOSEF • *Department of Clinical Microbiology and Immunology, School of Medicine, Tel Aviv University, Tel Aviv, Israel*
JUNJIE ZHANG • *Center for Phage Technology, Department of Biochemistry and Biophysics, Texas A&M University, College Station, TX, USA*

RUI ZHANG • *Department of Clinical Laboratory, Beijing Chao-yang Hospital, Capital Medical University, Beijing, China*
JING ZHAO • *State Key Laboratory of Tribology in Advanced Equipment, Tsinghua University, Beijing, China; Department of Clinical Laboratory, Beijing Chao-yang Hospital, Capital Medical University, Beijing, China*
XINLEI ZHUANG • *School of Pharmaceutical Sciences, Zhejiang University, Hangzhou, China*
MICHELLE ZORAWIK • *Department of Microbiology, Immunology & Molecular Genetics, University of California, Los Angeles, CA, USA; Keck School of Medicine, University of Southern California, Los Angeles, CA, USA*

Part I

Phage Display: Selection and Evolution

Chapter 1

Protocols for Building and Producing High Diversity Peptide Phage Display Libraries

Ricardo Jose Giordano and Lilian Costa Alecrim

Abstract

Phage display is an important technology to study protein-protein interaction and protein evolution, with applications in basic science and applied biotechnology, such as drug discovery and the development of targeted therapies. However, in order to be successful during a phage display screening, it is paramount to have good phage libraries. Here, we described detailed procedures to generate peptide phage display libraries with high diversity and billions of transformants.

Key words Phage display library, High diversity, Saturation mutagenesis, Electroporation

1 Introduction

Phage display technology was developed in 1985 by George Smith, initially as a method to identify and map antigens and epitopes [1]. However, very rapidly, phage display proved to be a robust and flexible technology, which can be utilized for the identification of peptides and proteins (such as antibodies) that bind to virtually any biological target [2–4]. In 2018, George Smith and *Sir* Gregory Winter were awarded the Nobel Prize in Chemistry for the development of such elegant technology [5].

In a nutshell, a phage library is a collection of phage particles each containing (or displaying) a different peptide or protein fused to a surface coat protein [2]. Thus, in order to be successful when performing a phage display screening (or biopanning), it is necessary to have good high diversity phage libraries. A well-built library allows for the display of billions of different peptides [6, 7]. To give an idea regarding the diversity of phage libraries, a single high diversity library contains more peptides than all known chemicals and biochemicals studied by researchers in the history of science, and cataloged by the Chemical Abstract Society (a division of the American Chemical Society). Furthermore, several phage libraries

Huan Peng et al. (eds.), *Phage Engineering and Analysis: Methods and Protocols*,
Methods in Molecular Biology, vol. 2793, https://doi.org/10.1007/978-1-0716-3708-2_1,

may be built, each displaying peptides in different formats (i.e., linear or cyclic, short, or longer peptides), increasing even further the molecular diversity which may be explored by researchers.

Building phage display libraries is a relatively straightforward process but hard work. Although it relies on basic molecular biology cloning technology, in order to produce good high-quality libraries with high diversity (>10^9 peptides), one must pay attention to every detail to assure that each step of the process is carried to its highest efficiency. Here, we will share some of our experience building phage libraries [6–8]. First, all reagents need to be produced with the highest grade: vectors should be purified using CsCl gradient; insert should be digested and purified using ion-exchange columns (Qiagen). Finally, freshly made electrocompetent cells are the best option to obtain the highest number of transformants. Nevertheless, there are several types of vector and modes to display the peptide or protein on the surface of bacteriophage. In this chapter, we describe procedures using the fUSE55 vector for displaying peptides on the coat protein III. So, this protocol may also be seen as a guideline, to be adjusted accordingly, depending on your resources, phage display vector, and type of library being built.

2 Materials

Water (molecular biology grade).

dNTP 10 mM.

DTT 100 mM.

$MgCl_2$ 100 mM.

Isopropanol.

Ethanol 70%.

Qiagen Midi and Maxi Prep kits, Qiagen QIAquick Nucleotide Removal kit, and extra QIAGEN-tips 100 and 500.

2.1 Phage Vectors

There are several vectors available for building phage libraries [2, 9]. Here, we describe procedures to build libraries using the fUSE55 vector, which allows for the display of up to five peptides fused to the minor coat protein III. However, these procedures can be easily adapted to other phage display vectors by switching to the recommended bacterial cell lines, insert design, and corresponding restriction enzymes. The fUSE55 is an fd-tet-based bacteriophage vector developed by Dr. George Smith [2]. Most of the methods described here were based and adapted from protocols that were previously organized and distributed by Dr. George Smith.

2.2 Enzymes

BglI and indicated 10× buffer.

DNA Polymerase I, large (Klenow) fragment and indicated 10× buffer.

T4 DNA ligase and corresponding buffer.

2.3 Culture Media and Buffers

1. LB (Luria Bertani) media: Add 10 g of tryptone, 5 g of yeast extract, and 10 g NaCl. Add all the solid ingredients to ~900 mL of ddH_2O and stir to dissolve. Adjust pH 7 with NaOH, complete to 1000 mL and autoclave.
2. LB tet/strp: Add 2 mL of the tetracycline stock (20 mg/mL) and 2 mL of the streptomycin stock (25 mg/mL) to 1 L of LB. The final concentration should be, respectively, 40 μg/mL and 50 μg/mL. Keep at 4 °C in the dark.
3. SOC media: SOC medium is identical to SOB medium, except that it contains 20 mM glucose. To prepare 1 L of SOB, dissolve the following reagents in ~800 mL of deionized water: 20 g tryptone, 0.584 g NaCl, 0.186 g KCl, and 2.4 g $MgSO_4$. Adjust the pH of the medium to 7.0 with NaOH. Adjust the volume of the solution to 1 L with deionized H_2O and autoclave. After the SOB medium has been autoclaved, allow it to cool to 50 °C or less and add 20 mL of filter-sterilized 20% glucose solution.
4. LB tet/strp plates: Prepare 1 L of LB, adjust pH, and add 15 g of bacto-agar and autoclave. Let the bottles cool down to 50 °C and add the antibiotics: 500 μL of the tetracycline stock (final concentration 10 μg/mL) and 2 mL of streptomycin stock solution (final concentration 50 μg/mL). Pour the plates, let them cool down and keep at 4 °C in the dark.
5. STE buffer: 100 mM NaCl, 10 mM Tris-Cl, 1 mM EDTA-$2H_2O$, pH 8.0. Add 5.84 g of NaCl, 1.21 g of Tris base, 0.37 g of EDTA to ~900 mL of ddH_2O and stir to dissolve. Adjust pH 8.0, complete to 1000 mL and autoclave.
6. PEG/NaCl: Dissolve 100 g PEG 8000 and 116.9 g NaCl in 500 mL ddH_2O. Add the remaining water to 1 L, stir until completely homogenized and autoclave.
7. PBS: 1.37 M NaCl, 27 mM KCl, 100 mM Na_2HPO_4, and 18 mM KH_2PO_4. Dissolve the following reagents in 800 mL of ddH_2O: 8 g of NaCl, 0.2 g of KCl, 1.44 g of Na_2HPO_4, and 0.24 g of KH_2PO_4. Adjust the pH to 7.4 with HCl, add ddH_2O, and then autoclave.
8. 0.5 M EDTA: Add 18.61 g of disodium EDTA-$2H_2O$ to 80 mL of H_2O and stir. Adjust the pH to 8.0 with NaOH. The disodium salt of EDTA will not go into solution until the pH of the solution is adjusted to ~8.0. Once fully dissolved, top up the solution to 100 mL using distilled water.

9. TAE: First, prepare a concentrated 50× stock solution of TAE buffer. To do this, dissolve 242 g of Tris base in 750 mL of deionized water. Add 57.1 mL of acetic acid and 100 mL of 0.5 M EDTA (pH 8.0), and adjust the volume to 1 L by adding water. The final pH of the 50× TAE buffer should be about 8.5. To make the 1× TAE working buffer, add 49 parts of deionized water to 1 part of 50× TAE buffer.
10. TBE: First, prepare a concentrated 10× stock solution of TBE buffer. Add 108 g of Tris base, 55 g of boric acid, and 7.44 g of disodium EDTA-$2H_2O$ in 800 mL of deionized water. Dissolve and complete to 1000 mL.
11. Qiagen buffers recipes (*see* Table 1).

2.4 0.8% Agarose Gel

Measure 0.4 g of agarose (final agarose concentration in gel should be 0.8%) and mix agarose powder with 50 mL 1× TAE in a microwavable flask. Microwave for 1–3 min until the agarose is completely dissolved. Let agarose solution cool down to about 50 °C (about when you can comfortably keep your hand on the flask) and add ethidium bromide to a final concentration 0.5 μg/mL. Pour into gel dock with comb and allow to solidify.

2.5 DNA Polyacrylamide Gel Electrophoresis

Assemble the glass plates with spacers in gel caster. To prepare the gel solution, add 8 mL of 30% acrylamide/*bis*-acrylamide (29:1), 1.2 mL of TBE 10×, 200 μL of ammonium persulfate (10% w/v), 10 μL of TEMED (N,N,N′,N′-tetramethylethylenediamine), and 2.59 mL of dH_2O. Work quickly after addition of TEMED to complete the gel before the acrylamide polymerizes. Immediately insert the appropriate comb into the gel, being careful not to allow air bubbles to become trapped under the teeth. Allow the acrylamide to polymerize for 30 min at room temperature. When polymerized, remove gels from gel caster, and insert gels into gel box. Add TBE 1× buffer and carefully pull the combs from the polymerized gel.

3 Methods

All procedure may be performed at room temperature unless indicated otherwise.

3.1 Phage Display Vector Preparation

The fUSE55 vector can be amplified using most commonly available laboratory *E. coli* strains. However, we recommend using strain MC1061, which is a high-efficiency transformation cell, streptomycin resistant, and F minus. The use of a female uninfectible host strain prevents the growth of replicating forms that reverted the frameshift in gene III. In order to obtain sufficient amounts (~1 mg) of fUSE55 to build a high diversity library, we

Table 1
Qiagen buffer recipes

Buffer	Composition	Preparation
P1	50 mM Tris-HCl, pH 8.0, 10 mM EDTA, 100 μg/mL RNase A	Dissolve 6.06 g Tris base, 3.72 g EDTA-$2H_2O$ in 800 mL dH_2O. Adjust the pH to 8.0 with HCl. Adjust the volume to 1 L with dH_2O and add 100 mg RNase A
P2	200 mM NaOH, 1% SDS	Dissolve 8.09 g of NaOH in 950 mL dH_2O, 50 mL 20% SDS solution. The final volume should be 1 L
P3	3.0 M potassium acetate pH 5.5	Dissolve 294.5 g potassium acetate in 500 mL dH_2O. Adjust the pH to 5.5 with glacial acetic acid (about 110 mL). Adjust the volume to 1 L with dH_2O
QBT	750 mM NaCl, 50 mM MOPS, pH 7.0, 15% isopropanol, 0.15% Triton X-100	Dissolve 43.83 g NaCl, 10.46 g MOPS (free acid) in 800 mL dH_2O. Adjust the pH to 7.0. Add 150 mL pure isopropanol and 15 mL 10% Triton X-100 solution. Adjust the volume to 1 L with dH_2O
QC	1.0 M NaCl, 50 mM MOPS, pH 7.0, 15% isopropanol	Dissolve 58.44 g NaCl and 10.46 g MOPS (free acid) in 800 mL dH_2O. Adjust the pH to 7.0. Add 150 mL pure isopropanol. Adjust the volume to 1 L with dH_2O
QF	1.25 M NaCl, 50 mM Tris-HCl, pH 8.5, 15% isopropanol	Dissolve 73.05 g NaCl and 6.06 g Tris base in 800 mL dH_2O and adjust the pH to 8.5 with HCl. Add 150 mL pure isopropanol. Adjust the volume to 1 L with dH_2O
TE	10 mM Tris-HCl, 1 mM EDTA, pH 8	Dissolve 1.21 g Tris base, 0.37 g EDTA-$2H_2O$ in 800 mL dH_2O. Adjust the pH to 8.0 with HCl and adjust the volume to 1 L with dH_2O
PNI	40% (v/v) 5 M guanidinium chloride, 60% (v/v) isopropanol	First, prepare 5 M guanidinium chloride. To do this, dissolve 477.5 g of guanidinium chloride in 800 mL dH_2O. Dissolve and adjust the volume to 1 L with dH_2O. In 600 mL of isopropanol, add 400 mL of 5 M guanidinium chloride
PE	10 mM Tris-HCl, 80% ethanol, pH 7.5	Dissolve 1.21 g Tris base in 800 mL 100% ethanol and 100 mL dH_2O. Adjust the pH to 7.5 with HCl and adjust the volume to 1 L with dH_2O
EB	10 mM Tris-HCl, pH 8.5	Dissolve 1.21 g Tris base in 800 mL dH_2O. Adjust the pH to 8.5 with HCl and adjust the volume to 1 liter with dH_2O

recommend growing enough MC1061 cell culture transformed with the fUSE55 vector. Because fUSE55 behaves as a very low copy plasmid, in our hands, we needed approximately 12 L of saturated LB media culture (supplemented with 40 μg/mL tetracycline and 50 μg/mL streptomycin) to generate enough vector to produce one phage library. After the overnight culture, bacterial cells should then be carefully washed once with STE buffer (100 mM NaCl, 10 mM Tris-Cl, pH 8.0, and 0.1 mM EDTA) to eliminate the single-stranded form of fUSE55, present due to the production of viral particles that remain attached to the surface of the bacteria. Bacterial pellets may then be kept frozen (−20 °C) while waiting processing for DNA vector purification.

Purification of the double-stranded form of fUSE55 should be performed using commercially available plasmid prep kits. For the Qiagen Max Prep, we modified the manufacturer's protocol: for every 2 L of culture, we used 100 mL of solutions P1, P2, and P3 (composition indicated in the Materials section), which was then loaded into one QIAGEN-tip 500. To obtain a vector with high purity, free of contaminants (i.e., genomic DNA), it is necessary to carry out two consecutive maxi-preparations (load the first purified vector into a second QIAGEN-tip 500) followed by equilibrium centrifugation with cesium chloride [10]. The quality of the DNA should be assessed by electrophoresis in a 0.8% agarose gel stained with ethidium bromide (0.5 μg/mL) and quantified by spectrophotometry (readings at wavelengths of 260 nm and 280 nm) using a Nanodrop equipment (Thermo Scientific) (ratio 260 nm/280 nm ~1.8).

3.2 fUSE55 Restriction Digestion

Once a suitable amount of vector has been produced, it needs to be linearized using the *BglI* restriction enzyme (recognition site: GCCNNNN/NGGC). There are two adjacent *BglI* sites in the fUSE55 vector. Because the *BglI* enzyme recognizes a degenerate site and the two *BglI* sites in fUSE55 do not share the same exact sequence (Figs. 1 and 2), there is no need to dephosphorylate the vector: upon digestion and purification using the QIAGEN-tip 500, the small fragment stuffer is lost, and the vector is no longer capable of self-ligation.

The amount of vector used to build libraries may vary slightly. For instance, a random hexapeptide library (X6, X = any amino acid) has a maximum diversity of 64 million peptides; thus, a library with 10^9 transformants will most likely encode at least 1 copy of each possible peptide (*see* **Note 1**). However, for an octapeptide library (X8) with a theoretical diversity of 6×10^{10}, even a library with 10^{10} transformants will not encode all possible octapeptides. Because the number of transformants in the library will be proportional to the amount of vector, the protocol also lists relations (enzyme/μg of DNA, etc.) which can be useful to scale up the reaction accordingly. Below, we will reproduce the protocol used to build a CX8C library with approximately 10^{10} transformants. For that, we started with 1 mg of fUSE55 and 5 μg of insert. These steps should be taken as guidelines and optimized accordingly to your laboratory conditions and the kind of library that is being built.

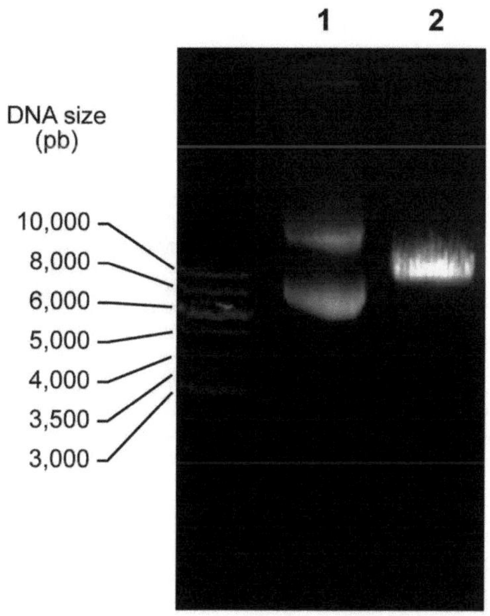

Fig. 1 fUSE55 vector before and after *BglI* digestion. Agarose gel (0.8%) electrophoresis analysis of ~1 μg of CsCl purified fUSE55 vector before [1] and after *BglI* digestion [2]. DNA was stained with ethidium bromide

1. Prepare the fUSE55 digestion following the enzyme manufacturer's recommendations and buffer. In our condition, we used 5 U of enzyme per μg of DNA, added in two consecutive additions (2500 U/4 h + 2500 U/overnight) to completely linearize 1 mg of fUSE55 in a 15 mL reaction (Table 2). Upon completion, take a 1 μg sample of and analyze the DNA by electrophoresis in agarose gel (0.8%). A single ~10 kbp DNA band should be visible (Fig. 1).

Next Day

2. Equilibrate three QIAGEN-tip 500 columns with Qiagen QBT buffer (20 mL each column).
3. Divide the fUSE55 digestion into 3 tubes (50 mL), 5 mL in each tube. Add 45 mL Qiagen QBT buffer.
4. Load each of the diluted fUSE55 *BglII* digestion (50 mL) into one of the QIAGEN-tip 500, and let the reaction pass through the columns.
5. Wash twice each QIAGEN-tip 500 with Qiagen QC buffer (2 × 30 mL).
6. Elute the linearized fUSE55 vector with warm Qiagen QF buffer (15 mL).

   Very important: To increase DNA recovery, warm the Qiagen QF buffer to 50 °C.

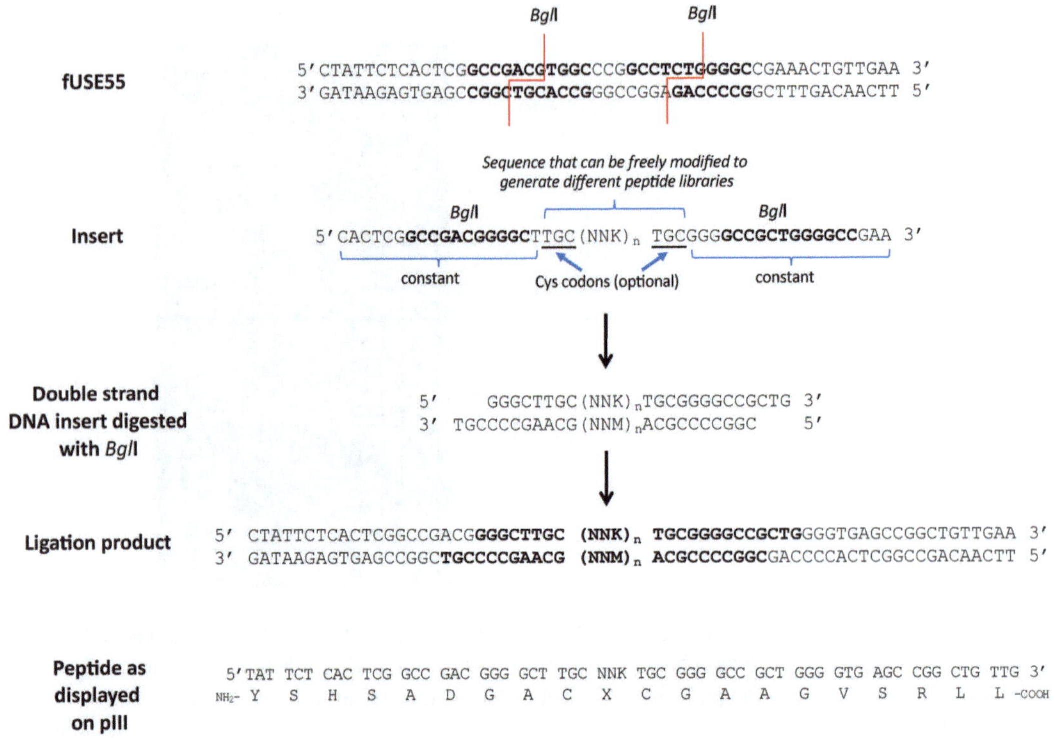

Fig. 2 Insert design for peptide library construction. The fUSE55 cloning site with two adjacent BglI restriction enzyme sites is shown above. Next, a single-strand oligonucleotide used for library construction of a CX8C peptide library is shown. The constant sequences containing the BglI restriction enzyme sites, flanking the custom-designed part of the oligonucleotide, are indicated (including, in this case, two Cys codons flanking eight NNK sequences to generate the random octapeptides). Once the insert is converted to double-stranded DNA and digested with BglI enzyme, it is ready to be cloned into the BglI-linearized fUSE5 vector. Finally, a fragment of the pIII protein containing the displayed peptide is shown below

Table 2
fUSE55 digestion

Component	Amount
10× buffer (supplied with the enzyme)	1.5 mL
fUSE55 vector	1 mg
Water *q.s.*	15 mL
*Bgl*I (10,000 U/ml)	0.5 mL (5000 U)[a]

[a]First add 0.25 mL followed by another 0.25 mL, 4–6 h later. Leave overnight at 37 °C

7. Precipitate the eluted digested fUSE55 by adding 10.5 mL (0.7 volumes) of room temperature isopropanol to the eluted DNA. Mix and centrifuge immediately at ≥15,000 × *g* for 30 min at 4 °C. (Use centrifuge tubes that can withstand centrifugations at ≥15,000 × *g*.)

8. Carefully decant the supernatant and wash the fUSE55 pellet with room temperature ethanol 70%. Centrifuge at ≥15,000 × *g* for 10 min at 4 °C.
9. Carefully decant the supernatant and air-dry the fUSE55 pellet for a few minutes (1–2 min). *Attention*: Do not overdry the DNA or it will extremely difficult to dissolve it.
10. Add 1 mL TE buffer and redissolve the linearized fUSE55 by washing carefully the walls of the tube. It helps to incubate at 37 °C for 5–10 min with agitation.
11. Analyze and quantify the eluted DNA by electrophoresis in agarose gel (0.8%) and Nanodrop. A single ~10 kbp DNA band should be visible in the gel (Fig. 1). Reserve the vector, which can be kept at −20 °C until use. Expect DNA recovery to be around 40%.

3.3 Insert Design and Preparation

Custom-synthetic oligonucleotides should be designed and synthesized (Thermo Fisher, 50 nmol scale) to include random nucleotides at specific positions following the $(NNK)_n$ pattern, in which N represents any of the four nucleotides (G, T, A, or C); K represents only G or T; and "n" indicates the number of codons desired (hence, amino acids to be displayed) (*see* **Notes 2** and **3**). For example, to generate a phage library that displays cyclic octapeptides (CX8C, X = any amino acid and C = Cys), one should first-include a Cys codon (TGC or TGT followed by 8 NNK trinucleotides plus another TGC or TGT codon) (Figs. 2 and 4) (*see* **Note 4**).

The use of the NNK motif reduces the number of randomly generated *stop* codons from 3 to 1 while still encoding all 20 amino acids. The sequences of the degenerate oligonucleotides are flanked by restriction sites recognized by the *Bgl*I enzyme. An oligonucleotide complementary to the 3′ region of the degenerate oligonucleotides should also be synthesized in order to produce double-stranded DNA inserts using the DNA polymerase I large (Klenow) fragment. There are excellent reviews regarding the different kind of libraries that can be built and how to design the oligonucleotides, accordingly [2, 9]. As an example, we show the sequences of oligonucleotides used to produce the X6 and CX8C phage display libraries and the antisense oligo used to generate the double-stranded DNA insert (Fig. 3) [6, 7].

To calculate the amount of oligonucleotide that you need, use the formula (*see* **Note 5**). Here, we describe procedures to produce approximately 43 μg of insert starting from 1 nmol of each oligonucleotide listed above (CX8C + antisense).

1. Resuspend oligos in molecular biology grade water (100 pmol/μL).

Library sense oligos:

CX8C: 5'CACTCGGCCGACGGGGCTTGCNNKNNKNNKNNKNNKNNKNNKNNKTGCGGGGCCGCTGGGGCCGAA 3'
X6: 5'CACTCGGCCGACGGGGCTNNKNNKNNKNNKNNKNNKGGGGCCGCTGGGGCCGAA 3'

Library antisense:

5' TTCGGCCCCAGCGGC 3'

Fig. 3 Oligonucleotides to produce the CX8C and X6 phage display libraries and antisense oligonucleotide

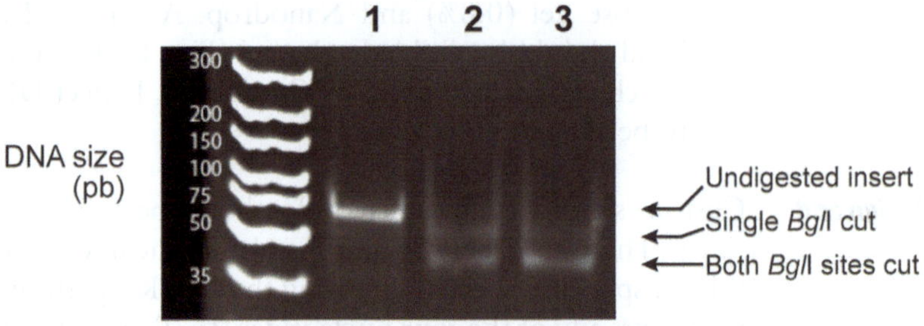

Fig. 4 Insert preparation. Sense and antisense oligos were used to generate the double-stranded DNA insert for the CX8C library production (1). The DNA insert (60 ng) was then digested with BglI restriction enzyme in two conditions: a single addition of 1 U of enzyme and digestion overnight (2) or following two additions of enzyme (0.5 U/8 h + 0.5 U/overnight) (3). Arrows indicate when the insert that were not digested, partially digested (at a single BglI site) or fully digested (both 5′ and 3′ BglI sites)

2. In a 500 μL tube, add 10 μL of each oligo (sense and the desired antisense oligonucleotide) (1 nmol each), 10 μL of *10× reaction buffer* for Klenow fragment, 2.5 μL 10 mM dNTP, and 66 μL of water (final volume of 100 μL).
3. Heat the solution first to 50 °C (15 min), then to 37 °C (15 min).
4. Add 20 U of DNA Polymerase I, Large (Klenow) Fragment (2 μL, 10,000 U/mL), and incubate at 37 °C for 1 h.
5. Stop the reaction by heating to 75 °C for 10 min.
6. Purify the double-stranded product using the QIAquick Nucleotide Removal Kit (Qiagen), following the manufacturer's instruction.

   Important: Pay attention to the cartridge capacity (10 μg). In this case, follow the protocol for small DNA fragments (<100 pb) and add ~10 volumes of Qiagen Buffer PNI (1.1 mL) to the Klenow reaction for final volume of 1.2 mL. Distribute the solution into 4 QIAquick spin column (0.3 mL each), wash with 750 μL Buffer PE, and elute double-stranded DNA insert with buffer EB (50 μL). Combine the eluates (200 μL), and remove 2 × 1 μL sample (undigested DNA).

Before proceeding to the next step (digestion with restriction enzyme *BglI*), analyze and quantify one of the samples of the eluted DNA by polyacrylamide gel electrophoresis (20% gel/TBE buffer) (Fig. 4).

3.4 Insert Digestion with BglI

1. Prepare the reaction to digestion the double-stranded DNA insert following the enzyme manufacturer's recommendation with regard to buffer and reaction conditions. In our conditions, to completely linearize the insert, we need up to 20 U of enzyme per μg of DNA added in two consecutive additions (10 U/4–6 h + 10 U/overnight) (Table 3).

Next Day

Purify the digested insert using the QIAquick Nucleotide Removal Kit as described above.

2. Analyze and quantify the eluted DNA by electrophoresis in acrylamide gel (TBE buffer, 20% acrylamide) and Nanodrop. A single DNA band should be visible in the gel (Fig. 4). Reserve the insert, which can be kept at −20 °C until use.

3.5 Library Ligation

The final step is the ligation of the restricted double-stranded DNA insert to the linearized fUSE55 vector. Here, it is important to set up pilot reactions to determine the best ratio (vector to insert). In the pilot reactions (10–20 μL in volume), insert and vector concentrations should mirror exactly the final scaled-up ligation. For vector circularization, concentrations of DNA should be maintained within the 1–10 μg/mL total DNA (linker + insert). So, starting with 1 mg of fUSE55, after *BglI* digestion and purification, expect to have approximately 300–400 μg of linearized vector and approximately 10–20 μg of digested insert. fUSE55 is 9200 bp and insert is 45 bp, a ~200:1 (m/m) relation. Hence, fUSE55 is usually the limiting factor, and you should set up your ligation based on the optimum vector to insert ratio (as determined by the pilot ligations) and the amount of linearized fUSE55 vector available (Table 4).

Table 3
Insert digestion

Component	Amount
10× buffer (supplied with the enzyme)	0.1 mL
DNA insert	20 μg
Water *q.s.*	1 mL
BglI (10,000 U/ml)	40 μL (400 U)[a]

[a]First add 20 μL followed by another 20 μL, 4–6 h later. Leave overnight at 37 °C

Table 4
Ligation condition

Component	Amount
Tris-Cl, pH 7.5	50 mM
$MgCl_2$	10 mM
DTT	10 mM
ATP	1 mM
T4 DNA ligase[a]	2000 CEU

Incubate overnight at 8–10 °C (refrigerator)
[a]Check the units from the enzyme vendor; in case it uses Weiss units, make the appropriate adjustments (one Weiss unit = ~200 cohesive end units)

Next Day

1. For 100 μg total DNA (fUSE55 + insert), the ligation volume will be 10 mL (final DNA concentration of 10 μg/mL).
2. Divide the fUSE55 ligation into two tubes (50 mL), 5 mL in each tube. Add 45 mL Qiagen QBT buffer to each tube. You should end up with 100 mL solution.
3. Equilibrate one QIAGEN-tip 100 column with Qiagen QBT buffer (5 mL).
4. Apply the diluted ligation (100 mL) into the QIAGEN-tip 100 and let the ligation pass through the column.
5. Wash twice the QIAGEN-tip 100 with Qiagen QC buffer (2 × 30 mL).
6. Elute the fUSE55-insert ligation product with warm Qiagen QF buffer (15 mL).

   Very important: To increase DNA recovery, warm the Qiagen QF elution buffer to 50 °C.
7. Precipitate the eluted fUSE55-insert ligation product by adding 10.5 mL (0.7 volumes) of room temperature isopropanol to the eluted DNA. Mix and centrifuge immediately at ≥15,000 × *g* for 30 min at 4 °C. (Use centrifuge tubes that can withstand centrifugations at ≥15,000 × *g*.)
8. Carefully decant the supernatant and wash the fUSE55 pellet with room temperature ethanol 70%. Centrifuge at ≥15,000 × *g* for 10 min at 4 °C.
9. Carefully decant the supernatant and air-dry the fUSE55 pellet for a few minutes (1–2 min). *Attention*: Do not overdry the DNA or it will be extremely difficult to dissolve it.
10. Add 50 μL water and redissolve the fUSE550-insert ligation product by washing carefully the walls of the tube. It helps to

incubate at 37 °C for 5–10 min with agitation. Aim for a 1–2 mg/mL total DNA final concentration.

11. Analyze and quantify the eluted DNA by electrophoresis in agarose gel (0.8%) and Nanodrop, and proceed to the electroporation step. The ligation product may be kept frozen (−20 °C) until use.

3.6 Large-Scale Electroporation

We used freshly prepared *E. coli* MC1061 cells to produce the library and electroporation to transform the ligation and produce the virion. Cells were made competent following standard protocols [11]. Freshly prepared cells have higher efficiency but due to lot-to-lot efficiency variation, it is recommended to "practice" and optimize the protocol to assure good reproducibility in generating electrocompetent cells. For instance, to improve efficiency, maintain stock cells in minimal media and streak fresh plates prior to inoculating the cell cultures to produce the electrocompetent cells. The protocol yields ~1 mL of competent cells for each 500 mL of culture, so scale up accordingly depending on your ligation yield. Aliquot cells into 1.5 mL tube and keep them on ice at all times.

1. Estimate the number of electroporations that will be performed. For 100 μg of total DNA ligation, we recommend 7–10 electroporations.
2. For each electroporation, add 10 mL SOC media supplemented with 0.2 μg/mL tetracycline in a 250 or 500 mL flask. (For instance, if ten electroporations will be performed, add 100 mL of SOC media to a 500 mL flask.) Keep in the incubator at 37 °C, 200 rpm rotation.
3. Set up the electroporator (Gene Pulser) to 2500 V, capacitor 25 μF, and resistance 400 Ω.
4. Add up to 10 μL (10–15 μg of DNA) of the purified ligation to one aliquot (200 μL) of freshly prepared ice-cold *E. coli* MC1061 cells.
5. Transfer to an ice-cold electroporation cuvette (0.2 cm gap) being extra careful to avoid air bubbles (which may cause arcing).
6. Pulse once your electroporation.
7. Immediately, remove the cuvette from the electroporator and add 1 mL SOC media. Gently but surely, resuspend and homogenize cells and transfer to the flask with warm (37 °C) SOC low-tet media. Keep the flask in the incubated at 37 °C with 200 rpm agitation while the remaining electroporation is performed.
8. Check the electroporation parameters: a time pulse close to 5 ms is expected.

9. Repeat the same procedure until all DNA has been electroporated.
10. Remove a 10 μL aliquot from the SOC media and reserve. This is important to determine the number of transformants and, therefore the diversity, the number of "unique" peptides in our library.
11. Culture the SOC media for 30–40 min at 37 °C with 200 rpm rotation. Transfer 50 mL aliquots of the SOC media to new 2 L flasks with 500 mL of LB supplemented with 40 μg/mL tetracycline and culture at 37 °C (250 rpm) overnight. You should have 20 flasks with a total of 10 L of LB culture.

3.7 Determining the Diversity of the Phage Library

The diversity of the library should be determined using the small (10 μL) aliquot that was removed from the SOC media once all electroporations have been performed. It is important to remove the aliquot before 30 min of culture to avoid bacterial cell growth, which may inadvertently augment the number of transformants. The numbers of transformant are calculated by colony counting. Plate serial dilutions (10^{-1}, 10^{-2}, 10^{-3}, etc. up to 10^{-6}) of the SOC media onto LB/tet/strp agar plates. For example, a library with 10^9 transformants (in 100 mL of SOC media, as described above) will have 10,000,000 transforming units (TU) per mL of culture or 10^5 TU/μL.

IMPORTANT: Once the library is transferred to the LB media and cultured overnight, it is no longer possible to know the number of transformants since phage slows but do not inhibit bacterial growth.

1. Label six 1.5 mL Eppendorf tube 10^{-1}, 10^{-2}, 10^{-3}, etc. up to 10^{-6}.
2. Add 90 μL of LB to each tube.
3. Make the serial dilutions by transferring the 10 μL aliquot of SOC media to the first tube (10^{-1}), homogenizing, and transferring 10 μL to the next tube.
4. Plate 10 μL of each tube onto three individual LB tet/strp agar plates (triplicates).
5. Next day, count the colonies in each plate (when possible) and calculate the number of transformants.

3.8 Producing the Phage Peptide Library

Next Day

Bacteriophage is secreted to the media during cell culture. Here, we describe procedures to produce the final phage library by the PEG/NaCl method [6]. It is not necessary to purify virion from all 10 L of culture. Each 500 mL will yield approximately 500 μL of library. Of course, phage library is hard to produce and a precious reagent, so decide wisely how much library you want to produce and how much you want to discard. Below, we illustrate phage purification from 300 mL of culture. Scale up accordingly to your needs and production capacity.

1. Transfer 300 mL of the bacterial cell culture to 400 mL centrifuge tubes. Make sure the O-ring in the lead is intact and not bended; otherwise, liquid will leak during centrifugation.
2. Centrifuge at 8000 × *g* for 15 min and transfer the supernatant to a new 500 mL.
3. Add 45 mL PEG/NaCl solution. Close with the lead and homogenize by inverting the tube several times. Keep on ice for 30 min.
4. Centrifuge at 10,000 × *g* for 30 min, discard the supernatant, and leave the tube in an inverted position for a few minutes to collect and discard the remaining liquid from the wall of the tube.
5. Centrifuge at 10,000 × *g* for 5 min and discard any remaining liquid.
6. Resuspend the phage pellet with 10 mL of PBS. It helps to incubate the tube at 37 °C with rotation (200 rpm).
7. Transfer the solution to a 50 mL tube and centrifuge at 10,000 × *g* for 10 min.
8. Transfer the phage solution to a new 50 mL centrifuge tube, and add 1.5 mL of PEG/NaCl solution. Close and homogenize by inverting the tube several times. Leave on ice for 30 min.
9. Centrifuge at 10,000 × *g* for 30 min, discard the supernatant, and leave the tube in an inverted position for a few minutes to collect and discard the remaining liquid from the wall of the tube.
10. Centrifuge at 10,000 × *g* for 5 min and discard any remaining liquid.
11. Resuspend the phage pellet with 0.3–0.5 mL of sterile PBS. It helps to incubate the tube at 37 °C with rotation (200 rpm).
12. Transfer the phage solution to a 1.5 mL Eppendorf tube and centrifuge at 15,000 × *g* for 10 min. Transfer supernatant to a new 1.5 mL Eppendorf tube.
13. Keep the phage library at 4 °C.

4 Notes

1. There are two key concepts: the theoretical diversity and the actual number of unique peptides encoded by a phage library. The first can be calculated based on the insert design. For a

random heptapeptide (X7) library, there are $20^7 = 1.28 \times 10^9$ (1.28 billion) heptapeptides that can be created. On the other hand, the actual number of peptides that a library contains will be the number of transformant obtained during the final stage of production. Considering the empirical rule "68-95-99.7," an X7 phage library with a billion transformants will encode ~68% of all possible heptapeptides, while a library with 10 billion transformants should have close to 95% of them.

2. The use of NNN to generate randomized codons is not advised due to the high chance of introducing a stop codon (3 in 64 codons) compared with the NNK or NNS alternatives (1 in 32 codons). Alternatively, NDT, DBK, or NRT will encode a more limited number of amino acids but with all biophysical properties (anionic, cationic, aliphatic hydrophobic, aromatic hydrophobic, hydrophilic and small) with the advantage of no stop codons. Finally, specific amino acids may be fixed with the display peptide, for instance, a tyrosine to mimic a phosphorylation site, or a glutamic/aspartic acid to mimic phosphoserine/threonine.
3. New emerging technologies can substitute the use of degenerate oligonucleotide synthesis (NNK), such as the trimer (codon) phosphoramidites to produce oligonucleotides of mixed sequences without stop codons.
4. Protein III, used for display, is processed in the *E. coli* membrane periplasmic space, an environment that allows for S-S disulfide-bridge formation.
5. To convert DNA μg to pmol and vice versa, use the formula below:

$$\text{pmol of DNA} = \mu\text{g of DNA} \times (\text{pmol}/660\ \text{pg}) \times \left(10^6\ \text{pg}/1\ \mu\text{g}\right) \times (1/N)$$

in which

pmol = amount of DNA in pmols

N = DNA length in bp

Acknowledgement

This work was supported by Sao Paulo State Agency (FAPESP) grant 2019/25.828-8 to RJG.

References

1. Smith GP (1985) Filamentous fusion phage: novel expression vectors that display cloned antigens on the virion surface. Science 228(4705):1315–1317
2. Smith GP, Petrenko VA (1997) Phage display. Chem Rev 97(2):391–410
3. Winter G, Griffiths AD, Hawkins RE et al (1994) Making antibodies by phage display technology. Annu Rev Immunol 12:433–455
4. Ozawa MG, Zurita AJ, Dias-Neto E et al (2008) Beyond receptor expression levels: the relevance of target accessibility in ligand-directed pharmacodelivery systems. Trends Cardiovasc Med 18(4):126–132
5. Barderas R, Benito-Peña E (2019) The 2018 Nobel Prize in chemistry: phage display of peptides and antibodies. Anal Bioanal Chem 411(12):2475–2479
6. Michaloski JS, Redondo AR, Leila S, Magalhães LS et al (2016) Discovery of pan-VEGF inhibitory peptides directed to the extracellular ligand-binding domains of the VEGF receptors. Sci Adv 2:e1600611
7. Tang FHF, Staquicini FI, Teixeira AAR et al (2018) A ligand motif enables differential vascular targeting of endothelial junctions between brain and retina. Proc Natl Acad Sci USA 116(6):2300–2305
8. Teixeira AAR, Carnero LR, Kuramoto A et al (2021) A refined genome phage display methodology delineates the human antibody response in patients with Chagas disease. iScience 24(6):102540
9. Barbas CFIII, Burton DR, Scott JK et al (2001) Phage display: a laboratory manual. Cold Spring Harbor Laboratory, Cold Spring Harbor
10. Maniatis T, Fritsch EF, Sam-brock J (1982) Molecular cloning: a laboratory manual. Cold Spring Harbor Laboratory, Cold Spring Harbor, pp 86–96
11. Seidman C E, Struhl K, Sheen J et al (2001) Introduction of plasmid DNA into cells. Curr Protoc Mol Biol Chapter 1:Unit1.8

Chapter 2

Generation of a Naïve Human scFv Phage Display Library and Panning Selection

Brenda Pei Chui Song, Jing Yi Lai, and Theam Soon Lim

Abstract

Phage display antibody libraries have been successfully used as the essential tool to produce monoclonal antibodies against a plethora of targets ranging from diseases to native biologically important proteins as well as small molecules. It is well documented that diverse antibody genes are the major genetic source for the construction of a high-quality antibody library and selection of high-affinity antibodies. Naïve antibody libraries are derived using the IgM repertoire of healthy donors obtained from B-cells isolated from human peripheral blood mononuclear cell (PBMC). Single-chain fragment variable (scFv) is a routinely used format due to its smaller size and preference for phage display. The process involves the use of a two-step cloning method for library construction. The protocol also covers the biopanning process for target positive clone selection.

Key words Naïve antibody repertoires, Antibody libraries, Combinatorial, scFv, Human, Phage display, Phagemid, Monoclonal antibodies

1 Introduction

Antibody phage display technology represents an alternative technique to the conventional hybridoma technology for monoclonal antibody generation [1, 2]. Antibody phage display involves the isolation of human-derived monoclonal antibodies from a collection of diverse Ig gene repertoire that is displayed on the surface of bacteriophages [3]. This was made possible with the first demonstration of peptides being physically displayed on the surface of bacteriophages by George P. Smith in 1985 [4]. Utilizing robust and stable filamentous phages, this versatile technology has been instrumental in the production of monoclonal antibodies which are extensively employed in basic research, diagnostics, and therapeutics. Its ability to rapidly generate and modify high-affinity monoclonal antibodies has rendered it indispensable in immunology till this day [5]. The prerequisite for a successful antibody phage

Huan Peng et al. (eds.), *Phage Engineering and Analysis: Methods and Protocols*,
Methods in Molecular Biology, vol. 2793, https://doi.org/10.1007/978-1-0716-3798-2_2,

display panning campaign lies in the construction of high-quality antibody gene libraries [6]. To construct a high-quality phage display antibody library, the source of antibody genes must be large and diverse enough to replicate and in some instance exceed the natural diversity of the antibody response in vivo [3].

The antibody phage display library consists of a large collection of antibody presenting phages with each phage particle displaying unique antibody proteins that are linked to their respective phage particles harboring the unique gene sequence [7]. This collection of phage particles displaying unique antibody variable domains in the form of a library would normally consist of 10^8–10^{11} unique clones depending on the type of libraries. The display of antibodies from the library involves having the antibody protein being fused to the bacteriophage coat protein by cloning the antibody gene fragment to the minor coat protein, pIII [7]. Generally, antibody libraries are classified into three categories, namely, the naïve, immune, and synthetic libraries [2, 4]. This classification is based on the source of antibody V gene used for its generation [8]. Naïve antibody libraries are derived from healthy donors focusing on the application of the IgM repertoire [2]. This allows naïve libraries to be unbiased toward a certain group of antigens, assuming there is no known infection at the point of collection or vaccination [1]. However, antibodies isolated from naïve antibody libraries usually exhibit lower affinities compared to immunized libraries, as they are yet to undergo in vivo affinity maturation [6]. Therefore, the size of the library and repertoire diversity are critically monitored for this case, where larger library sizes are preferred to ensure higher affinity monoclonal antibodies are able to be isolated [9]. A very large antibody library can be constructed using multiple "naïve" donors to increase the overall antibody repertoire of the library [9]. Therefore, a balance between the number of donors with cloning diversity is important for a successful antibody phage display library generation.

There are several common human antibody formats displayed using phage display. This includes the single-chain fragment variable (scFv), fragment antigen binding (Fab), and domain antibodies. Due to the limitation in *E. coli* folding machinery, antibody fragments such as scFv, Fab, VHH, and dAbs are used routinely for antibody phage display instead of the larger formats [10]. However, the scFv format is the most used format due to its ability to present the diversity of both variable heavy and light chain in an efficient display size range for bacteriophages. The scFv format is made up of the variable regions of heavy (VH) and light (VL) chains which are linked by a flexible peptide linker that can be expressed in a functional form in *E. coli* [10]. It is the preferred antibody format used for phage display as it is not prone to degradation and is easier to be expressed in its functional form [9–11].

Biopanning is a term used to describe the in vitro selection process using phage display. It is a repetitive process that functions to sieve positive clones from the diverse library and subsequently enrich the positive population [12]. The selection process involves several repetitive cycles of binding, washing, and amplification of the positive phage clones until a predominant population is present [12]. The selection process is based on affinity selection whereby strong affinity clones will bind to the target and are successfully rescued to the next round by a process of amplification. Therefore, the population of the target binding clone will be concentrated over several rounds of selection. This chapter describes the steps involved in the construction of a human naïve scFv antibody library and the subsequent biopanning process to identify target specific monoclonal antibodies. The protocol consists of a two-step cloning method to construct a diverse human scFv antibody phage display library. A standard biopanning protocol is described in detail for the selection process. It is our hope that the protocol will help new antibody developers to successfully develop and apply phage display for their antibody development programs.

2 Materials

2.1 Isolation of B-Cells

1. Ficoll-Paque™ PLUS (Cytiva, USA).
2. Phosphate-buffered saline (PBS), pH 7.4: 8 g NaCl, 0.2 g KCl, 1.44 g $Na_2HPO_4{\cdot}2H_2O$, 0.24 g KH_2PO_4 in 1 L dH_2O, autoclave and store at room temperature.
3. QIAamp® RNA Blood Mini Kit (QIAGEN, Germany).

2.2 First-Strand cDNA Synthesis

1. 300–500 ng RNA.
2. SuperScript First-Strand Synthesis System for RT-PCR (Invitrogen, USA).

2.3 Amplification of Heavy Chain (HC) and Light Chain (LC) Gene Repertoire

1. Vent® DNA Polymerase (NEB, USA) (*see* **Note 1**).
2. *Pfu* DNA Polymerase (Thermo Scientific, USA) (*see* **Note 1**).
3. Forward and reverse primers for first amplification of VH and VL (*see* Table 1).
4. Forward and reverse primers for second amplification of VH and VL with restriction endonuclease sites (*see* Table 2).
5. 10 mM dNTP mixture.
6. Agarose.
7. 10× TBE buffer, pH 8.0: 108 g Tris-HCl, 55 g boric acid, and 7.4 g EDTA in 1 L dH_2O, autoclave and store at room temperature.
8. QIAquick® Gel Extraction Kit (QIAGEN, Germany).

Table 1
Primers for first amplification of VH and VL gene repertoire

Primer name	Primer sequence
VH amplification	
VH1 Fw	5′- CAGGTCCAGCTKGTRCAGTCTGG -3′
VH157 Fw	5′- CAGGTGCAGCTGGTGSARTCTGG -3′
VH2 Fw	5′- CAGRTCACCTTGAAGGAGTCTG -3′
VH3 Fw	5′- GAGGTGCAGCTGKTGGAGWCY -3′
VH4 Fw	5′- CAGGTGCAGCTGCAGGAGTCSG -3′
VH4 DP63 Fw	5′- CAGGTGCAGCTACAGCAGTGGG -3′
VH6 Fw	5′- CAGGTACAGCTGCAGCAGTCA -3′
Human scFv IgM CH1 Rv	5′- AAGGGTTGGGGCGGATGCACT -3′
VL amplification	
Vλ1 Fw	5′- CAGTCTGTSBTGACGCAGCCGCC -3′
Vλ1459 Fw	5′- CAGCCTGTGCTGACTCARYC -3′
Vλ15910 Fw	5′- CAGCCWGKGCTGACTCAGCCMCC -3′
Vλ2 Fw	5′- CAGTCTGYYCTGAYTCAGCCT -3′
Vλ3 Fw	5′- TCCTATGWGCTGACWCAGCCAA -3′
Vλ3 DPL16 Fw	5′- TCCTCTGAGCTGASTCAGGASCC -3′
Vλ338 Fw	5′- TCCTATGAGCTGAYRCAGCYACC -3′
Vλ6 Fw	5′- AATTTTATGCTGACTCAGCCCC -3′
Vλ78 Fw	5′- CAGDCTGTGGTGACYCAGGAGCC -3′
Vκ1 Fw	5′- GACATCCRGDTGACCCAGTCTCC -3′
Vκ246 Fw	5′- GATATTGTGMTGACBCAGWCTCC -3′
Vκ3 Fw	5′- GAAATTGTRWTGACRCAGTCTCC -3′
Vκ5 Fw	5′- GAAACGACACTCACGCAGTCTC -3′
scFv Fab Lambda CL1 Rv	5′- TGAACATTCTGTAGGGGCCACTG -3′
scFv Fab Lambda CL2 Rv	5′- TGAACATTCCGTAGGGGCAACTG -3′
scFv Fab Kappa CL Rv	5′- ACACTCTCCCCTGTTGAAGCTCTT -3′

2.4 Two-Step Cloning

1. Antarctic phosphatase.
2. T4 DNA Ligase.
3. 3 M sodium acetate, pH 5.2: sodium acetate in 800 mL dH_2O; adjust pH to 5.2 with glacial acetic acid, top up to 1 L with dH_2O and store at room temperature.

Table 2
Primers with restriction endonuclease site for second amplification

Primer name	Primer sequence
VH amplification	
VH1 NcoI Fw	5′- CCC AGC CGG <u>CCA TGG</u> CC CAG GTC CAG CTK GTR CAG TCT GG -3′
VH157 NcoI Fw	5′- CCC AGC CGG <u>CCA TGG</u> CC CAG GTG CAG CTG GTG SAR TCT GG -3′
VH2 NcoI Fw	5′- CCC AGC CGG <u>CCA TGG</u> CC CAG RTC ACC TTG AAG GAG TCT G -3′
VH3 NcoI Fw	5′- CCC AGC CGG <u>CCA TGG</u> CC GAG GTG CAG CTG KTG GAG WCY -3′
VH4 NcoI Fw	5′- CCC AGC CGG <u>CCA TGG</u> CC CAG GTG CAG CTG CAG GAG TCS G -3′
VH4 DP63 NcoI Fw	5′- CCC AGC CGG <u>CCA TGG</u> CC CAG GTG CAG CTA CAG CAG TGG G -3′
VH6 NcoI Fw	5′- CCC AGC CGG <u>CCA TGG</u> CC CAG GTA CAG CTG CAG CAG TCA -3′
Human scFv IgM CH1 XhoI Rv	5′- ACCG <u>CTC GAG</u> AC AAG GGT TGG GGC GGA TGC ACT -3′
VL amplification	
Vλ1 SalI Fw	5′- TGT GAC AAA <u>GTC GAC</u> G CAG TCT GTS BTG ACG CAG CCG CC -3′
Vλ1459 SalI Fw	5′- TGT GAC AAA <u>GTC GAC</u> G CAG CCT GTG CTG ACT CAR YC-3′
Vλ15910 SalI Fw	5′- TGT GAC AAA <u>GTC GAC</u> G CAG CCW GKG CTG ACT CAG CCM CC -3′
Vλ2 SalI Fw	5′- TGT GAC AAA <u>GTC GAC</u> G CAG TCT GYY CTG AYT CAG CCT -3′
Vλ3 SalI Fw	5′- TGT GAC AAA <u>GTC GAC</u> G TCC TAT GWG CTG ACW CAG CCA A -3′
Vλ3 DPL16 SalI Fw	5′- TGT GAC AAA <u>GTC GAC</u> G TCC TCT GAG CTG AST CAG GAS CC -3′
Vλ338 SalI Fw	5′- TGT GAC AAA <u>GTC GAC</u> G TCC TAT GAG CTG AYR CAG CYA CC -3′
Vλ6 SalI Fw	5′- TGT GAC AAA <u>GTC GAC</u> G AAT TTT ATG CTG ACT CAG CCC C -3′
Vλ78 SalI Fw	5′ - TGT GAC AAA <u>GTC GAC</u> G CAG DCT GTG GTG ACY CAG GAG CC -3′
Vκ1 SalI Fw	5′- TGT GAC AAA <u>GTC GAC</u> G GAC ATC CRG DTG ACC CAG TCT CC -3′

(continued)

Table 2
(continued)

Primer name	Primer sequence
Vκ246 SalI Fw	5′- TGT GAC AAA GTC GAC G GAT ATT GTG MTG ACB CAG WCT CC -3′
Vκ3 SalI Fw	5′- TGT GAC AAA GTC GAC G GAA ATT GTR WTG ACR CAG TCT CC -3′
Vκ5 SalI Fw	5′- TGT GAC AAA GTC GAC G GAA ACG ACA CTC ACG CAG TCT C -3′
scFv Lambda NotI Rv	5′- ATG ATG ATG TGC GGC CGC AGA GGA SGG YGG GAA CAG AGT GAC -3′
scFv Kappa NotI Rv	5′- ATG ATG ATG TGC GGC CGC GAA GAC AGA TGG TGC AGC CAC AGT -3′

4. pLABEL phagemid vector.
5. QIAGEN Plasmid Maxi Kit (QIAGEN, Germany).
6. QIAquick® PCR Purification Kit (QIAGEN, Germany).
7. Ampicillin stock solution (50 mg/mL): 2.5 g ampicillin sodium salt in 50 mL of 50% (v/v) ethanol; filter-sterilize and store at −20 °C.
8. 40% (w/v) glucose: 40 g D-(+)-glucose in 100 mL of dH_2O; autoclave and store at 4 °C.
9. 2× YT medium: 16 g tryptone, 10 g yeast extract, and 5 g NaCl in 1 L dH_2O; autoclave and store at room temperature.
10. 2× YT agar: 16 g tryptone, 10 g yeast extract, 5 g NaCl, and 15.5 g agar in 1 L dH_2O; autoclave.
11. 2× YT-amp medium: 2× YT medium, 0.1 mg/ mL ampicillin, and 2% (v/v) glucose.
12. 2× YT-amp agar plate: 2× YT agar, 0.1 mg/mL ampicillin, and 2% (v/v) glucose.
13. 80% (v/v) glycerol: 40 mL glycerol in 10 mL dH_2O; autoclave and store at room temperature.
14. Sterile Petri Dish, 90 mm.
15. Nunc™ Square BioAssay Dish, 25 mm (Thermo Scientific, USA).
16. MicroPulser™ Electroporator (Bio-Rad, USA).

2.4.1 First Step Cloning (VL)

1. SalI.
2. NotI.

3. ElectroMAX™ DH10β Cells (Invitrogen, USA), genotype: F^{-} *mcrA* Δ(*mrr-hsdRMS-mcrBC*) Φ80*lacZ*Δ*M15* Δ*lacX74 recA1 endA1 araD139*Δ(*ara, leu*)*7697 galU galK* λ^{-} *rpsL nupG*.

2.4.2 Second Step Cloning (VH)

1. NcoI.
2. XhoI.
3. XL1-Blue MRF' Competent Cells (Agilent Technologies, USA), genotype: *recA1 endA1 gyrA96 thi-1 hsdR17 supE44 relA1 lac* [F′ *proAB lacI*^{q} *Z*Δ*M15* Tn*10* (Tet^{r})].

2.5 Colony PCR

1. DreamTaq DNA Polymerase (Thermo Scientific, USA).
2. Forward primer, LMB3: 5′ – CAG GAA ACA GCT ATG AC – 3′.
3. Reverse primer, pIII: 5′ – TTA GAT CGT TAC GCT AAC - 3′.

2.6 scFv Phage Library Packaging

1. 2× YT medium: 16 g tryptone, 10 g yeast extract, and 5 g NaCl in 1 L dH_2O; autoclave and store at room temperature.
2. 2× YT agar: 16 g tryptone, 10 g yeast extract, 5 g NaCl, and 15.5 g agar in 1 L dH_2O; autoclave.
3. 50 mg/mL ampicillin stock solution: 2.5 g ampicillin sodium salt in 50 mL of 50% (v/v) ethanol; filter-sterilize and store at −20 °C.
4. 30 mg/mL kanamycin stock solution: 1.5 g kanamycin sulfate in 50 mL of dH_2O; filter-sterilize and store at −20 °C.
5. 40% (w/v) glucose: 40 g D-(+)-glucose in 100 mL of dH_2O; autoclave and store at 4 °C.
6. 2× YT-amp medium: 2× YT medium, 0.1 mg/ mL ampicillin, and 2% (v/v) glucose.
7. 2× YT-amp/kan medium: 2× YT medium, 0.1 mg/mL ampicillin, and 0.06 mg/mL kanamycin.
8. 2× YT-amp agar: 2× YT agar, 0.1 mg/mL ampicillin, and 2% (v/v) glucose.
9. 2× YT-kan agar: 2× YT agar and 0.06 mg/mL kanamycin.
10. M13K07 helper phage (Invitrogen, USA).
11. PEG/NaCl solution: 200 g polyethylene glycol 6000 (PEG 6000) and 146 g NaCl in 1 L dH_2O; autoclave and store at room temperature.
12. Phosphate-buffered saline (PBS), pH 7.4: 8 g NaCl, 0.2 g KCl, 1.44 g $Na_2HPO_4{\cdot}2H_2O$, 0.24 g KH_2PO_4 in 1 L dH_2O; autoclave and store at room temperature.

2.7 Phage Titration

1. 2× YT medium: 16 g tryptone, 10 g yeast extract, and 5 g NaCl in 1 L dH_2O; autoclave and store at room temperature.
2. 2× YT agar: 16 g tryptone, 10 g yeast extract, 5 g NaCl, and 15.5 g agar in 1 L dH_2O; autoclave.
3. 2× YT-amp agar: 2× YT agar, 0.1 mg/mL ampicillin, and 2% (v/v) glucose.
4. 2× YT-kan agar: 2× YT agar and 0.06 mg/mL kanamycin.
5. Phosphate-buffered saline (PBS), pH 7.4: 8 g NaCl, 0.2 g KCl, 1.44 g $Na_2HPO_4{\cdot}2H_2O$, 0.24 g KH_2PO_4 in 1 L dH_2O; autoclave and store at room temperature.
6. TG1 Electrocompetent Cells, genotype: *supE thi-1 Δ(lac--proAB) Δ(mcrB-hsdSM)5(rK - mK -)* [F′ *traD36 proAB lacI^{q}Z ΔM15]*.

2.8 scFv Library Biopanning

2.8.1 scFv Selection

1. Corning® 96-well clear polystyrene high bind Stripwell™ microplate.
2. Coating buffer, pH 9.6: 1.59 g Na_2CO_3 and 2.93 g $NaHCO_3$ in 1 L dH_2O; store at 4 °C.
3. Phosphate-buffered saline (PBS), pH 7.4: 8 g NaCl, 0.2 g KCl, 1.44 g $Na_2HPO_4{\cdot}2H_2O$, 0.24 g KH_2PO_4 in 1 L dH_2O; autoclave and store at room temperature.
4. PBST: 1 mL Tween 20 in 1 L PBS.
5. PTM blocking buffer: 2 g skim milk in 100 mL PBST (*see* **Note 2**).
6. Trypsin (10 μg/mL): 10 μg trypsin in 1 mL dH_2O; store at −20 °C.
7. Ampicillin stock solution (50 mg/mL): 2.5 g ampicillin sodium salt in 50 mL of 50% (v/v) ethanol; filter-sterilize and store at −20 °C.
8. Kanamycin stock solution (30 mg/mL): 1.5 g kanamycin sulfate in 50 mL of dH_2O; filter-sterilize and store at −20 °C.
9. 40% (w/v) glucose: 40 g D-(+)-glucose in 100 mL of dH_2O; autoclave and store at 4 °C.
10. 2× YT medium, pH 7.2: 16 g tryptone, 10 g yeast extract, and 5 g NaCl in 1 L dH_2O; autoclave and store at room temperature.
11. 10× amp: 1 mg/mL ampicillin and 20% (v/v) glucose in PBS.
12. 2× YT-amp: 2× YT medium with 0.1 mg/mL ampicillin and 2% (v/v) glucose.
13. 2× YT-amp/kan: 2× YT medium with 0.1 mg/mL ampicillin and 0.06 mg/mL kanamycin.

14. XL1-Blue Competent Cells (Agilent Technologies, USA), genotype: *recA1 endA1 gyrA96 thi-1 hsdR17 supE44 relA1 lac* [F′ *proAB lacI*^{q} *Z*Δ*M15* Tn*10* (Tet^r)].

2.8.2 Polyclonal and Monoclonal Phage ELISA

1. Bovine serum albumin (BSA) Fraction V.
2. PBST: 1 mL Tween 20 in 1 L PBS.
3. PTM blocking buffer: 2 g skim milk in 100 mL PBST (*see* **Note 2**).
4. Anti-M13 horseradish peroxidase (HRP): Prepare at 1:5000 dilution in PTM blocking buffer.
5. 2,2′-azino-bis(3-ethylbenzothiazoline-6-sulphonic acid) (ABTS) Tablets.
6. ABTS buffer.

3 Methods

3.1 Total Lymphocyte Isolation

The antibody repertoire used for the construction of a human naïve library is obtained from blood samples collected from healthy donors. A set of inclusion and exclusion criteria for donors are included in the notes (*see* **Note 3**). Changes to the criteria can be made depending on your specific requirements:

1. EDTA-treated fresh blood is diluted with PBS at a ratio of 1:1.
2. Layer 10 mL of whole blood on top of 7.5 mL of Ficoll-Paque™ PLUS (*see* **Note 4**) with care, and centrifuge at 1200 × *g* for 15 min (min) at 18 °C.
3. Aspirate lymphocytes layered between Ficoll-Paque™ PLUS and plasma carefully, and remove to a sterile 50 mL polypropylene tube.
4. The isolated lymphocytes are diluted with PBS to a final volume of 50 mL and centrifuge at 250 × *g* for 10 min at 18 °C.
5. The supernatant is removed using a clean filter tip.
6. Repeat the washing step with PBS once.
7. The isolated lymphocytes are then subjected to RNA extraction using a total RNA extraction kit based on the manufacturer's protocols (*see* **Note 5**).
8. Keep extracted total RNA at −80 °C until use. The quality of RNA can be determined before use.

3.2 First-Strand cDNA Synthesis

1. First-strand cDNA is synthesized using SuperScript™ II Reverse Transcriptase according to the manufacturer's procedure.
2. Store the cDNA at −20 °C until use.

3.3 Amplification of Heavy Chain and Light Chain Gene Repertoire

The scFv fragment consists of the variable region of heavy (VH) and light (VL) chain linked by a flexible peptide linker. The cDNA is used as the template to amplify both the VH and VL genes for library cloning. The diverse V genes available for both VH and VL means different primers are used for the repertoire amplification. A set of primers consisting of seven VH primers, nine Vλ primers, and four Vκ primers are used for this purpose (*see* Table 1).

1. Each PCR reaction is prepared consisting of 200 ng of cDNA, 200 μM dNTPs, 0.2 μM Fw and Rv primers (Table 1), 2 μL of *Pfu* buffer with $MgSO_4$, and 0.5 U *Pfu* DNA Polymerase. The reaction is topped up with dH_2O to a final volume of 20 μL.
2. The specific PCR program used is as follows: initial denaturation at 95 °C (2 min), 30 cycles of amplification with denaturation at 95 °C (30 s per cycle), annealing at 55 °C or 62 °C (30 s), elongation at 72 °C (45 s), and a final elongation at 72 °C (5 min) (*see* **Note 6**).
3. The amplified PCR product is separated on 1.2% TBE agarose gel using gel electrophoresis at 110 V for 50 min. Excise the corresponding band (~400 bp for VH and 650 bp for LC) and gel extract using QIAquick® Gel Extraction Kit according to the manufacturer's protocol (*see* **Note 7**).
4. Determine the concentration of each purified PCR product and continue to second amplification. The second amplification will introduce the restriction endonuclease site. Alternatively, store the DNA at −20 °C until use.
5. For the second amplification, each PCR reaction is prepared with 20 μg of the purified VH or LC, 200 μM dNTPs, 0.2 μM Fw and Rv primers (Table 2), 2 μL of ThermoPol® buffer, and 0.4 U Vent® DNA Polymerase. Top up the reaction to 20 μL using dH_2O. At this stage, the LC will be amplified with the Rv primers to shorten the fragment to yield only the VL domain.
6. The PCR program is as follows: initial denaturation at 95 °C (2 min), 30 cycles of amplification with denaturation at 95 °C (30 s per cycle), annealing at 55 °C or 62 °C (30 s), elongation at 72 °C (45 s), and a final elongation at 72 °C (5 min) (*see* **Note 6**).
7. Separate the amplified PCR product on 1% TBE agarose gel using gel electrophoresis at 110 V for 50 min. Excise the band (~400 bp for both VH and VL) and extract using QIAquick® Gel Extraction Kit according to the manufacturer's protocol.
8. The purified DNA products are then pooled according to the subfamilies in a set of five donors each. Determine the concentration and store the DNA at −20 °C until use.

3.4 Two-Step Cloning

A two-step cloning strategy is used for the library cloning. The first cloning step will generate a mini library containing the VL (λ and κ) repertoire using DH10β cells, whereas the second cloning step will introduce the VH repertoire to the VL-mini library. This will result in a final scFv library using XL1-Blue cells. The schematic diagram of the scFv library construction process is outlined in Fig. 1.

3.4.1 First Step Cloning (VL)

1. Double digest the pLABEL vector and pooled Vλ or Vκ with SalI and NotI overnight at 37 °C. Heat-inactivate the reaction at 65 °C for 20 min (*see* **Note 8**).
2. Add 5 U of alkaline phosphatase only to the double digested pLABEL vector and incubate at 37 °C for 1 hour (h). Heat-inactivate the reaction at 80 °C for 2 min.
3. Separate the digested vector on 1% TBE agarose gel using gel electrophoresis at 110 V for 50 min. Excise the band of interest, and extract the DNA using QIAquick® Gel Extraction Kit according to the manufacturer's protocol.
4. The digested Vλ and Vκ fragments are purified using QIAquick® PCR Purification Kit according to the manufacturer's protocol.
5. Ligate digested pLABEL and digested Vλ or Vκ pools at 1: 2 ratio using T4 DNA Ligase (*see* **Note 9**). Incubate the ligation reaction overnight at 16 °C. Heat-inactivate the reaction at 65 °C for 10 min.
6. Precipitate the ligated DNA using 2.5 volume of ethanol and 0.1 volume of 3 M sodium acetate. Incubate the DNA in - 80 °C for 1 h or alternatively flash frozen in liquid nitrogen, followed by centrifugation for 20 min at 14,000 × *g*.
7. Wash the DNA pellet with 500 μL of 70% (v/v) ethanol and centrifuge again at 14,000 × *g* for 20 min.
8. Dissolve the DNA pellet with 4 μL dH_2O per ligation reaction.
9. Thaw DH10β cells on ice for 2 min and mix with 2 μL DNA.
10. Transfer the mixture to a pre-chilled 0.1 cm electroporation cuvette. Transform the mixture at 1.7 kV using an electroporator.
11. Resuspend the electroporated mixture with 1 mL of pre-warmed 2× YT medium, and transfer the suspension to a 1.5 mL microcentrifuge tube. Incubate the cell suspension for 1 h at 37 °C and 700 rpm.
12. Pool the cell suspension according to subfamilies and take 10 μL of cells to determine the cloning efficiency. Dilute the 10 μL cells in 90 μL 2× YT-amp and plate out on 2× YT-amp agar plates in 90 mm petri dish.

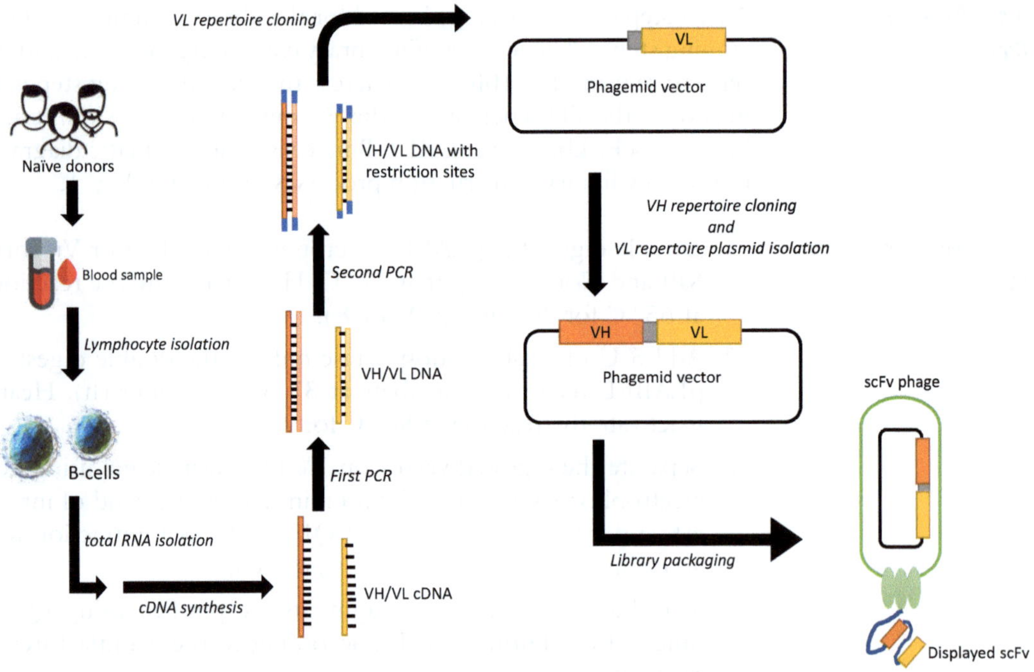

Fig. 1 A schematic diagram of the scFv naïve library construction process. Total RNA is isolated from total lymphocytes of healthy donors. cDNA is synthesized via reverse transcription. First PCR amplification is performed to produce the VH and VL repertoire. A second PCR amplification is down to introduce restriction enzyme sites to the VH and VL repertoire. A two-step cloning is done by cloning the VL repertoire first into the phagemid vector, followed by the cloning of VH repertoire. The phagemid vector containing the V genes is transformed and pooled to generate a glycerol stock of the scFv naïve phage library

13. Plate out the remaining pooled cell suspension on 30 BioAssay Dish with 2× YT-amp agar. Incubate the agar plates overnight at 37 °C.
14. Scrape the colonies on the BioAssay Dish with 2× YT-amp.
15. Estimate the library diversity by titrating the scraped library stock (*see* Subheading 3.4.3).
16. Prepare glycerol stocks of the VL-mini library (*see* Subheading 3.4.4).
17. Perform a colony PCR using colonies on the 90 mm petri dish to confirm successful cloning of the VL repertoire (*see* Subheading 3.5) (*see* **Note 10**).

3.4.2 Second Step Cloning (VH)

1. Culture a tube of VL-mini library glycerol stock overnight in 500 mL of 2× YT-amp at 37 °C with shaking at 200 rpm.
2. Extract the plasmid using QIAGEN Plasmid Maxi Kit according to the manufacturer's protocol.

3. Double digest the VL-mini library and VH with NcoI and XhoI overnight at 37 °C. Heat-inactivate the reaction at 65 °C for 20 min (*see* **Note 8**).
4. Perform the digestion and ligation as in the first step cloning procedure (*see* Section 3.4.1).
5. Transform the scFv library into XL1-Blue MRF' cells.
6. Pool the cell suspension according to subfamilies and take 10 μL of cells to determine the cloning efficiency. Dilute 10 μL of cells in 90 μL 2× YT-amp and plate out on 2× YT-amp agar plates in 90 mm petri dish.
7. Plate out the remaining pooled cell suspension on 40 BioAssay Dish with 2× YT-amp agar. Incubate the agar plates overnight at 37 °C.
8. Scrape the colonies on BioAssay Dish with 2× YT-amp.
9. Estimate the library diversity by titrating the scraped library stock (*see* Subheading 3.4.3) (*see* **Note 11**).
10. Prepare glycerol stock of the scFv library (*see* Subheading 3.4.4).
11. Perform a colony PCR using colonies on the 90 mm petri dish to confirm successful cloning of the VH repertoire (*see* Section 3.5) (*see* **Note 10**).

3.4.3 Library Size Estimation

1. Take 10 μL of cells from the scraped library suspension to perform a tenfold serial dilution to 10^{-13}.
2. Spot 10 μL of the dilution on the 2× YT-amp agar plate and incubate overnight at 37 °C.
3. Count the library size based on the formula:

$$\text{Library size} = \frac{\text{Number of colonies} \times \text{Dilution factor}}{\text{Total volume}}$$

3.4.4 Preparation of Bacteria Library Stock

1. Pellet down the colonies scraped from the BioAssay Dish at 4500 × *g*, 10 min.
2. Resuspend the pellet with adequate amount of 2× YT-amp, and add 80% (v/v) glycerol to the cell suspension to make a 20% glycerol library stock (*see* **Note 12**).
3. Aliquot the final library suspension into 2 mL cryogenic vial and keep in −80 °C until use.

3.5 Colony PCR

1. Pick single colonies randomly and resuspend in 10 μL dH_2O.
2. Set up the PCR reaction using 2 μL of colony supernatant, 200 μM dNTPs, 0.5 μM LMB3 Fw and pIII Rv primers, 1× DreamTaq buffer, and 0.5 U DreamTaq DNA Polymerase. Top up the reaction to 20 μL with dH_2O.

3. Set the PCR program as follows: initial denaturation at 95 °C (5 min), 20 cycles of amplification with denaturation at 95 °C (30 s/cycle), annealing at 55 °C (30 s), elongation at 72 °C (2 min), and a final elongation at 72 °C (5 min).
4. Separate the amplified PCR product on 1% TBE agarose gel using gel electrophoresis run at 110 V for 50 min (*see* **Notes 10** and **13**).
5. Send the colonies with the correct band size for sequencing.
6. Analyze the DNA sequences for V,D,J gene usage and mutations using IMGT/V-quest (www.imgt.org/IMGT_vquest/vquest) (*see* **Note 14**).

3.6 scFv Phage Library Packaging

1. Thaw a tube of the library glycerol stock and inoculate into 500 mL of 2× YT-amp. Culture at 37 °C with shaking at 200 rpm until OD_{600} ~0.5.
2. Infect 250 mL of the culture with 10^{12} M13KO7 helper phage, following the multiplicity of infection (MOI) of 1:20. Incubate the culture static at 37 °C for 30 min.
3. Pellet down the culture at 3500 × *g* for 10 min. Remove the supernatant (*see* **Note 15**).
4. Resuspend the pellet with 300 mL 2× YT-amp/kan. Culture overnight at 30 °C, 200 rpm.
5. Pellet down the cells at 10,000 × *g* for 10 min. Transfer the phage-containing supernatant to a new 50 mL polypropylene tube.
6. Add 1/5 volume of PEG/NaCl solution to 4/5 volume of supernatant. Mix well and incubate on ice for 1 h (*see* **Note 16**).
7. Centrifuge the mixture at 10,000 × g for 30 min at 12 °C, remove the supernatant, and resuspend the phage pellet in 8 mL PBS.
8. Add 2 mL PEG/NaCl solution, mix well and incubate on ice for 20 min.
9. Centrifuge the mixture at 10,000 × g for 30 min at 12 °C. Remove the supernatant, short spin briefly and pipette out the remaining PEG/NaCl solution.
10. Resuspend the phage pellet in 2 mL PBS.
11. Centrifuge at 10,000 × *g* for 10 min to remove remaining bacterial cells from the phage precipitation. Repeat centrifugation until no bacteria pellet is observed.
12. Take 10 μL of the phage to check the titer of the packaged scFv library (*see* Subheading 3.7).
13. Store the scFv antibody library at 4 °C until use (*see* **Note 17**).

3.7 Phage Titration

1. Dilute 10 μL of scFv phage library in 90 μL of PBS and perform a tenfold serial dilution.
2. Add 100 μL of TG1 or XL1-Blue MRF' cells at OD_{600} ~0.5 to the diluted phage.
3. Spot 10 μL of the infected cells on 2× YT-amp agar plate and 2× YT-kan agar plate. Incubate overnight at 37 °C (*see* **Note 18**).
4. Calculate the phage titer using the formula:

$$\text{Phage titer (cfu/mL)} = \frac{\text{Number of colonies} \times \text{Dilution factor}}{\text{Total volume spotted}}$$

3.8 scFv Library Biopanning

The collection of displayed antibody fragments can be screened by biopanning. The number of biopanning rounds depends on the phage discovery rates. The rate will increase with each subsequent round of biopanning. Generally, 3–4 rounds are sufficient to produce a suitably enriched subpopulation of phage-displaying proteins. The schematic representation of the biopanning procedure is shown in Fig. 2.

3.8.1 scFv Selection

1. Coat a microtiter well with 1–10 μg of antigen using coating buffer at a final volume of 100 μL. Reduce the amount of antigen for subsequent biopanning rounds to increase the selection stringency. Incubate overnight at 4 °C (*see* **Note 19**).
2. Wash the antigen-coated well 3× with 300 μL PBST using a ELISA plate washer (*see* **Note 20**).
3. Block the antigen-coated well with 300 μL of PTM blocking buffer for 1 h at room temperature with constant shaking. The wells are sealed with a breathable tape. The wells are then washed with 300 μL PBST for 3×.
4. Simultaneously, dilute 10^{11} phage particles in PTM blocking buffer at final volume of 200 μL, and incubate for 1 h on a blocked microtiter well for pre-absorption (*see* **Note 21**).
5. Transfer 100 μL of the pre-absorbed phage to the antigen-coated well, and incubate for 2 h at room temperature with constant shaking at 700 rpm.
6. Remove unbound and unspecific phage particles by washing 10× with PBST. An additional ten wash steps are introduced for every subsequent biopanning round.
7. Elute the bound phages with 100 μL of 10 μg/mL trypsin, and incubate static at 37 °C for 30 min (*see* **Note 22**).
8. Transfer the eluted phages to a new 2 mL microcentrifuge tube, and add 100 μL of XL1-Blue MRF' cells at OD_{600} ~0.5. Incubate the cells static at 37 °C for 30 min followed by 30 min of incubation with shaking at 700 rpm.

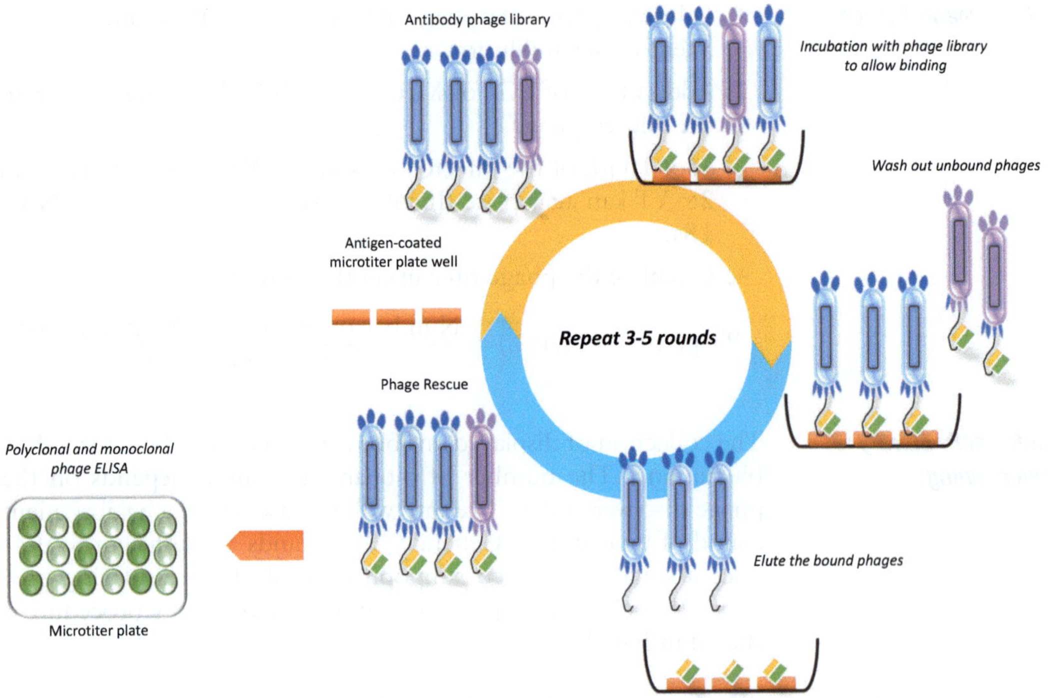

Fig. 2 A schematic representation of biopanning process. Biopanning involves 3–5 consecutive rounds of binding, washing, and elution of phages for rescue. ELISA is carried out to determine the enrichment of unique phage clones

9. Take 10 μL of infected culture to determine the rescued phage titer (*see* Subheading 3.7).
10. Add 20 μL of 10× amp to the remaining culture and culture overnight at 37 °C, 700 rpm.
11. Inoculate 190 μL of 2× YT-amp with 10 μL of the overnight culture. Culture at 37 °C with shaking at 700 rpm for 2.5 h.
12. Mix the remaining overnight culture with 65 μL of 80% (v/v) glycerol, and store the glycerol stock in −80 °C.
13. Infect the culture with 10^{11} M13KO7 helper phage, and incubate static at 37 °C for 30 min.
14. Pellet down the culture at 3500 × *g* for 10 min. Resuspend the cell pellet with 230 μL of 2× YT-amp/kan. Culture overnight at 30 °C with shaking at 700 rpm.
15. Pellet down the overnight culture at 3500 × *g* for 10 min.
16. Transfer the supernatant to new clean tube. Take 100 μL of the phage for subsequent round of biopanning and keep the remaining phage at 4 °C for polyclonal ELISA. The amplified phage titer from each round of biopanning is determined (*see* Subheading 3.7) (*see* **Note 23**).

3.8.2 Polyclonal and Monoclonal Phage ELISA

In a successful biopanning campaign, an obvious enrichment is observed in polyclonal ELISA of subsequent biopanning rounds. It is crucial to choose the best biopanning round in order to isolate different monoclonal antibodies with good specificity and diversity:

1. Coat appropriate number of microtiter wells with 1–10 μg of antigen using coating buffer at a final volume of 100 μL. Incubate overnight at 4 °C. Concurrently, coat equal number of empty wells with BSA at the same amount as the negative controls to observe for nonspecific binders.
2. Wash the coated wells 3× with PBST using ELISA plate washer (*see* **Note 20**).
3. Block the coated wells with 300 μL of PTM blocking buffer for 1 h at room temperature shaking at 700 rpm, and then wash 3× with PBST.
4. Dilute 50 μL phage with 50 μL PTM blocking buffer; add the 100 μL mixture to corresponding antigen-coated wells and the control wells. Incubate for 2 h at room temperature with shaking at 700 rpm.
5. Wash the wells 3× with PBST.
6. Add 100 μL of anti-M13-HRP to each antigen-coated wells and control wells. Incubate for 1 h at room temperature with shaking at 700 rpm.
7. Wash the wells 3× with PBST.
8. Develop the wells with 100 μL ABTS solution for 30 min in the dark. Measure the readings at 405 nm using a microtiter plate spectrophotometer.
9. After determining the biopanning round for monoclonal selection, take 10 μL of the remaining phage to perform a tenfold serial dilution and infect 100 μL of XL1-Blue MRF' cells of OD_{600} ~0.5 at 37 °C, static. Plate out the infected cell on 2× YT-amp agar plate.
10. Pick 92 monoclonal antibody clones randomly to inoculate 200 μL of 2× YT-amp. Culture overnight at 37 °C, 1250 rpm.
11. Package the monoclonal scFv antibody as described in **steps 11–15** in Subheading 3.8.1.
12. Perform monoclonal phage ELISA using the procedures as described in **steps 1–8** in this section.
13. Identify positive clones with good signal/noise ratio for colony PCR (*see* Subheading 3.5) and send for DNA sequencing analysis.

4 Notes

1. Other high fidelity proof-reading polymerase can be used as an alternative.
2. Bovine serum albumin (BSA), casein, ovalbumin, gelatin, and any other suitable blocking agents can be utilized as an alternative. It is required for the blocking agent to be made fresh each time to prevent microbial contamination.
3. Sampling requirement:
   (a) Inclusion criteria: Donors with healthy backgrounds, received standard common vaccination, no infection in the past 6 months prior to collection date, and not on immunosuppressors or other medications.
   (b) Exclusion criteria: Donors with family backgrounds of severe illnesses, suffering from autoimmune disorder, having ongoing infection, experiencing symptoms of infection or on medication including antibiotics and immunosuppressors within a month from date of blood collection.
4. Care must be taken to ensure that the blood layer is not mixed with the Ficoll-Paque™ PLUS solution to prevent aggregation of erythrocytes which can reduce the yield of lymphocytes as lymphocytes trapped in the aggregate sediment with erythrocytes to the bottom of tube.
5. Integrity and concentration of the total RNA extracted from lymphocytes can be analyzed using Bioanalyzer Instrument (Agilent, USA).
6. The VH and VL repertoire is amplified from each donor independently to avoid bias and loss of repertoire caused by sample pooling. This step ensures better repertoire diversity.
7. VH fragment is about 400 base pair (bp); however, the rearrangement of V-genes might introduce bands of varied sizes approximately 400 bp. The targeted band must be excised carefully.
8. The number of digestion reaction can be increased as needed to achieve enough DNA for the following cloning procedure. In case of different restriction endonucleases with incompatible reaction buffers are used, a sequential digestion may be required as digestion efficiency is affected by the salt content in the buffer.
9. A naïve library size for the heavy chain repertoire should reach about 10^7–10^9 or higher as the variable heavy (VH) chain is the predominant region for antigen binding. To achieve a highly diverse library, multiple ligation reactions may be required.

10. The expected band size of colony PCR is approximately 750 bp for the first step cloning and 1200 bp for the second step cloning.
11. The final naïve library size should be between 10^9 and 10^{12} or higher.
12. The library stock is prepared to ensure when the tube of stock is thawed into 500 mL of 2× YT-amp during library packaging, it will have a starting OD_{600} ~0.1.
13. It is preferred to have inserts for both VL-mini library and full scFv library with a cloning efficiency of more than 80%.
14. Deep sequencing can also be used to determine the diversity of the library repertoire.
15. It is required to pellet down the cells and to remove the glucose-containing medium in order to discard excess glucose which will suppress the expression of the scFv::pIII fusion protein.
16. The enhancement of the number of phage particles precipitated by the PEG/NaCl solution can be done by extending the ice-cold incubation period to 2 h.
17. Store phage preparation for no longer than 4 weeks at 4 °C. Ensure the phage to be packaged freshly to avoid loss of binding brought on by proteolysis of displayed antibodies.
18. Ampicillin agar plate should have the colony numbers of at least two times higher than on kanamycin plate to indicate that less helper phage is being produced alongside the phage library. Additionally, the colony can be used to estimate the size of phage library before biopanning.
19. SDS-PAGE and Western blot are used to determine the purity of the antigens before proceeding to biopanning to ensure total isolation of specific scFv antibody binders.
20. The wells can be washed alternatively using squirt bottle filled with PBST followed by shaking to remove the washing buffer. Tap them several times on dry paper towel to remove residue. Care must be taken to ensure the bubbles during washing are not carried over to the next step.
21. The first biopanning round should employ 100 times as much phage as the total library size. For a library size of 10^9, a total of 10^{11} phage particles should be used.
22. Alternatively, other elution buffers or methods can be employed at this stage. Competitive elution can also be carried out by competing with free antigens. Acid-based elution using glycine-HCL, pH 2.2 can also be used. The best elution method would be dependent on the requirements and preference of each user.

23. To determine the phage recovery for each biopanning round, the rescued phage titer is divided by the input phage titer. The gradual increase of phage recovery should be observed over each cycle of biopanning to indicate the success enrichment of binding clones.

Acknowledgements

This work was supported by a Universiti Sains Malaysia, Special (Matching) Short-Term Grant with Project No: 304. CIPPM.6315708.

References

1. Yeoh SG, Sum JS et al (2019) Potential of phage display antibody technology for cardiovascular disease immunotherapy. J Cardiovasc Transl Res 15(2):360–380
2. Guliy OI, Evstigneeva SS, Dykman LA (2023) Recombinant antibodies by phage display for bioanalytical applications. Biosens Bioelectron 222:114909
3. Chan SK, Lim TS (2017) Immune human antibody libraries for infectious diseases. Adv Exp Med Biol 1053:61–78
4. Ledsgaard L et al (2022) Advances in antibody phage display technology. Drug Discov Today 27(8):2151–2169
5. Bashir S, Paeshuyse J (2020) Construction of antibody phage libraries and their application in veterinary immunovirology. Antibodies 9(2):1–13
6. Mahdavi SZB et al (2022) An overview on display systems (phage, bacterial, and yeast display) for production of anticancer antibodies; advantages and disadvantages. Int J Biol Macromol 208:421–442
7. Lai JY, Lim TS (2020) Infectious disease antibodies for biomedical applications: a mini review of immune antibody phage library repertoire. Int J Biol Macromol 163:640–648
8. Valldorf B et al (2021) Antibody display technologies: selecting the cream of the crop. Biol Chem 403(5–6):455–477
9. Ponsel D et al (2011) High affinity, developability and functional size: the holy grail of combinatorial antibody library generation. Molecules 16(5):3675–3700
10. Manoutcharian K, Gevorkian G (2023) Shark VNAR phage display libraries: an alternative source for therapeutic and diagnostic recombinant antibody fragments. Fish Shellfish Immunol 138:108808
11. Bashir S, Paeshuyse J (2020) Construction of antibody phage libraries and their application in veterinary immunovirology. Antibodies (Basel) 9(2):21
12. Mohd Ali MR et al (2021) Development of monoclonal antibodies against recombinant LipL21 protein of pathogenic Leptospira through phage display technology. Int J Biol Macromol 168:289–300

Chapter 3

Structure-Guided and Phage-Assisted Evolution of Therapeutic Antibodies to Reverse On-Target Point Mutation-Mediated Resistance

Xinlei Zhuang, Shuqing Chen, and Liqiang Pan

Abstract

Resistance to therapeutic antibodies caused by on-target point mutations is a major obstacle in anticancer therapy, creating an "unmet clinical need." To tackle this problem, researchers are developing new generations of antibody drugs that can overcome the resistance mechanisms of existing agents. We have previously reported a structure-guided and phage-assisted evolution (SGAPAE) approach to evolve cetuximab, a therapeutic antibody, to effectively reverse the resistance driven by $EGFR^{S492R}$ or $EGFR^{G465R}$ mutations, without changing the binding epitope or compromising the antibody efficacy. In this protocol, we provide detailed instructions on how to use the SGAPAE approach to evolve cetuximab, which can also be applied to other therapeutic antibodies for reversing on-target point mutation-mediated resistance. The protocol consists of four steps: structure preparation, computational prediction, phage display library construction, and antibody candidate selection.

Key words On-target point mutation, Therapeutic antibody resistance, SGAPEA, Computational prediction, Phage display

1 Introduction

Antibody therapy has become a standard treatment for various cancers in the past 20 years [1]. However, the effectiveness of therapeutic antibodies is often compromised by the emergence of drug resistance [2–4]. One of the most detrimental resistance mechanisms is the on-target point mutation, which prevents antibody binding [5] and limits further applications such as combination therapy. As therapeutic antibodies become more widely used, such resistance issues are also more common. For example, some colorectal cancers acquire resistance to cetuximab by mutating the epidermal growth factor receptor (EGFR) ectodomain within the cetuximab epitope (e.g., $EGFR^{S492R}$, $EGFR^{G465R}$) [6, 7]; some cancers show primary resistance to pertuzumab by mutating the

Huan Peng et al. (eds.), *Phage Engineering and Analysis: Methods and Protocols*,
Methods in Molecular Biology, vol. 2793, https://doi.org/10.1007/978-1-0716-3798-2_3,

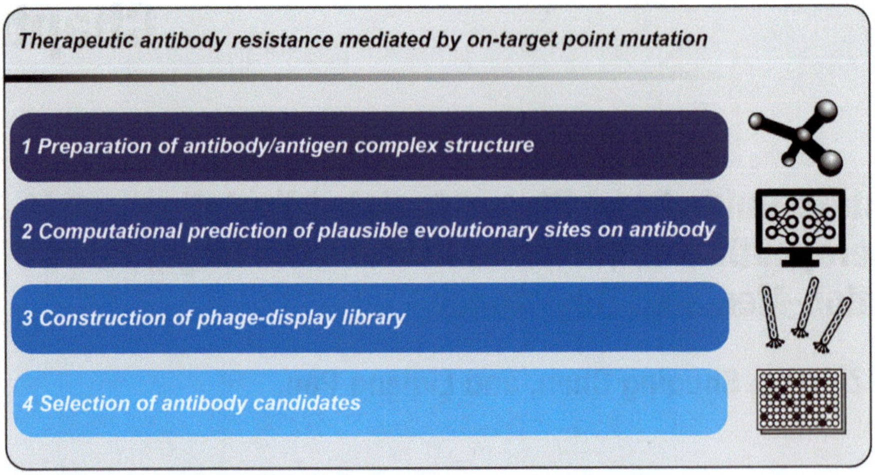

Fig. 1 Flow diagram of structure-guided and phage-assisted evolution (SGAPAE) of therapeutic antibodies to overcome on-target point mutation-mediated resistance

human epidermal growth factor receptor 2 (HER2) extracellular domain (e.g., $HER2^{S310F}$, $HER2^{S310Y}$) [8, 9]. Therefore, developing an efficient strategy to overcome the antibody drug resistance mediated by on-target point mutation is of great importance.

To overcome the on-target point mutation-driven resistance to therapeutic antibodies, current strategies mainly rely on new antibody combinations that target nonoverlapping epitopes, e.g., anti-EGFR antibodies MM-151 and Sym004 [10, 11], which are still in clinical phase I/II [12, 13]. However, developing a new antibody drug is often time-consuming and costly, taking up to a decade and a billion dollars. If new antibodies with excellent properties, druggability, and target functions can be obtained by minor engineering based on existing therapeutic antibodies, the development process of new antibody drugs can be greatly accelerated. Therefore, we have developed a structure-guided and phage-assisted evolution (SGAPAE) approach (Fig. 1) for efficient evolution of therapeutic antibodies to reverse on-target point mutation-mediated resistance [14].

Here, as an example, we applied SGAPAE approach to evolve cetuximab to reverse resistance mediated by $EGFR^{S492R}$ or $EGFR^{G465R}$, the two most common antibody-resistance point mutations in cetuximab treatment (Fig. 2). We used the Rosetta platform [15, 16] to calculate the energy difference between the bound and unbound states of the interface and identify key interface residues that are essential for reestablishing the $EGFR^{S492R}$/cetuximab or $EGFR^{G465R}$/cetuximab interaction by scanning all residues on the interface. We then developed a semi-rationally designed library of cetuximab mutants with a restricted epitope. The focused library enabled us to efficiently identify cetuximab variants with minimal mutations that can reverse $EGFR^{S492R}$- or

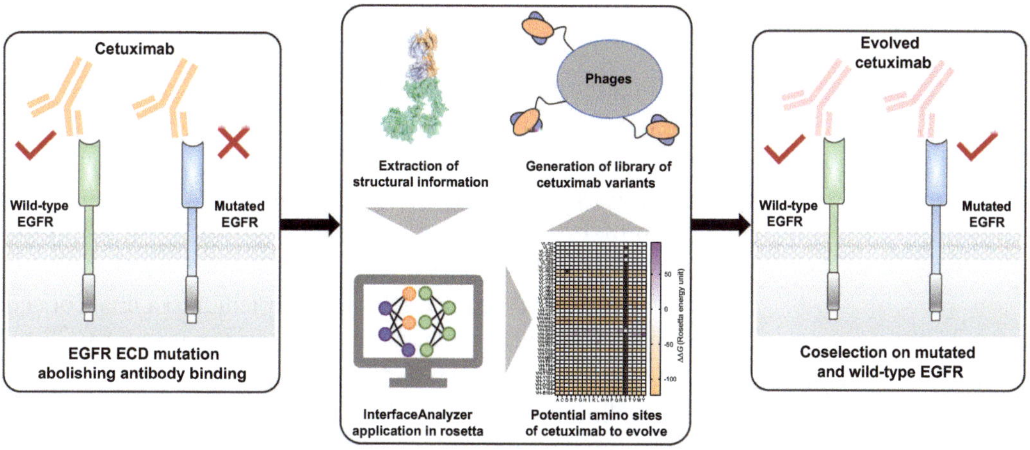

Fig. 2 The application of the SGAPAE approach for evolving cetuximab to overcome resistance mediated by point mutations in the EGFR extracellular domain

$EGFR^{G465R}$-driven resistance to cetuximab while largely preserving the optimized properties and druggability of cetuximab. The size of the antibody mutant library was reduced significantly to 10^6~10^7 phages due to the limited key residues for randomization, which showed an advantage over conventional phage display libraries with a size of 10^9~10^{12} [17–19] phages in biased screening processes. The diversity of the semi-rationally designed antibody mutant library was confined within the binding epitope of the parental antibody structure, providing a clear evolutionary direction toward antigen and retaining a highly similar binding site in the selected antibody variants. Combined with structural information and computational predictions, our SGAPAE approach could enhance the screening efficiency for effective antibody variants. The binding epitope of the parental antibody is usually related to antibody efficacy. With the epitope-restricted antibody mutant phage display library, we discovered that minimal or even single point mutations in the antibody CDR could fully reverse on-target point mutation-mediated antibody drug resistance without changing the binding epitope or weakening antibody efficacy. Moreover, the evolved antibody variants with one or two amino acid substitutions in the CDR would not affect the rest of the antibody domains, inheriting the optimized properties of therapeutic antibody to the largest possible extent. The SGAPAE approach thus provides an efficient and feasible strategy to overcome on-target point mutation-mediated antibody drug resistance while maintaining the druggability of the parental antibodies, and should be a promising approach for the evolution of therapeutic antibodies.

2 Materials

1. $MgCl_2$ solution: Dissolve $MgCl_2$ to a final concentration of 1 M in water and sterilize using a 0.22-μm syringe filter. Store at 4 °C.
2. Ampicillin stock solution:Dissolve ampicillin to a final concentration of 100 mg/mL in water and sterilize using a 0.22-μm syringe filter. Store at −20 °C.
3. Kanamycin stock solution:Dissolve ampicillin to a final concentration of 50 mg/mL in water and sterilize using a 0.22-μm syringe filter. Store at −20 °C.
4. Tetracycline stock solution:Dissolve tetracycline to a final concentration of 50 mg/mL in water and sterilize using a 0.22-μm syringe filter. Store in the dark at −20 °C.
5. SOC liquid media: To 600 mL of water, add tryptone powder (20 g), yeast extract powder (5 g), NaCl (0.5 g), and KCl (0.186 g), and then autoclave to sterilize at 121.0 °C. To 400 mL of water, add 3.96 g glucose and then autoclave to sterilize at 121.0 °C. Allow to cool completely. Mix together above two solutions and add 1 mL of 1 M $MgCl_2$ solution to make complete SOC liquid media. Store in the dark at 4 °C.
6. 2×YT liquid media: Add tryptone powder, yeast extract powder, and NaCl to a final concentration of 16 g/L, 10 g/L, and 5 g/L, respectively, in water. Mix to dissolve and then autoclave to sterilize at 121.0 °C. Store at 4 °C.
7. 2×YT+G liquid media: To 600 mL of water, add tryptone powder (16 g), yeast extract powder (10 g), and NaCl (5 g), and then autoclave to sterilize at 121.0 °C. To 400 mL of water, add 25 g glucose and then autoclave to sterilize at 121.0 °C. Allow to cool completely. Mix together above two solutions to make complete 2× YT+G liquid media. Store at 4 °C.
8. 2×YT+tet liquid media: Mix together 2×YT liquid media and 0.1% (vol/vol) tetracycline stock solution to make complete 2×YT+tet liquid media. Prepare when using.
9. 2×YT+tet+K liquid media: Mix together 2×YT liquid media, 0.1% (vol/vol) tetracycline stock solution and 0.1% (vol/vol) kanamycin stock solution to make complete 2×YT+tet+K liquid media. Prepare when using.
10. 2×YT+tet+K+A liquid media: Mix together 2×YT liquid media, 0.1% (vol/vol) tetracycline stock solution, 0.1% (vol/-vol) kanamycin stock solution, and 0.1% (vol/vol) ampicillin stock solution to make complete 2×YT+tet+K+A liquid media. Prepare when using.

11. 2×YT+G+K liquid media: Mix together 2×YT+G liquid media and 0.1% (vol/vol) kanamycin stock solution to make complete 2×YT+G+K liquid media. Prepare when using.
12. 2×YT+G+K+A liquid media: Mix together 2×YT+G liquid media, 0.1% (vol/vol) kanamycin stock solution and 0.1% (vol/vol) ampicillin stock solution to make complete 2×YT +G+ K+A liquid media. Prepare when using.
13. 2×YT agar: Add tryptone powder, yeast extract powder, NaCl, and agar to a final concentration of 16 g/L, 10 g/L, 5 g/L, and 1.5% (wt/vol), respectively, in water. Mix to dissolve and then autoclave to sterilize at 121.0 °C. Allow to cool appropriately. Pour into Petri dishes and allow to solidify. Store at 4 °C.
14. 2×YT+tet agar:Add tryptone powder, yeast extract powder, NaCl, and agar to a final concentration of 16 g/L, 10 g/L, 5 g/L, and 1.5% (wt/vol), respectively, in water. Mix to dissolve and then autoclave to sterilize at 121.0 °C. Allow to cool appropriately. Mix together above solution and 0.1% (vol/vol) tetracycline stock solution, and pour into Petri dishes and allow to solidify. Store in the dark at 4 °C.
15. 2×YT+G agar:To 60 mL of water, add tryptone powder (1.6 g), yeast extract powder (1.0 g), NaCl (0.5 g), and agar (1.5 g), and then autoclave to sterilize at 121.0 °C. To 40 mL of water, add 2.5 g glucose and then autoclave to sterilize at 121.0 °C. Allow to cool appropriately. Mix together above two solutions and pour into Petri dishes and allow to solidify. Store at 4 °C.
16. 2×YT+G+A agar: To 60 mL of water, add tryptone powder (1.6 g), yeast extract powder (1.0 g), NaCl (0.5 g), and agar (1.5 g), and then autoclave to sterilize at 121.0 °C. To 40 mL of water, add 2.5 g glucose and then autoclave to sterilize at 121.0 °C. Allow to cool appropriately. Mix together above two solutions and 0.1% (vol/vol) ampicillin stock solution and pour into Petri dishes and allow to solidify. Store at 4 °C.
17. 2×YT+G+A+tet agar: To 60 mL of water, add tryptone powder (1.6 g), yeast extract powder (1.0 g), NaCl (0.5 g), and agar (1.5 g), and then autoclave to sterilize at 121.0 °C. To 40 mL of water, add 2.5 g glucose and then autoclave to sterilize at 121.0 °C. Allow to cool appropriately. Mix together above two solutions, 0.1% (vol/vol) ampicillin stock solution, 0.1% (vol/vol) tetracycline stock solution, and pour into Petri dishes and allow to solidify. Store in the dark at 4 °C.
18. 2×YT+G+A+K agar: To 60 mL of water, add tryptone powder (1.6 g), yeast extract powder (1.0 g), NaCl (0.5 g), and agar (1.5 g), and then autoclave to sterilize at 121.0 °C. To 40 mL of water, add 2.5 g glucose and then autoclave to sterilize at 121.0 °C. Allow to cool appropriately. Mix together above two

solutions, 0.1% (vol/vol) ampicillin stock solution and 0.1% (vol/vol) kanamycin stock solution, and pour into Petri dishes and allow to solidify. Store at 4 °C.

19. PEG–NaCl solution: Add PEG-8000 and NaCl to a final concentration of 200 g/L and 146.2 g/L in water, and autoclave to sterilize at 121.0 °C. Store in the dark at 4 °C.
20. Elution buffer (pH 2.2): Dissolve glycine to a final concentration of 15 g/L in water. Adjust pH and sterilize using a 0.22-μ m syringe filter. Store at 4 °C.
21. Neutralization buffer (pH 8.0): Dissolve Tris-HCl to a final concentration of 240 g/L in water. Adjust pH and sterilize using a 0.22-μm syringe filter. Store at 4 °C.
22. PBS (pH 7.4): Add NaCl, KCl, $Na_2HPO_4 \cdot 12H_2O$, KH_2PO_4 to water at final concentrations of 8 g/L, 0.2 g/L, 2.89 g/L, and 0.24 g/L, respectively. Store at room temperature (RT).
23. Trypsin–EDTA solution (pH 7.2): Dissolve trypsin and EDTA to a final concentration of 2.5 g/L and 0.2 g/L in PBS. Adjust pH and sterilize using a 0.22-μm syringe filter. Store at 4 °C.
24. PBST: Mix together PBS, 0.1% (vol/vol) TWEEN-20 to make complete PBST. Prepare when using.
25. Milk–PBST: Mix together PBST, 5% (wt/vol) nonfat milk powder to make complete Milk–PBST. Prepare when using.
26. ELISA coating buffer (pH 9.6): Add Na_2CO_3 and $NaHCO_3$ to water at final concentrations of 1.59 g/L and 2.93 g/L, respectively. Store at room temperature.
27. ELISA substrate buffer (pH 4.5–5.0): Add $Na_2HPO_4 \cdot 12H_2O$ and citric acid monohydrate to water at final concentrations of 17.5 g/L and 5.5 g/L, respectively. Store at RT.
28. TMB solution: Dissolve TMB to a final concentration of 6 mg/mL in DMSO. Mix together 250 μL above solution, 20 mL ELISA substrate buffer, and 4 μL 30% H_2O_2 to make complete TMB solution. Prepare in the dark when using.
29. XL-1 blue *E. coli*.
30. Electrocompetent XL-1 blue *E. coli*.
31. Helper-Phage M13KO7.
32. NIH/3T3 cell line.
33. WT-EGFR-NIH/3T3 cell line.
34. S492R-EGFR-NIH/3T3 cell line.
35. G465R-EGFR-NIH/3T3 cell line.
36. WT-EGFR-ECD-Fc protein.
37. S492R-EGFR-ECD-Fc protein.
38. G465R-EGFR-ECD-Fc protein.

39. DNA Gel Extraction Kit.
40. PrimeSTAR® HS DNA Polymerase.
41. Restriction endonucleases: *Not*I; *Sfi*I.
42. 10×CutSmart buffer.
43. T4 DNA Ligase.
44. 10×T4 DNA Ligation Buffer.
45. HRP anti-E tag antibody.
46. Syringe filter, PES, 33 mm (0.22 μm).
47. 1-L Erlenmeyer flask.
48. 50-mL polypropylene tube.
49. 10-mL centrifuge tube.
50. 2-mL EP tube.
51. 96-well microtiter plate.

3 Methods

3.1 Structure Preparation

1. Find the crystal structure of wild-type EGFR/Fab complex for cetuximab from Protein Data Bank (PDB code: 1YY9).
2. Remove waters and other het residues (e.g., NDG, NAG, BMA) in the crystal structure, and add missing loops and hydrogen atoms for the cleaned structure of antigen–antibody complex.
3. Use the *Relax* application in Rosetta for initial refinement.
4. Select the relaxed structure with the lowest energy for the subsequent mutant modeling.

3.2 Computational Prediction

1. Predict structural models for each of $EGFR^{S492R}$/cetuximab and $EGFR^{G465R}$/cetuximab complexes based on the officially provided "calculate_protein_protein_ddg" demo from the Rosetta package.
2. Calculate corresponding $\Delta\Delta G$ via a *RosettaScripts* protocol using the ref2015 energy function.
3. Determine the difference of binding energy (dG_separated) between mutant and wild type of EGFR via *InterfaceAnalyzer* application.
4. Cluster all predicted structures of $EGFR^{S492R}$/cetuximab and $EGFR^{G465R}$/cetuximab according to the interface residues between antigen and antibody.
5. Select three representative structures for each of $EGFR^{S492R}$/cetuximab and $EGFR^{G465R}$/cetuximab complex.

6. Perform a systematic single point mutant scan for cetuximab's amino acids within 5 Å from any residue of EGFR by using Rosetta *pmut_scan* application.
7. Calculate the binding energies of all $EGFR^{mutant}$/$cetuximab^{mutant}$ and $EGFR^{mutant}$/$cetuximab^{wild\text{-}type}$ complexes by using *InterfaceAnalyzer* application.
8. Identify potential effective cetuximab variants against EGFR mutants based on the change in binding energy. A negative value of $\Delta\Delta G$ indicates that the cetuximab variant is likely to be effective against EGFR mutants.
9. Identify potential "replaceable" residues for directed evolution of cetuximab toward binding with EGFR mutants.

3.3 Construction of the Phage Display Library

1. Use oligonucleotide primers with NNS (N=A, T, C or G; S=C or G) codons to replace the amino acid at the "replaceable" residues of cetuximab with 19 other amino acids (*see* **Note 1**).
2. Amplify the gene fragments of Ctx-scFv variants by PCR, and separate DNA by agarose gel electrophoresis. Cut the target DNA and purify the samples using the DNA Gel Extraction Kit (*see* **Note 2**).
3. Digest pCANTAB-5E vector and gene fragments of Ctx-scFv variants with *Sfi* I and *Not* I according to the manufacturer's instructions. Separate DNA by agarose gel electrophoresis and cut scFv insert and pCANTAB-5E vector DNA. Extract them using the DNA Gel Extraction Kit.
4. Ligate the DNA of insert and vector in a 1:4 ratio using a commercially available ligation kit (NEB T4 Ligase Kit).
5. Remove ions from the ligation reaction by dialysis membrane and deionized water (typically 4–8 h) (*see* **Note 3**).
6. For transformation into XL-1 blue *E. coli*, thaw electrocompetent bacteria on ice and mix with the ligation reaction containing 100-ng DNA.
7. Transfer the 80 μL mix to a prechilled 0.1-cm electroporation cuvette. Dry the electrode of the cuvette with a tissue paper.
8. Perform a 1.8-kV pulse using an electroporator. Immediately, add 1 mL of 37 °C pre-warmed SOC liquid medium, transfer the suspension to a 2-mL tube and shake for 1 h at 220 rpm and 37 °C.
9. To determine the amount of transformants, use 10 μL of the transformation, and perform a dilution series down to 10^{-6} dilution (six tenfold serial dilutions). Plate out each dilution on 2×YT+G+A+tet agar plates and incubate overnight at 37 °C.
10. Plate out the remaining 990 μL on 2×YT+G+A+tet agar plates, and incubate overnight at 37 °C.

11. Calculate the amount of transformants, which should be $>1 \times 10^9$ cfu (*see* **Note 4**).
12. Pick 10–20 clones and inoculate 5 mL of 2×YT+G+A+tet liquid media and grow for 6–8 h, prepare DNA and confirm the successful cloning by DNA-sequence analysis.
13. Float off the colonies on the 2×YT+G+A+tet agar plates with 20 mL of 2×YT liquid medium using a Drigalski spatula. Use 800 μL of bacteria solution and 200 μL of glycerol for glycerol stock. Make 25 glycerol stocks and store at −80 °C.
14. To package the library, inoculate 400 mL of 2×YT+G+A+tet liquid medium in a 1-L Erlenmeyer flask with 1 mL of scFv gene library stock. Grow at 220 rpm at 37 °C until an OD_{600}~0.5 is reached.
15. Infect 25 mL of bacteria culture (~1.25×10^{10} cells) with 1.25×10^{11} cfu of the M13KO7 helper phage at an moi of 1:10. Incubate for 30 min without shaking and then for another 30 min with 220 rpm at 37 °C.
16. To remove the glucose that represses the lac promoter of pCANTAB-5E and thus the scFv–pIII fusion protein expression, harvest the cells by centrifugation for 10 min at 3000*g* in 50-mL polypropylene tubes.
17. Resuspend the pellet in 400 mL of 2×YT+A+K+tet liquid medium in a 1-L Erlenmeyer flask. Produce scFv-phage overnight at 220 rpm and 30 °C.
18. Pellet the bacteria by centrifugation for 10 min at 10,000*g*.
19. Precipitate the phage from the supernatant by adding 1/5 volume of PEG–NaCl solution. Incubate for 1 h at 4 °C with gentle shaking, followed by centrifugation for 1 h at 5000*g*.
20. Discard the supernatant, and put the open tubes upside down on tissue paper. Let the viscous PEG–NaCl solution move out completely. Resuspend the phage pellet in 1 mL of ice-cold PBS. Store the packaged Ctx-scFv variants phage library at 4 °C.

3.4 Preparation of Cells (See Note 5)

1. Depletion should be performed on NIH/3T3 cells, and coselection on WT-EGFR-NIH/3T3 cells, S492R-EGFR-NIH/3T3 cells, and G465R-EGFR-NIH/3T3 cells (ideally: the same cell line, with or without the antigen transfection).
2. Adherent cells are enzymatically detached with Trypsin–EDTA solution to obtain a single-cell suspension. Keep trypsin incubation as short as possible. Add medium containing 10% (vol/vol) FBS to inhibit trypsin and to prevent further proteolytic degradation of surface molecules. Alternatively, use a cell scraper to detach cells from the cell culture flask.

3. Count the cells and assess their viability using trypan blue exclusion staining.
4. Centrifuge cell suspension for 5 min at 300*g*, 4 °C.
5. Resuspend cells in 10 mL of ice-cold PBS.
6. Centrifuge for 5 min at 300*g*, 4 °C.
7. Use 1~5 × 10^7 cells of each cell type for the next step.

3.5 Depletion of Nonspecific Phage

1. Use aliquots of the Ctx-scFv variant phage library. The phage number should be 100-fold higher than the library diversity. Incubate the phage in 1 mL of 2% BSA–PBS for blocking on a rotator for 1 h at 4 °C. This represents the input of your selection.
2. Block the NIH/3T3 cells by incubating them with 5 mL of 2% BSA–PBS on a rotator for 1 h at 4 °C.
3. Centrifuge the cell suspension at 300*g* and 4 °C for 5 min. Discard the supernatant and add the phage solution to the cells for depletion. Incubate on a rotator for 2 h at 4 °C.
4. Pellet the cell-phage suspension at 300*g* and 4 °C for 5 min, and transfer the supernatant to a new tube.

3.6 Coselection of Specific Phage

1. Block the WT-EGFR-NIH/3T3 cells, S492R-EGFR-NIH/3T3 cells, and G465R-EGFR-NIH/3T3 cells, respectively, by incubating them with 5 mL of 2% BSA–PBS on a rotator for 1 h at 4 °C.
2. Centrifuge cell suspension for 5 min at 300*g*, 4 °C, and add the depleted phage library (supernatant of the depletion) to the S492R-EGFR-NIH/3T3 cells and G465R-EGFR-NIH/3T3 cells. Incubate on a rotator for 2 h at 4 °C.
3. Pellet cell-phage suspension for 5 min at 300*g*, 4 °C and remove the supernatant. Wash cells with 1 mL of PBS.
4. Centrifuge for 5 min at 300*g*, 4 °C.
5. Repeat this washing procedure (**steps 3** and **4**) ten times.
6. Elute phage by resuspending cells in 500 μL of elution buffer on a rotator for 15 min at RT.
7. Centrifuge for 5 min at 300*g*, 4 °C and transfer the supernatant to a 1.5-mL tube and neutralize by adding 100 μL of neutralization buffer quickly.
8. Transfer neutralized phage to the WT-EGFR-NIH/3T3 cells blocked in **step 1**. Incubate on a rotator for 2 h at 4 °C.
9. Repeat **steps 3–7**.

3.7 Infecting XL-1 Blue E. coli with the Selected Phage for Next Panning Round

1. Inoculate 50 mL of 2×YT+G+tet liquid medium with XL-1-blue *E. coli*, and incubate until OD_{600} ~0.5 at 37 °C, 220 rpm.
2. Pour 100 μL of eluted phage into a 10-mL tube.
3. Add 100 μL of XL-1-blue *E. coli* at OD_{600} ~0.5.
4. Incubate for 30 min at 37 °C without shaking, and subsequently for 30 min at 37 °C, 220 rpm.
5. Add 1000 μL of 2×YT+G+tet liquid medium and 120 μL of ampicillin stock solution. Incubate for 1 h at 37 °C, 220 rpm (OD_{600} should reach 0.4–0.5).
6. Infect cells with M13KO7 helper phage in a tenfold surplus.
7. Incubate for 30 min at 37 °C without shaking, and subsequently for 30 min at 37 °C, 220 rpm.
8. Centrifuge for 10 min at 3000*g*, RT. Remove the supernatant completely.
9. Add 1 mL of 2×YT+A+K+tet liquid medium. Incubate overnight at 30 °C, 220 rpm.
10. Pellet bacteria for 10 min at 3000*g*, RT.
11. Transfer the supernatant into a fresh tube and run the selection cycle (*see* Subheadings 3.4–3.6) for 2–3 rounds.

3.8 Phage Titration

1. Inoculate 5 mL of 2×YT+G+tet liquid medium with XL-1 blue *E. coli*, and incubate overnight at 37 °C, 220 rpm.
2. Inoculate 50 mL of 2×YT+G+tet liquid medium with 500 μL of the overnight culture, and incubate at 37 °C, 220 rpm until OD_{600} ~0.5 is reached.
3. Add 5 μL of output to 495 μL of PBS. This is the 10^{-2} dilution of phage. Make serial dilution of phage until 10^{-8} for the output.
4. Use 10 μL of the dilutions to infect 50 μL of XL-1 blue *E. coli* at OD_{600} ~0.5 in a tube.
5. Incubate without shaking at 37 °C for 30 min.
6. Plate each dilution of the bacteria suspension on 2×YT+G+A +tet agar plates.
7. Grow overnight at 37 °C.
8. Count the colonies and calculate the cfu titer according to the dilution.

3.9 Production of Soluble scFv

1. Fill 96 tubes with 150 μL of 2×YT+G+A+tet liquid medium.
2. Pick 92 clones with sterile tips from the plate created in Subheading 3.8, **step 7** and inoculate each tube. Also inoculate two tubes with a positive control (e.g., XL-1 blue-pCANTAB-5E-Ctx-scFv). Keep two tubes without clones as negative control.

3. Incubate in a shaker overnight at 37 °C, 220 rpm.
4. Fill 96 new tubes with 190 μL of 2×YT+G+A+tet liquid medium.
5. Transfer 10 μL of each overnight culture into the corresponding new tube. Incubate for 2 h at 37 °C, 220 rpm.
6. Pellet cultures for 10 min at 3000*g*, RT. Remove the supernatant by turning over the tube and carefully beating out the liquid (alternatively, remove the supernatant carefully by pipetting).
7. Resuspend the pellet in 200 μL of buffered 2×YT+A+tet containing 100 μM IPTG (IPTG will induce expression of the pCANTAB-5E lac promoter).
8. Incubate overnight at 30 °C, 220 rpm.
9. Pellet cultures for 10 min at 3000*g*, RT. Transfer the supernatant into fresh tube, used for ELISA or other analytical methods.

3.10 Analysis of Soluble scFv by ELISA

1. To analyze the antigen specificity of the monoclonal soluble scFv, coat each well with 100 μL of 1 μg/mL antigen (WT-EGFR-ECD-Fc, S492R-EGFR-ECD-Fc, or G465R-EGFR-ECD-Fc, respectively, diluted in ELISA coating buffer), and incubate overnight at 4 °C.
2. Wash the coated wells three times with PBST.
3. Block the antigen-coated wells with Milk–PBST for 2 h at 37 °C. Make sure the wells are completely filled.
4. Remove Milk–PBST. Add 50 μL of fresh Milk–PBST and 50 μL of supernatant containing soluble scFv generated in Subheading 3.9, **step 9**, to each well. Incubate for 1 h at 37 °C.
5. Wash the coated wells three times with PBST.
6. Add 100 μL/well of HRP-anti-E tag antibody solution (diluted appropriately in Milk–PBST) and incubate for 1 h.
7. Wash the coated wells three times with PBST.
8. Add 100 μL of TMB solution to each well and incubate for 10–30 min.
9. Stop the color reaction by adding 100 μL of 2 M sulfuric acid solution per well. The color changes from blue to yellow.
10. Measure the absorbance at 450 nm using an ELISA reader, and identify positive candidates with a signal on both wild-type and mutant EGFR.
11. Sequence the DNA of the selected scFv for identification of unique clones using the oligonucleotide primers pCANTAB-5E-R1 and pCANTAB-5E-R2.

4 Notes

1. Avoid NNS codons within the primer overlap regions.
2. If the PCR reactions contain unwanted by products, perform gel extraction throughout the protocol with the DNA Gel Extraction Kit.
3. Note that removing ions from the ligation reaction is important, because ions can cause short circuits and explosions in subsequent electrotransformation.
4. Make sure not to lose library diversity due to low transformation efficiencies.
5. To avoid the internalization and loss of target antigen during the depletion, selection, or screening, perform all the procedures involving cells at 4 °C.

References

1. Lyu X, Zhao Q, Hui J, Wang T et al (2022) The global landscape of approved antibody therapies. Antib Ther 5(4):233–257
2. Misale S, Di Nicolantonio F, Sartore-Bianchi A et al (2014) Resistance to anti-EGFR therapy in colorectal cancer: from heterogeneity to convergent evolution. Cancer Discov 4(11): 1269–1280
3. Schlam I, Tarantino P, Tolaney SM (2022) Overcoming resistance to HER2-directed therapies in breast cancer. Cancers (Basel) 14(16):3996
4. Torka P, Barth M, Ferdman R et al (2019) Mechanisms of resistance to monoclonal antibodies (mAbs) in lymphoid malignancies. Curr Hematol Malig Rep 14(5):426–438
5. Arena S, Bellosillo B, Siravegna G et al (2015) Emergence of multiple EGFR extracellular mutations during cetuximab treatment in colorectal cancer. Clin Cancer Res 21(9): 2157–2166
6. Montagut C, Dalmases A, Bellosillo B et al (2012) Identification of a mutation in the extracellular domain of the epidermal growth factor receptor conferring cetuximab resistance in colorectal cancer. Nat Med 18(2):221–223
7. Newhall K, Price T, Peeters M et al (2014) Frequency of S492R mutations in the epidermal growth factor receptor: analysis of plasma dna from metastatic colorectal cancer patients treated with panitumumab or cetuximab monotherapy. Ann Oncol 25:i109
8. Diwanji D, Trenker R, Thaker TM et al (2021) Structures of the HER2-HER3-NRG1beta complex reveal a dynamic dimer interface. Nature 600(7888):339–343
9. Shin JW, Kim S, Ha S et al (2019) The HER2 S310F mutant can form an active heterodimer with the EGFR, which can be inhibited by cetuximab but not by trastuzumab as well as pertuzumab. Biomolecules 9(10):629
10. Arena S, Siravegna G, Mussolin B et al (2016) MM-151 overcomes acquired resistance to cetuximab and panitumumab in colorectal cancers harboring EGFR extracellular domain mutations. Sci Transl Med 8(324):314r–324r
11. Sanchez-Martin FJ, Bellosillo B, Gelabert-Baldrich M et al (2016) The first-in-class anti-EGFR antibody mixture Sym004 overcomes cetuximab resistance mediated by EGFR extracellular domain mutations in colorectal cancer. Clin Cancer Res 22(13):3260–3267
12. Harb W, Lieu C, Beeram M et al (2014) A first-in-human study evaluating the safety and pharmacology of MM-151, a novel oligoclonal anti-EGFR antibody combination in patients with refractory solid tumors. Ann Oncol 25: v152
13. Montagut C, Argiles G, Ciardiello F et al (2018) Efficacy of Sym004 in patients with metastatic colorectal cancer with acquired resistance to anti-EGFR therapy and molecularly selected by circulating tumor DNA analyses: a phase 2 randomized clinical trial. JAMA Oncol 4(4):e175245

14. Zhuang X, Wang Z, Fan J et al (2022) Structure-guided and phage-assisted evolution of a therapeutic anti-EGFR antibody to reverse acquired resistance. Nat Commun 13(1):4431
15. Fleishman SJ, Leaver-Fay A, Corn JE et al (2011) Rosettascripts: a scripting language interface to the rosetta macromolecular modeling suite. PLoS One 6(6):e20161
16. Leaver-Fay A, Tyka M, Lewis SM et al (2011) Rosetta3: an object-oriented software suite for the simulation and design of macromolecules. Methods Enzymol 487:545–574
17. Alfaleh MA, Alsaab HO, Mahmoud AB et al (2020) Phage display derived monoclonal antibodies: from bench to bedside. Front Immunol 11:1986
18. Ferrari D, Garrapa V, Locatelli M (2020) A novel nanobody scaffold optimized for bacterial expression and suitable for the construction of ribosome display libraries. Mol Biotechnol 62(1):43–55
19. Frenzel A, Schirrmann T, Hust M (2016) Phage display-derived human antibodies in clinical development and therapy. MAbs 8(7): 1177–1194

Chapter 4

Combining a Base Deaminase Mutator with Phage-Assisted Evolution

Maryam Ali, Raneem Akel, Maria J. Botero, and Jumi A. Shin

Abstract

Phage-assisted evolution has emerged as a powerful technique for improving a protein's function by using mutagenesis and selective pressure. However, mutations typically occur throughout the host's genome and are not limited to the gene-of-interest (GOI): these undesirable genomic mutations can yield host cells that circumvent the system's selective pressure. Our system targets mutations specifically toward the GOI by combining T7 targeted mutagenesis and phage-assisted evolution. This system improves the structure and function of proteins by accumulating favorable mutations that can change its binding affinity, specificity, and activity.

Key words Mutagenesis, M13 bacteriophage, Cytidine deaminase, Selective pressure, T7 promoter and terminator

1 Introduction

Phage-assisted evolution was developed by Liu and coworkers by looping the four steps of directed evolution into the bacteriophage life cycle [1]. These steps include mutagenesis, gene expression, selection, and replication. By doing so, minimal intervention is required, as all four steps are carried out within bacterial cells via the M13 filamentous bacteriophage machinery.

In this evolution system, the gene-of-interest (GOI)—which expresses the protein-of-interest (POI)—lies on the selection phage (SP), which is a modified version of the M13 bacteriophage that lacks the gIII required for phage propagation. The gIII is instead present on the accessory plasmid (AP), downstream of the DNA target site for the POI. The POI is fused to an RNA ω (omega) subunit; therefore, upon POI binding to its target site, gIII is then expressed and phages are propagated. The mutation plasmid (MP) contains the proteins required for targeted mutation of the

Huan Peng et al. (eds.), *Phage Engineering and Analysis: Methods and Protocols*,
Methods in Molecular Biology, vol. 2793, https://doi.org/10.1007/978-1-0716-3798-2_4,

GOI. With this setup, we have used phage-assisted evolution to generate protein variants with improved structure and function [2, 3].

In our system, mutations are restricted to the GOI by the T7 promoter and T7 terminator upstream and downstream of the GOI, respectively (Fig. 1). The MP carries a T7 RNA polymerase (T7 RNAP) fused to a mutation-inducing protein, such as cytidine deaminase [5]. Upon induction of mutagenesis, T7 RNAP specifically recognizes the T7 promoter upstream of the GOI and begins transcription, positioning cytidine deaminase near the GOI to induce cytosine-to-thymine mutations. Upon reaching the T7 terminator, the fusion protein is released, thereby terminating mutations. Thus, mutations will only occur in the GOI. To minimize cytidine deaminase activity outside of the GOI, we used a T7 terminator with 98% efficiency, thereby limiting off-target mutations [6].

Before starting phage-assisted evolution, the SP, AP, and appropriate MP must be acquired or properly designed [7]. Directed evolution is achieved by running passages with the required SP, AP, and MP transformed into bacterial cells (Fig. 2). The mutations obtained can be tracked using DNA Sanger sequencing and analyzed using a DNA alignment software, such as Clustal Omega. A high phage activity (or phage titer), calculated by using plaque assays, must be maintained throughout the experiment; this procedure is described below (Subheadings 3.5 and 3.6). Optionally, "drift" passages that do not impose selective pressure can be run to allow mutations to accumulate in the GOI before introducing selective pressure. Phage-assisted evolution can therefore be utilized to select for protein variants with the desired outcome and has been successfully applied to protein evolution [8].

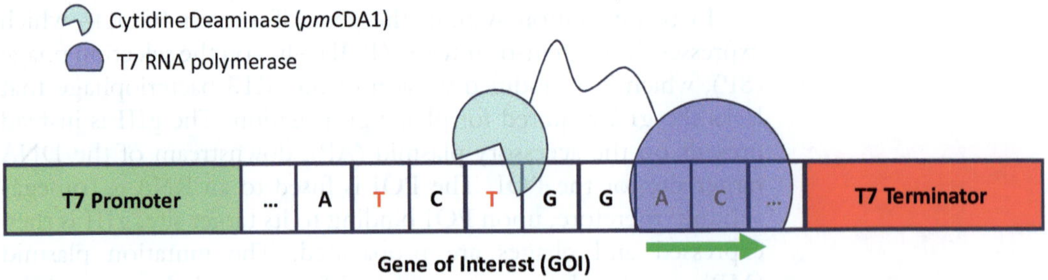

Fig. 1 Targeted mutation of the gene of interest. T7 RNA polymerase (T7 RNAP, purple) transcribes the gene of interest along the direction of the arrow starting at the T7 promoter (green). Cytidine deaminase (*pm*CDA1, light blue) is fused to T7 RNAP. During transcription, T7 RNAP brings cytidine deaminase to the GOI and induces cytidine-to-thymine mutations (indicated by red Ts). Upon reaching the T7 terminator (red), the T7RNAP-cytidine deaminase fusion releases from DNA [4]

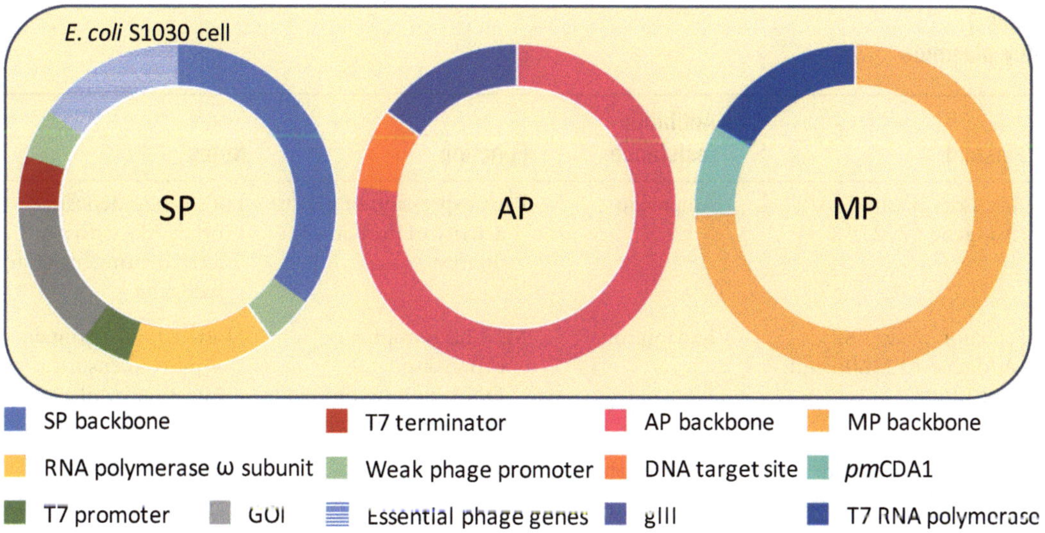

Fig. 2 M13 bacteriophage evolution system. Bacterial host cells must carry three plasmids: selection phage (SP), accessory plasmid (AP), and mutagenesis plasmid (MP) [2]. The white lines represent the start and end of fusion genes in a clockwise direction. SP contains the GOI fused to RNA polymerase ω subunit; the T7 promoter and T7 terminator lie upstream and downstream of the GOI, respectively. SP also contains the weak phage promoter that transcribes the essential phage genes. AP contains the DNA target site and gIII, which is the phage tail gene, necessary for proper phage propagation and infection. MP contains the *pm*CDA1-T7 RNAP fusion complex that acts on the SP to mutate the GOI [4]

2 Materials

2.1 Preparing the Plasmids

1. Obtain plasmids from Addgene, Table 1. Use sterile dd H_2O in all experiments.
2. Site-directed mutagenesis (SDM; Q5 Site-Directed Mutagenesis Kit, NEB). Can use other methods for SDM.
3. Cloning:
    - Appropriate restriction enzymes and buffers.
    - T4 DNA ligase and 10× T4 DNA ligase buffer.
    - Gel extraction kit.

2.2 Running Passages/Cycles

1. *E. coli* strains (Table 2):
    - S1030 (Addgene bacterial strain #105062). Used as the host strain for selection.
    - S1059 (Addgene plasmid #79219). Used for activity-independent plaque assays and for constitutive propagation of SP.
    - Grow cells in the appropriate media with antibiotics at 37 °C with shaking at 250 rpm overnight (12–18 h).

Table 1
Key plasmids

Plasmid	Antibiotic resistance	Function	Notes
Accessory plasmid (AP) (Addgene #79219)	Ampicillin	Pairs expression of gIII to activity of protein-of-interest	Liu's group deposited other APs with altered stringencies at Addgene
Selection phage (SP) (Addgene #122606 and #122605)	Kanamycin	M13 bacteriophage expressing DNA-binding protein in lieu of gIII protein	Host cell must contain F-pilus to ensure infection by phage [9]
Mutagenesis plasmid: eMutaT7-optimized ribosome binding site (Addgene #187621)	Chloramphenicol	Mutates GOI	T7 RNA polymerase fused to a deaminase. Induced by arabinose

Table 2
***E. coli* strains**

E. coli strain	Antibiotic resistance	Plasmid for transformation (s)	Notes
S1030 (Addgene #105063)	Streptomycin and tetracycline	1. Accessory plasmid 2. Mutagenesis plasmid	Activity-dependent Host for mutagenesis and selection passages
S1059 (Addgene plasmid #79219)	Ampicillin	Mutagenesis plasmid (for drift)	Activity-independent Host for plaque assays, recovery, and drift passages Carries AP with constitutively active gIII, allowing SP propagation in drift passages

2. 2× YT growth media supplemented with the appropriate antibiotics:
   - Dissolve 31 g 2× YT into 1 L H_2O and autoclave to sterilize.
   - After cooling to room temperature, add 1 mL appropriate antibiotic(s).
   - Store at 4 °C. Keep away from light.
3. 0.22 μm sterile filters for isolating purified bacteriophages.
4. 10% L-arabinose:

- Dissolve L-arabinose to 10% final concentration in water and filter-sterilize.
- Store at room temperature for up to 3 months.

5. Antibiotics. Store aliquots at −20 °C for up to 6 months:
   - Ampicillin (1000× stock). 30 mg/mL solution in water, filter-sterilized.
   - Kanamycin (1000× stock). 30 mg/mL solution in water, filter-sterilized.
   - Chloramphenicol (1000× stock). 25 mg/mL solution in anhydrous ethanol.
   - Streptomycin (1000× stock). 50 mg/mL solution in water, filter-sterilized.
   - Tetracycline (1000× stock). 10 μg/mL solution in anhydrous ethanol. Store in the dark.

2.3 Plaque Assay [2]

1. 2× YT top agar:
   - Add 3.1 g 2× YT media powder to 0.75 g agar. Dissolve in 100 mL dd H_2O.
   - Autoclave to sterilize and store at room temperature in the dark.
2. LB agar plates:
   - Add 3.12 g of Lysogeny broth (LB) to 1.88 g agar and suspend in 125 mL dd H_2O.
   - Autoclave to dissolve and sterilize.
   - Cool to 55 °C; then pour into sterile petri dishes.
   - Once fully solidified, wrap with parafilm and store at 4 °C for up to 1 month.
3. 55 °C water bath to maintain molten top agar.

2.4 Sequencing and Analysis

1. Method for plasmid purification, such as QIAprep Spin Miniprep kit (Qiagen).
2. Multiple sequence alignment tool, such as Clustal Omega, to monitor mutations in the genetic sequence.

3 Methods

3.1 Preparing the SP

1. Clone GOI into SP downstream of RNA ω subunit.
2. Clone T7 promoter and T7 terminator directly upstream and downstream of the GOI, respectively:

- Terminator selection:
  - Native T7 terminator [10].
  - Optimized T7 terminator [6]:
    - Can choose appropriate terminator based on efficiency requirements [6].

3. Ensure all essential phage genes are being transcribed. The presence of the T7 terminator may stop transcription of essential phage genes required for phage assembly, maturation, and infection [11]:
   - In the SP, there is a weak phage gene promoter upstream of the RNA ω subunit. This weak promoter transcribes essential phage genes gVI, gI, and gXI (Fig. 2, essential phage genes). These three genes are necessary for phage assembly and maturation. Inserting a highly efficient T7 terminator prevents transcription of these genes by the weak promoter; thus, low phage propagation results. To circumvent this problem, we inserted another weak phage promoter after the T7 terminator to allow expression of these essential phage genes.

3.2 Preparing the AP

1. Clone target site of POI upstream of gIII.
2. Check for autoactivation using the bacterial-one-hybrid system [2, 12].

3.3 Preparing the MP

We used Kim's eMutaT7-optimized ribosome binding site (RBS) as the mutation vector [4, 5], with some changes based on what we required for our system:

1. Cloned in *pm*CDA1 gene (expressed cytidine deaminase) with bacterially optimized codons (Addgene #167974) [13]:
   (a) Performed SDM to introduce new restriction sites in both plasmids.
   (b) Used these new restriction sites to clone bacterially optimized *pm*CDA1 into the eMutaT7 backbone.
2. Used SDM to inactivate the f1 *ori* in eMutaT7 to prevent competition with SP using SDM [14, 15].

3.4 Passages

Selection pressure refers to Darwinian selection, where phages expressing a functional POI survive and propagate. Stringency refers to the strength of the selection, where higher stringency only enables phages expressing highly functional POIs to propagate; lower stringency enables any phage expressing a functional protein to propagate, regardless of the functionality level. Higher stringency conditions lead to a smaller library of variants.

Table 3
Effects of various factors on stringency of experiments

Factors	Stringency
Dilution factor	Fewer phages = higher stringency
Remove drift	Smaller gene pool = higher stringency
Starting titer	Infection starting at lower titer = higher stringency
Incubation time	Less incubation time = higher stringency

We recommend starting with less stringent conditions to build a library of variants and then increase stringency to select the most advantageous mutations. Stringency levels can be altered according to Table 3.

- Transform SP containing the GOI into chemically competent S1059 cells. Rest on ice for 15 min and then heat shock at 42 °C for 1 min. Immediately after heat shock, incubate the cells in 10 mL 2× YT media overnight.
- The next day, centrifuge for 10 min and filter-sterilize the supernatant to remove residual cells. Store at 4 °C as is or at −20 °C as glycerol stocks (*see* Notes). This is "Passage 0." Perform a plaque assay to ensure a high phage titer before beginning phage-assisted evolution.
- Culture preparation:
  (a) Transform the AP into chemically competent S1030 cells using a standard transformation protocol.
  (b) Make S1030-AP-competent cells for selection passages and S1059-competent cells for drift passages and plaque assays.
  (c) Transform MP into S1030-AP and into S1059-competent cells using a standard transformation protocol.
  (d) Set up overnight cultures using colonies from the transformation in 2× YT media and appropriate antibiotics.
  (e) The next day, subculture cells in a 1:10 ratio and grow in a shaker at 37 °C at 250 rpm until OD_{600} 0.4–0.6 (~2 h). These cells can be stored at 4 °C for 1 week.
  (f) Take 2 mL cell culture and add 10% L-arabinose to 0.2% final concentration. Based on the desired stringency, add phages from the most recent phage passage to 0.1–10% final concentration. Incubate for 4–18 h at 37 °C at 250 rpm in a shaker.
  (g) Following incubation, centrifuge for 10 min and filter-sterilize the supernatant to remove residual cells. Store at

4 °C as is or at −20 °C as glycerol stocks (*see* **Notes**). Use this purified phage population to start the next passage or recovery.

(h) Determine the titer of the purified phage solution by an activity-independent plaque assay (Subheading 3.4).

3.5 Plaque Assay [2]

We use plaque assays to determine phage titer (also referred to as "phage activity"). A phage titer of 10^6–10^9 plaque-forming units/mL must be maintained throughout the experiment. Carry out all following steps under sterile conditions:

1. Grow activity-independent S1059 cells to OD_{600} 0.6–0.9. These cells can be stored at 4 °C for up to 1 week.
2. Warm a water bath to 55 °C. Warm LB agar plates in a 37 °C incubator.
3. Melt 2× YT top agar in the microwave and aliquot 4–5 mL molten top agar into culture tubes. Warm top agar enough until it is melted homogeneously but be careful not to burn. Place culture tubes in the water bath.
4. Make serial dilutions (10^{-1}–10^{-6}) of the passage whose phage activity is to be determined. Add 180 μL sterile water to each sterile microcentrifuge tube. Add 20 μL phages from passage to the first dilution. Mix and add 20 μL dilution 1 into dilution 2 and so on. Repeat mixing and dilution until the desired final dilution.
5. Set up new sterile microcentrifuge tubes for each dilution that is to be plated. Transfer 30 μL from each serial dilution into the corresponding new tube.
6. Aliquot 270 μL S1059 cells into tubes containing 30 μL serial dilution (another tenfold dilution). Mix and let tubes sit for 10–15 min to allow phage particles to infect the host cells. *Do not exceed 15 min in total*; 10 min is generally recommended.
7. After 10–15 min, plate the infected cells by adding 300 μL phage-cell mixture into a tube of molten agar; gently invert to mix to avoid bubbles. Plate onto the corresponding warmed LB plate. Repeat for remaining dilutions.
8. Following a 24-h incubation period, calculate the phage titer by counting the number of clear plaques formed on the LB agar plate and using the following formula:

$$\text{Phage titre (PFU/mL)} = \frac{\text{Plaque forming units}}{\text{Volume of phage} \times \text{Dilution factor}}$$

where the dilution factor is the plate with the highest number of plaques that you can clearly quantify (e.g., $10^{-4} = 0.0001$), volume of phage is typically 0.003 mL, and the plaque-forming units (PFU) are the plaques counted on that dilution plate.

9. Clonal plaques can be isolated to check for mutations via Sanger sequencing (Subheading 3.7).

3.6 Recovery Phase

If the phage titer is low, run a recovery phase to maintain a titer between 10^6 and 10^9 PFU/mL:

1. In a sterile flask, incubate a tube of S1059-competent cells in 25 mL 2× YT media with antibiotics for 2–3 h in a 37 °C shaker at 250 rpm.
2. Add fresh 25 mL 2× YT media with antibiotics and up to 1 mL from the most recent phage culture. Re-incubate overnight at 250 rpm.
3. The following day, centrifuge for 30 min and filter-sterilize the phage solution. If the supernatant is still cloudy, repeat centrifugation for 10–15 min. Store this supernatant at 4 °C as is or at −20 °C as glycerol stocks.
4. Perform a plaque assay to check the titer of recovery. If the titer is within appropriate range (10^6–10^9 PFU/mL), then it is acceptable to use this passage to start the next passage.

3.7 Result Analysis

Sequence after the plaque assay to check for mutations in the passages:

1. Use a pipet tip to gently poke an isolated plaque and transfer it to a culture tube containing 6 mL 2× YT media. Grow in a 37 °C shaker at 250 rpm overnight (12–18 h).
2. Purify plasmid and choose appropriate primer for Sanger sequencing.
3. Compare the original genetic sequence with the passage(s) sent for sequencing using a multiple gene alignment tool to check for mutations.

4 Notes

- All steps involving phages and cells should be performed under sterile conditions.
- Any surface or equipment that comes into contact with phages should be disinfected using a 2% bleach solution, as phages are highly infectious. Alternatively, equipment can be placed under UV light for 20 min for disinfection.
- Filtered phages can be stored indefinitely at 4 °C or at −20 °C as glycerol stocks [16]. However, phage titers will decrease gradually over time:

– Making glycerol stocks of phage solution: mix purified phage solution with a sterile solution containing 50% glycerol/50% water in a 1:1 ratio. Store at −20 °C and ensure that the solution does not freeze.

Acknowledgement

We are grateful for assistance from Afnan Khan and Montdher Hussain and for support from NSERC and the UTM Office of Research.

References

1. Esvelt KM, Carlson JC, Liu DR (2011) A system for the continuous directed evolution of biomolecules. Nature 472:499–505
2. Popa SC, Inamoto I, Thuronyi BW, Shin JA (2020) Phage assisted continuous evolution (PACE): a guide focused on evolving protein-DNA interactions. ACS Omega 5:26957–26966
3. Xie VC, Styles MJ, Dickinson BC (2022) Methods for the directed evolution of biomolecular interactions. Trends Biochem Sci 47: 403–416
4. Seo D, Koh B, Eom G-E, Kim HW, Kim S (2023) A dual gene-specific mutator system installs all transition mutations at similar frequencies in vivo. Nucleic Acids Res 51(10): e59–e59
5. Park H, Kim S (2021) Gene-specific mutagenesis enables rapid continuous evolution of enzymes in vivo. Nucl Acid Res 49:e32
6. Calvopino-Chavez DG, Gardner MA, Griffitts JS (2022) Engineering efficient termination of bacteriophage T7 RNA polymerase transcription. G3 Genes Genomes Genet 12:jkac070
7. Thuronyi BW, Koblan LW, Levy JM, Yeh W-H, Zheng C, Newby GA, Wilson C, Bhaumik M, Shubina-Oleinik O, Holt JR, Liu DR (2019) Continuous evolution of base editors with expanded target compatibility and improved activity. Nat Biotechnol 37:1070–1079
8. Inamoto I, Sheoran I, Popa SC, Hussain M, Shin JA (2020) Combining rational design and continuous evolution on minimalist proteins that target DNA. ACS Chem Biol 16:35–44
9. Rakonjac J, Bennett NJ, Spagnuolo J, Gagic D, Russel M (2011) Filamentous bacteriophage: biology, phage display and nanotechnology applications. Curr Iss Mol Biol 13:51–76
10. Dunn JJ, Studier FW, Gottesman M (1983) Complete nucleotide sequence of bacteriophage T7 DNA and the locations of T7 genetic elements. J Mol Biol 166(4):477–535
11. Smeal SW, Schmitt MA, Rodrigues Pereira R, Prasa A, Fisk JD (2017) Simulation of the M13 life cycle I: assembly of a genetically-structured deterministic chemical kinetic simulation. Virology 500:259–274
12. Meng X, Brodsky MH, Wolfe SA (2005) A bacterial one-hybrid system for determining the DNA-binding specificity of transcription factors. Nat Biotechnol 23:988–994
13. Álvarez B, Mencía M, de Lorenzo V, Fernández LÁ (2020) In vivo diversification of target genomic sites using processive base deaminase fusions blocked by dCas9. Nat Commun 11(1):6436. https://doi.org/10.1038/s41467-020-20230-z
14. Dotto GP, Horiuchi K, Zinder ND (1984) The functional origin of bacteriophage f1 DNA replication: its signals and domains. J Mol Biol 172(4):507–521
15. Sathiamoorthy S, Shin JA (2012) Boundaries of the origin of replication: creation of a pET-28a-derived vector with p15A copy control allowing compatible coexistence with pET vectors. PLoS One 7:e47259
16. Miller SM, Wang T, Liu DR (2020) Phage-assisted continuous and non-continuous evolution. Nat Protoc 15(12):4101–4127

Chapter 5

Understanding the Structural Requirements of Peptide–Protein Interaction and Applications for Peptidomimetic Development

Angy Liseth Davalos Macias, Lilian Costa Alecrim, Fabio C. L. Almeida, and Ricardo Jose Giordano

Abstract

Protein–protein interaction is at the heart of most biological processes, and small peptides that bind to protein binding sites are resourceful tools to explore and understand the structural requirements for these interactions. In that sense, phage display is a well-suited technology to study protein–protein interactions, as it allows for unbiased screening of billions of peptides in search for those that interact with a protein binding domain. Here, we will illustrate how two distinct but complementary approaches, phage display and nuclear magnetic resonance (NMR), can be utilized to unveil structural details of peptide–protein interaction. Finally, knowledge derived from phage mutagenesis and NMR studies can be streamlined for quick peptidomimetic design and synthesis using the retroinversion approach to validate using in vitro and in vivo assays the therapeutic potential of peptides identified by phage display.

Key words Phage display, Peptide and protein interaction, Phage mutagenesis, Peptidomimetic design, Nuclear magnetic resonance, NMR

1 Introduction

The importance of peptides in biology is undeniable: they may function as hormones, cytokines, and neurotransmitters, modulating almost all processes in the human body, from digestion and homeostasis to the immune system and our brain. Likewise, peptide versatility in research is almost endless. With the advances in biotechnology, peptides that bind to virtually any protein target can be isolated and readily synthesized. These peptides can then be used to study the role of such proteins in different cellular processes. Considering that proteins are key molecules in life but, most importantly, that they comprise the overwhelming majority of targets for all the drugs in use today [1], peptides that modulate protein function are important leads for drug development

Huan Peng et al. (eds.), *Phage Engineering and Analysis: Methods and Protocols*,
Methods in Molecular Biology, vol. 2703, https://doi.org/10.1007/978-1-0716-3798-2_5,

[2, 3]. Therefore, understanding protein–peptide structure and function is an important step for rational development of novel therapeutics.

Although protein binding peptides may be designed based on available structure of protein complexes [4], this approach is limited to those protein complexes with structural information available to researchers. On the other hand, combinatorial approaches do not have such bottlenecks and can be used to identify novel biologically active peptides, which bind to virtually any protein or protein complex [5, 6].

One of these approaches is phage display, a formidable platform for peptide discovery [6]. Phage libraries often display billions of peptides which can be readily screened in an unbiased manner against any biological target. Peptides identified by phage display frequently interact with biologically relevant binding sites found on proteins and, as a result, behave as agonists or antagonists in biological processes [6–9]. Such biologically active peptides are valuable assets to study protein–protein interaction and, in particular, for the development of new therapeutic drugs [8–11]. For that, it is paramount to understand the structural requirements for the interaction of the peptide with its target protein.

In that sense, phage display is not only a discovery technique but also a versatile tool to explore how peptides interact with their target. For instance, phage mutagenesis can be used to identify key residues necessary to peptide binding [7, 12]. This is often achieved by alanine scanning in which each amino acid is substituted for alanine followed by a quantitative phage binding assay. Other substitutions can also be tested, such as serine and threonine with glutamic or aspartic acid to mimic phosphorylation sites, or cysteine with serine/threonine to evaluate potential disulfide bond formations.

Finally, in more sophisticated settings, NMR methods can be utilized not only to identify key amino acids important for protein binding but also to obtain valuable information regarding the peptide conformation in solution and its structural dynamics [7, 13]. These complementary approaches may indicate that as little as three residues within what would already be considered a short peptide with 8–10 amino acids are sufficient for peptide binding [8, 9]. Once these key residues have been mapped, bioactive peptidomimetics may be readily designed and synthesized using the retroinversion approach [14, 15]. Using this approach, a VEGF receptor binding peptide was converted into peptidomimetic compounds with antiangiogenic activity with potential use for the treatment of retinopathy [8, 11]. Here, we will describe methods for phage mutagenesis, phage binding assay, and nuclear magnetic resonance (NMR) methods that can be used to identify and pinpoint key amino acid residues necessary for peptide–protein interaction and how such information can then be used to design peptidomimetic molecules.

2 Materials

Alanine-Scanning Mutagenesis

Water (molecular biology grade).

Qiagen PCR purification kit.

Competent *E. coli* MC1061, DH5α, or XL-1blue XTR strain.

E. coli K91blu/kan (Available from Cell Origins, Cat. # COBS0001—https://www.cellorigins.com/k91blukan).

Phage PCR primers: forward 5′-AGCAAGCTGATAAACCGATACAATT -3′ and reverse 5′- CCCTCATAGTTAGCGTAACGATCT-3′.

2.1 Vector

fUSE5 (available from Cell Origins, Cat. # COV0002—https://www.cellorigins.com/fuse5).

2.2 Enzymes

*Sfi*I enzyme with corresponding 10× buffer.

T4 DNA ligase and corresponding buffer.

Platinum Taq Polymerase and corresponding buffer.

2.3 Culture Media and Buffers

1. Annealing buffer: 10 mM Tris–HCl pH 8.0, containing 100 mM NaCl and 1 mM EDTA. Dissolve 1.21 g Tris base, 5.84 g NaCl, and 0.37 g EDTA-2H_2O in 800 mL dH_2O. Adjust the pH to 8.0 with HCl. Adjust the volume to 1 liter with dH_2O.
2. Phosphate-buffered saline (PBS): 1.37 M NaCl, 27 mM KCl, 100 mM Na_2HPO_4, and 18 mM KH_2PO_4. Dissolve the following reagents in 800 mL of ddH_2O: 8 g of NaCl, 0.2 g of KCl, 1.44 g of Na_2HPO_4, and 0.24 g of KH_2PO_4. Adjust the pH to 7.4 with HCl, add ddH_2O, and then autoclave.
3. TAE: First, prepare a concentrated 50× stock solution of TAE buffer. To do this, dissolve 242 g of Tris base in 750 mL of deionized water. Add 57.1 mL of acetic acid and 100 mL of 0.5 M EDTA (pH 8.0), and adjust the volume to 1 L by adding water. The final pH of the 50× TAE buffer should be about 8.5. To make the 1× TAE working buffer, add 49 parts of deionized water to 1 part of 50× TAE buffer.
4. PEG/NaCl: Dissolve 100 g PEG 8000 and 116.9 g NaCl in 500 mL ddH_2O. Add the remaining water to 1 L, stir until completely homogenized and autoclave.

5. LB (Luria Bertani) media: Add 10 g of tryptone, 5 g of yeast extract, and 10 g NaCl. Add all the solid ingredients to ~900 mL of ddH_2O and stir to dissolve. Adjust pH 7 with NaOH, complete to 1000 mL and autoclave.
6. LB kan/tet: Add 4 mL of the tetracycline stock (final concentration 40 μg/mL) and 1 mL of the kanamycin stock solution (final concentration 100 μg/mL) to 1 L of LB. Keep at 4 °C in the dark.
7. Terrific broth: Add 12 g of tryptone, 24 g of yeast extract, and 4 mL of glycerol. Add all the solid ingredients to ~900 mL of ddH_2O and stir until homogenized. Filter and autoclave for 50 min.
8. TB supplement: Add 23.1 g of KH_2PO_4, 125.4 g of K_2HPO_4. Add all the solid ingredients to ~900 mL of ddH_2O and stir to dissolve. Complete to 1000 mL and filter through 0.22 μm filter.
9. TB sup/kan: Add to 450 mL of terrific broth media 50 mL of TB supplement and 1 mL of kanamycin stock solution (final concentration 50 μg/mL).
10. SOC media: SOC medium is identical to SOB medium, except that it contains 20 mM glucose. To prepare 1 liter of SOB, dissolve the following in ~800 mL of deionized water: 20 g tryptone, 0.584 g NaCl, 0.186 g KCl, and 2.4 g $MgSO_4$. Adjust the pH of the medium to 7.0 with NaOH. Adjust the volume of the solution to 1 L with deionized H_2O and autoclave. After the SOB medium has been autoclaved, allow it to cool to 50 °C or less and add 20 mL of filter-sterilized 20% glucose solution.
11. LB kan/tet agar plates: Prepare 1 L of LB, adjust pH, and add 15 g of bacto-agar and autoclave. Let the bottles cool down to 50 °C and add the antibiotic: 4 mL of the tetracycline stock (final concentration 40 μg/mL) and 2 mL of kanamycin stock solution (final concentration 100 μg/mL). Pour the plates, let them cool down and keep at 4 °C in the dark.

2.4 0.8% and 1.2% Agarose Gel

Measure 0.4 g of agarose (final agarose concentration in gel should be 0.8%), and mix agarose powder with 50 mL 1× TAE in a microwavable flask. Microwave for 1–3 min until the agarose is completely dissolved. Let agarose solution cool down to about 50 °C (about when you can comfortably keep your hand on the flask), and add ethidium bromide to a final concentration 0.5 μg/mL. Pour into gel dock with comb and allow to solidify. To prepare 1.2% agarose gel, it is necessary to have 0.6 g of agarose and repeat the procedure.

3 Methods

Phage Mutagenesis (Alanine Scanning)

Phage binding assay is a versatile tool to understand the importance of individual residues in peptide–receptor interaction. The idea is to generate individual phage mutants, each containing one (or more) residues mutated to a neutral amino acid. The most common example is alanine mutagenesis, in which each amino acid in the peptide is individually changed to alanine (a naturally occurring Ala residue may be changed to Gly). However, other interesting changes involve serine or threonine to glutamic or aspartic acids to mimic phosphorylation sites, or cysteine to serine to evaluate disulfide bond importance, among others. Once each individual phage is produced and sequenced to confirm the mutation, they can then be tested by phage binding assay to assess the impact each mutation has on ligand interaction. First, we describe procedures to generate alanine-scanning mutagenesis using the Fd-tet-based vector fUSE5. Next, we describe the protocol for phage binding assay to test each mutated phage.

3.1 Oligonucleotide Design

For every amino acid of interest, two single-stranded oligonucleotides with complementary sequences should be designed (Fig. 1). To facilitate cloning, once the oligonucleotides are annealed, they will generate a double-stranded DNA insert with protuberant ends compatible with fUSE5 vector cut with the *Sfi*I enzyme (Fig. 1). To design the oligonucleotides, use the following template:

IL11 receptor binding phage. Sequence for the DNA insert and encoded peptide:

```
        C    G    R    R    A    G    G    S    C
     5'-TGT  GGG  CGG  AGG  GCG  GGC  GGT  TCG  TGT -3'
```

Gly to Ala:

```
CARRAGGSC-f  5'     GGGCT  TGT  GCG  CGG  AGG  GCG  GGC  GGT  TCG  TGT  GGGGCCGCTG 3'
CARRAGGSC-r  3' TGCCCCGA   ACA  CGC  GCC  TCC  CGC  CCG  CCA  AGC  ACA  CCCCGGC    5'
```

Arg to Ala:

```
CGRAAGGSC-f  5'     GGGCT  TGT  GGG  CGG  GCG  GCG  GGC  GGT  TCG  TGT  GGGGCCGCTG 3'
CGRAAGGSC-r  3' TGCCCCGA   ACA  CCC  GCC  CGC  CGC  CCG  CCA  AGC  ACA  CCCCGGC    5'
```

Fig. 1 Example of oligonucleotide design for phage alanine-scanning mutagenesis. The interleukin-11 receptor binding peptide CGRRAGGSC and its corresponding DNA-encoded sequence are used to illustrate the design [7]. Two self-complementary oligonucleotides should be designed and ordered. Upon annealing, they will form a double-stranded DNA insert with protuberant ends compatible with the fUSE5 vector digested with the *Sfi*I restriction enzyme. Two pairs of oligonucleotides containing mutations (red) that will be translated as alanine in the final peptide are shown

Forward oligonucleotide: 5′-GGGCT (insert sequence) GGGGCCGCTG-3′

Reverse oligonucleotide: 5′- CGGCCCC (RC insert sequence*) AGCCCCGT-3′

*RC = reverse and complementary

3.2 Oligo Annealing

1. Resuspend oligos in molecular biology grade water (100 pmol/μL).
2. Mix equal volumes (5 μL) of the oligonucleotides in a microtube, and add 90 μL of annealing buffer to obtain an equimolar mixture of both oligonucleotides (10 pmol/μL).
3. Incubate the microtube with oligonucleotide solution at 100 °C for 5 min.
4. Allow the microtube with oligonucleotide solution to slowly cool to room temperature.

Once annealed, the oligos will leave a 3′-protuberating end for ligating into *Sfi*I-digested fUSE5 vectors (Fig. 1).

3.3 fUSE5 Restriction Digestion

The *fUSE5* vector needs to be linearized using the *Sfi*I restriction enzyme (recognition site: GGCCNNNN/NGGCC) (note that *Sfi*I enzyme requires digestion at 50 °C). There are two adjacent *Sfi*I sites in fUSE5 which do not share the same exact sequence. Thus, there is no need to dephosphorylate the vector. Upon digestion and purification using Qiagen PCR purification kit, the small fragment stuffer is lost, and the vector is no longer capable of self-ligation. Furthermore, because the oligonucleotides are dephosphorylated, the protocol will not work if the vector is also dephosphorylated:

1. Prepare the fUSE5 digestion following the enzyme manufacturer buffer and recommendations. We suggest digesting 10 μg of vector (1 pmol) using 100 U of *Sfi*I enzyme (10 U of enzyme/μg of DNA) for 2–4 h. This should be enough to perform at least ten mutations.
2. Once the digestion is complete, take a 100 ng aliquot to run on a 0.8% agarose gel to confirm that the vector has been completely linearized (see Chapter 1).
3. Divide the reaction into two 1.5 mL tubes, and purify the DNA using two spin columns from the Qiagen PCR purification kit (or equivalent) following the manufacturer recommendations and buffers.

   Important: Elute the fUSE5 vector using warm (50 °C) elution buffer for maximum recovery. Expect up to 50% loss.
4. Run a small aliquot (1/10) of the eluted sample to confirm DNA presence. Using NanoDrop, calculate DNA concentration.

3.4 Ligation

fUSE5 is 9200 bp and insert is ~40–60 bp (depending on the length of the encoded peptide). This represents a ~200:1 (m/m) vector-to-insert relation.

1. Set up the ligation reactions following the T4 DNA ligase manufacturer's recommendations and buffer. We recommend using 100 ng of fUSE5 vector (0.02 pmol) with the corresponding amount of annealed oligonucleotides for a 1:5 molar ratio (0.1 pmol). Upon ligation, transform each reaction into 20 μL of electrocompetent *E. coli* MC1061 cells (streptomycin resistant).

   Note: Electrocompetent cells may be prepared using standard protocols (see Chapter 1) [16]. Alternatively, another F minus strain (such as DH10B) may be used.
2. For each electroporation, prepare a 1.5 mL tube with 0.8 mL SOC media supplemented with 0.2 μg/mL tetracycline (SOC low-tet). Keep in the incubator at 37 °C, 200 rpm rotation.
3. Set up the electroporator (Gene Pulser) to 1800 V, capacitor 25 μF, and resistance 200 Ω.
4. Add 1 μL (~10 ng of DNA) of the ligation to one aliquot (20 μL) of ice-cold *E. coli* MC1061 cells.
5. Transfer to an ice-cold electroporation cuvette (0.1 cm gap) being extra careful to avoid air bubbles (which may cause arcing).
6. Pulse once your electroporation.
7. Immediately, remove the cuvette from the electroporator and add 200 μL SOC media. Gently but surely, resuspend and homogenize cells and transfer to the 1.5 mL tube with warm (37 °C) SOC low-tet media. Keep the tube incubated at 37 °C with 200 rpm agitation for 1 h.
8. Check the electroporation parameters: a time pulse close to 5 ms is expected.
9. Repeat the same procedure until all ligations have been electroporated.
10. After 1 h at 37 °C, centrifuge the tubes with the transformed bacteria for 2 min (2000 × *g*) to pellet the cells.
11. Remove the supernatant and resuspend cells in ~120 μL of SOC media. Plate 100, 10, and 1 μL aliquots from the SOC media in LB-tet plates.

3.5 Colony PCR

Perform colony PCR with 3–5 individual colonies for each phage mutant produced and sequence by Sanger to confirm mutagenesis (see forward and reverse primer sequences in the Reagents and Material section).

1. In a 96-well microtiter plate (round bottom), add 100 μL sterile PBS containing 10% glycerol to each well (enough for the number of colonies that will be analyzed).
2. Pick individual colonies with a sterile toothpick (or a sterile micropipette tip), and transfer to each well containing 100 μL of sterile PBS 10% glycerol. Homogenize well to make sure all bacteria in the colony have been transferred to the liquid. The plate can then be kept at −20 °C for several months.
3. Take 1 μL of each individual diluted colony to perform phage PCR reactions (15 μL reaction; denaturation, 95 °C/2 min; followed by 35 cycles, 95 °C 10 s, 55 °C 15 s, 72 °C 30 s; Platinum Taq Polymerase, ThermoFisher).
4. Also perform a PCR reaction using 10 ng of fUSE5 vector (non-digested).
5. Run a 5 μL aliquot of each PCR reaction on agarose gel (1.2%) to confirm the presence of insert. The PCR product for the fUSE5 vector is 247 bp, and PCR product for the colony PCR should be 247 plus insert size.
6. Sequence by Sanger the products that are positive by PCR to confirm the presence of the mutagenesis. (Use the reverse PCR primer to perform the sequencing reaction.)

3.6 Producing Phage Mutants

Bacteriophage is secreted to the media during cell culture. Here, we describe procedures to generate phage particles by the PEG/NaCl method [17, 18]. In brief, bacterial cells infected with each phage mutant are cultured in LB media (supplemented with streptomycin and tetracycline), and the phage particles which are secreted to the media are precipitated by the addition of polyethylene glycol (PEG) and sodium chloride (NaCl). Then, phage is solubilized in phosphate-buffered solution, and bacterial debris are removed by centrifugation before phage is precipitated one more time. Alternatively, phage from 1 μL of frozen MC1061 colony may be used to infect *E. coli* K91BluKan cells (see below, phage binding assay), which can be cultured in LB-kan/tet to produce phage.

1. In 50 mL flasks containing 10 mL of LB-str/tet, inoculate 5 μL of an individual colony of MC1061 (previously diluted in PBS/glycerol) that is positive for one of the mutageneses; culture the cells overnight at 37 °C with agitation (300 rpm).

Next day:

2. Transfer the cell culture to a 15 mL conical sterile polypropylene tube, and centrifuge at 8000 × *g* for 15 min; transfer the supernatant to a new 15 mL conical sterile polypropylene tube.
3. Add 1.5 mL PEG/NaCl solution. Close with the lead and homogenize by inverting the tube several times. Keep on ice for 30 min.

4. Centrifuge at 10,000 × *g* for 30 min, discard the supernatant, and leave the tube in an inverted position for a few minutes to collect and discard the remaining liquid from the wall of the tube.
5. Centrifuge at 10,000 × *g* for 5 min and discard any remaining liquid.
6. Resuspend the phage pellet with 10 mL of PBS. It helps to incubate the tube at 37 °C with rotation (200 rpm).
7. Centrifuge at 10,000 × *g* for 10 min.
8. Transfer the phage solution to a new 15 mL centrifuge tube, and add 1.5 mL of PEG/NaCl solution. Close and homogenize by inverting the tube several times. Leave on ice for 30 min.
9. Centrifuge at 10,000 × *g* for 30 min, discard the supernatant, and leave the tube in an inverted position for a few minutes to collect and discard the remaining liquid from the wall of the tube.
10. Centrifuge at 10,000 × *g* for 5 min and discard any remaining liquid.
11. Resuspend the phage pellet with 100 μL of sterile PBS. It helps to incubate the tube at 37 °C with rotation (200 rpm).
12. Transfer the phage solution to a 1.5 mL polypropylene tube, and centrifuge at 15,000 × *g* for 10 min. Transfer supernatant to a new 1.5 mL polypropylene tube.
13. Keep each phage mutant at 4 °C.

3.7 Phage Titration

Before proceeding to the phage binding assay, it is important to determine the number of transforming units (TU) you have obtained for each phage mutant. Make serial dilutions of each phage, infect *log*-phase *E. coli* K91BluKan cells, and plate them in LB kan/tet agar plates. Next day, count the number of colonies and determine how many TU/mL each phage mutant has yielded. Expect phage yield to be in the range of 10^8–10^9 TU/μL. Make sure you also prepare and titer the control insertless phage (Fd-tet).

3.8 Quantitative Phage Binding Assay

To assess the impact that each amino acid substitution has in protein binding, individual phage mutant is incubated with the immobilized ligand, and the number of phage particle bound to it is quantified by colony count or quantitative qPhage [19].

1. In a 96-well microtiter plate (flat bottom), immobilize the protein of interest by coating wells with 100 μL of a protein solution (10 μg/mL) (overnight at 4 °C). Coat duplicate wells for each phage to be tested.

Next day:

2. Wash wells three times with 100 μL PBS and add 200 μL of PBS 3% BSA. Block wells for 2 h.

   Note: After 1-h blocking, you should start a bacterial culture for the recovery of the cell-bound phage. Inoculate *E. coli* K91BluKan bacteria in TB sup/kan and culture cells at 37 °C with vigorous agitation (300 rpm) until A_{600} is 1.5–2.2.
3. Dilute the corresponding phage in order to add 10^9 TU to each well in 100 μL of PBS 3% BSA. Incubate for 1–2 h. Use the original non-mutated phage as positive control and the Fd-tet phage as negative control (no binding).

   Note: Phage addition can be adjusted to 10^8 TU for peptides with high affinity for their specific target.
4. Once the *E. coli* K91BluKan bacteria have reached the desired growth, start washing the wells to remove unbound phage.
5. Wash well ten times with 200 μL of PBS.
6. Add 100 μL of *E. coli* K91BluKan bacteria to recover phage bound to the immobilized protein. Leave for 30 min at room temperature.
7. Transfer 100 μL of the bacterial culture from each well to individual 1.5 mL tubes containing 900 μL of LB-kan/tet (tube A). Homogenize well. Make another 1/10 dilution from each tube by transferring 100 μL to another 1.5 mL tube containing 900 of LB-kan/tet.
8. Plate in triplicate 100 μL from tube A, and 10 and 100 μL of tube B, onto LB-kan/tet agar plates.

   Note: To facilitate plating, use the glass bead method to evenly spread the dilutions onto each plate.
9. Next day, calculate phage binding to each well by colony count.

   Note: Estimate the effect of each amino acid mutation based on the number of phage bound to the target protein relative to the original peptide (full binding) and the background binding of the insertless phage (Fd-tet) (no binding).

3.9 Studying Peptide Receptor Interaction by NMR

As a starting point, it is interesting to first understand how the free peptide behaves in solution. Take into account that the exchange between two or more conformations may lead to the appearance of two or more resonance lines for the different conformers. Thus, it is important to determine if the observed resonances belong to a single or more than one conformation being sampled by the peptide. For this, the first step will be performing 1D ^{1}H NMR spectroscopy to obtain a general view of the peptide dynamics and compare the number of resonances lines observed in the spectra with those expected for a unique peptide conformation.

The second step is two assign each of the resonances observed in the ^{1}H NMR spectrum. For this, a combination of 2D ^{1}H NMR spectra, such as correlation spectroscopy (COSY), total correlation spectroscopy (TOCSY), and nuclear Overhauser effect spectroscopy (NOESY), are collected. The COSY and TOCSY correlate hydrogens that are separated by up to three covalent bonds, while NOESY correlates hydrogens through space, enabling the connection of adjacent amino acid residues and the "walking" through the peptide sequence. A detail description of the sequential assignment strategy was described by Wuthrich [20].

The 1D ^{1}H spectra may be sufficient to get an overall view on the peptide conformational flexibility; however, other types of NMR experiments such as chemical exchange saturation transfer (CEST) and Car–Purcell–Meiboom–Gill (CPMG) NMR methods could be useful to determine the existence of other conformational states and their exchange timescales. To facilitate the study, in cases where the peptide has multiple conformations, one can decrease the temperature of the NMR experiments. Working at lower temperatures tends to slow down the exchange rate and stabilize preferential conformations.

Once the peptide has been characterized in solution and its resonance signals have been identified for each residue, we can start testing the peptide interaction with its ligand. In this context, the exchange regime in the binding interaction is an important parameter to take into consideration, since it will affect the appearance of resonance lines in the spectra. For example, let's consider the exchange of a spin between the peptide-bound and peptide-free conformations. In the fast exchange regime ($k_{ex} \gg \Delta\omega/2\pi$), it will be observed a single resonance signal due to the averaging of the chemical shifts of both conformations. This is because the lifetime of the peptide-free and peptide-bound forms is too short for the observation of individual resonance signals, which leads to the collapse of both signals into a single peak (Fig. 2). On the contrary, in the slow exchange regime ($k_{ex} \ll \Delta \gg \Delta\omega/2\pi$), there is not averaging of the chemical shifts, and two resonance lines will be observed in the spectra, one for the free and other for the bound conformer. $\Delta\omega/2\pi$ is the difference between the chemical shifts of the free and bound conformation of the given spin, and k_{ex} is the exchange rate for the binding reaction. Both are expressed in the frequency unit s^{-1}.

Weak peptide–ligand binding interactions are generally observed in the fast exchange regime. In our experience, this is the preferred regime for peptides identified by phage display [7, 13]. Therefore, the addition of a small amount of receptor over the peptide concentration will lead to the formation of a peptide–receptor complex in fast exchange with the peptide-free form, and the NMR parameters will reflect the bound state even if

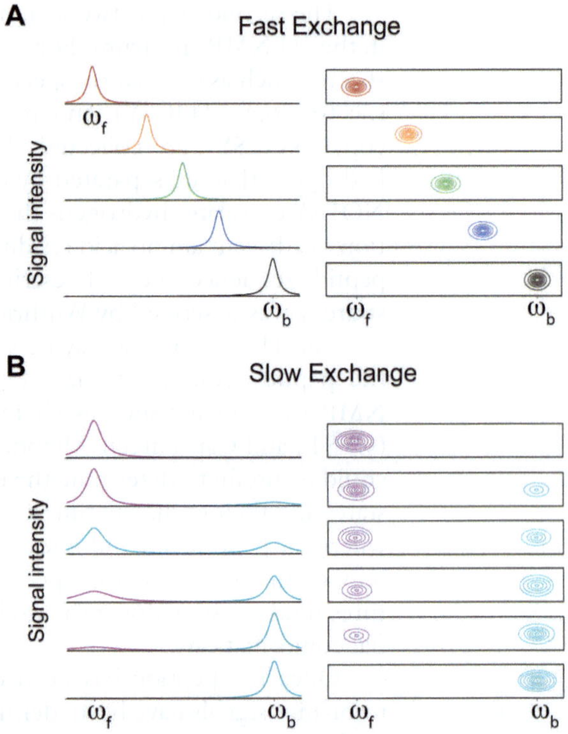

Fig. 2 Simulation of an NMR signal titration with increasing concentrations of ligand. The figure illustrates 1D and 2D spectrum simulations in the (**a**) fast exchange regime of the fully free peptide signal (ω_f) "walking toward" the fully ligand-bound peptide signal (ω_b) or in the slow exchange regime (**b**), where the intensity of the free peptide signal (ω_f) decreases while the ligand-bound peptide signal (ω_b) increases as the ligand is added to the system

an excess of free peptide is used [13]. Thus, chemical shift changes in the peptide can be tracked by performing NMR experiments with a large molar excess of peptide over ligand in the fast exchange scenario.

NMR Experiments

3.10 General Setup

1. Once your sample is place inside the spectrometer, adjust the working temperature and let it equilibrate.
2. Lock on according to your sample conditions. In general, the sample is buffered in a deuterated or non-protonated buffer, and it contains 5% of D_2O for lock purpose.
3. Perform the matching and tuning for each channel, followed by shimming.
4. Calibrate pulses and improve water suppression pulse length.

3.11 Setting Up a 1D ^1H Experiment

1. First start by acquiring a ^1H spectrum of your peptide at 25 °C, and determine if the number of observed resonance lines matches the expected ones.
2. If the observed resonance lines are greater in number than expected, acquire a series of one-dimensional ^1H spectrum of your peptide within a range of temperatures (e.g., from 0 °C to 45 °C) (see Fig. 1 in reference [13]).
3. Set a working temperature by comparing the recorded ^1H spectrum, and choose the one that better matches the number of resonance lines expected with improved signal to noise ratio.
4. Once the temperature has been set, acquire a ^1H spectrum of your peptide using 4096 data points with 256 scans, for example.
5. Identify the resonance signals for each amino acid present in your peptide (you will have to assign your peptide; see next session). Assign your peptide through the use of NOESY and TOCSY experiments.
6. Acquire a ^1H spectrum of your peptide at increasing concentrations (from 50 μM up to 1 mM), and determine the optimal concentration in which oligomeric states are not observed. To improve sensitivity, use the maximum concentration in which there is no peptide oligomerization. We have used peptide concentrations between 140 μM and 400 μM [7, 13].
7. Next, acquire the baseline ^1H spectrum of your peptide at the optimum temperature and peptide concentration.
8. Add a small amount of your receptor–peptide ligand (you may start with 1 μM of ligand concentration), and record the ^1H spectrum of your peptide.
9. Add more ligand to the system and continue acquiring the ^1H spectrum tracking for signal changes. We have observed interactions in fast exchange regime for peptide–receptor in the range of 1:80–1:200 (mol/mol).
10. Track the presence of shifts on the peptide resonance lines as more ligand is added to the sample. If changes are observed, the experiment may be repeated with increasing concentrations of our peptide (see Fig. 2 in references [7, 13]).

Resonance Assignment

If changes in the resonance lines are observed, it is then necessary to assign them to the corresponding amino acid in the peptide. These can be achieved using standard TOCSY and NOESY experiments. The TOCSY and NOESY experiments may be carried out and the resonances of the peptide assigned either free in solution or in the presence of the receptor [21–23]. These analyses may further circumvent eventual ambiguities observed in the ^1H spectrum and improve the identification of side-chain interactions [7, 13].

3.12 Setting Up 2D ^{1}H Experiments

The ^{1}H resonances of small peptides may be achieved through the use of homonuclear ^{1}H-^{1}H 2D experiments TOCSY and NOESY [21–23]:

1. Prepare you peptide sample at a concentration where it remains monomeric in solution.
2. Assign the ^{1}H resonances of the peptide using TOCSY and NOESY in the absence of ligand.
3. Gradually add small amounts of the ligand to the sample and record the spectra each time.
4. Finally, record TOCSY and NOESY spectra in the presence of an equimolar and/or saturating concentration of ligand and track changes in the chemical shift. If there is an interaction between the peptide and the ligand, we will observe chemical shift changes (rather small in general).

The protocols for TOCSY and NOESY experiments are described briefly as follows.

3.13 Setting Up a TOCSY Experiment

TOCSY (total correlation spectroscopy) is based on the magnetization transfer between spins that are coupled through-bond interactions. In the TOCSY spectrum, cross peaks are generated between all members of a coupled spin network, that is, between nuclei connected by a chain of couplings in addition to nuclei that are directly coupled [24]. Therefore, the TOCSY experiment contains all cross peaks among protons of the same amino acid residue [23]:

1. Set up a new experiment with the appropriate TOCSY pulse sequence [7, 13]. The TOCSY experiment utilizes a mixing time in which the net magnetizations of the spins spend most of their time during t_{mix} along the xy plane. This way they exchange phase information through spin–spin (T2) relaxation, which generates the TOCSY cross peaks [25].
2. Set the spin-lock mixing time using the MLEV-17 pulse sequence [26]. The average mixing time is around 70–80 ms, but it will depend on the spin's T2 value of your molecule and the size of the J-couplings. Large mixing times could cause a signal decrease due to T2 relaxation and overheating of the sample [25].
3. Use the WATERGATE technique for water suppression.
4. Collect the spectrum with 512 data points in F1 and 4096 data point in F2, with 32 scans (this is a suggestion and the number points and scans may be adjusted depending on your samples and/or experimental conditions).

3.14 Setting Up a NOESY Experiment

The NOESY (nuclear Overhauser effect spectroscopy) experiment relies on the coherence transfer between spins through dipolar coupling (through-space interactions), showing a spectrum with a series of cross peaks disposed on both sides of a diagonal, for spatially close protons [23].

NOESY explores the properties of the NOE (nuclear Overhauser enhancement) effect. The NOE is an extremely powerful tool in structural studies by NMR, since it is used to generate distance restrains for structural calculation. For a pair of spin that are dipolar-coupled, it turns out that their relaxations are not independent and experiment cross-relaxation, leading to the nuclear Overhauser effect (NOE) [27]. This means that if there is a dipolar relaxation between two spins, the behavior of the z-magnetization of one spin will be affected by the amount of z-magnetization of the other. At the same time, the cross-relaxation rate constant is proportional to the cross-peak volume in the NOESY spectrum [23, 24].

Therefore, we may use the NOESY to deduce the distance between the spin pair using the cross-peak volume, since there is a relationship between cross-peak volume/intensity and the spatial distance among the coupled spins [23]:

1. Set up a new experiment with the NOESY pulse sequence as shown in Fig. 2. The mixing time in NOESY allows the transference of magnetization via cross-relaxation or chemical exchange, and the final 90° pulse creates transverse magnetization which is detected.
2. Set the experiment mixing time. It will depend on the size of your molecule. For biomolecules it generally used a mixing time between 0.005 s and 0.2 s [28]. You may test different mixing times between 50 and 200 ms and according to your peptide characteristics.
3. Use the WATERGATE technique for water suppression.
4. Collect the spectrum with 512 data points in F1 and 4096 data point in F2, and 16–80 scans (this is a suggestion and the number point and scans may be adjusted depending on your samples and/or experimental conditions).

3.15 Peptide-Mimetic Design

Peptides are susceptible to protease degradation and can be difficult to study using in vitro assays and even more challenging when applied to in vivo models. On the other hand, the development of peptidomimetic compounds by medicinal chemistry or biophysical approaches is expensive and time-consuming [29]. Nevertheless, the primary sequence of a peptide is vital for the biological activity of the peptide, and the retroinversion approach takes advantage of this. These retroinverted peptides have proven to be valuable assets for drug development, including the development of biologically

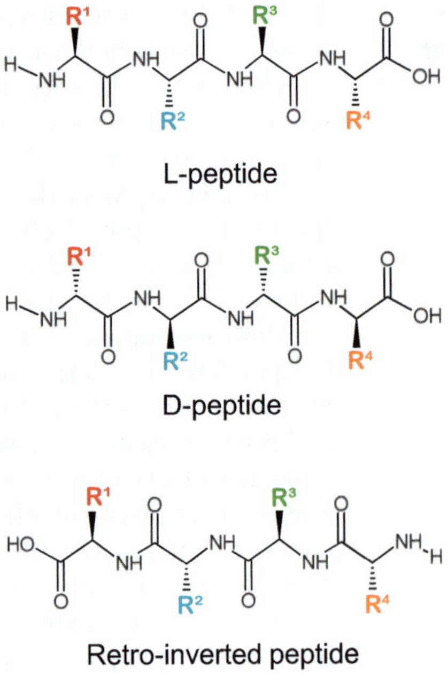

Fig. 3 The retroinversion approach for peptidomimetic design. The figure illustrates using the Fischer projection a biologically active tetrapeptide (L-amino acids). If the amino acids are switched to D-enantiomers, note that change in the spatial orientation of the side chains. However, by inverting at the same time the amino acid sequence, one generates a retroinverted peptidomimetic molecule that preserves the orientation of the side chains and is now resistant to protease degradation

active peptidomimetic version of peptide identified by phage display [8, 9]. The retroinversion approach consists in synthesizing a peptide by using D-amino acids (retro isomers) and inverting the primary sequence of the original peptide (inverted) (Fig. 3). This dual change in the peptide sequence and spatial conformation leads to a stable version of the peptide since biological systems are not well prepared to recognize D-amino acids and their bonds; at the same time, it maintains the spatial orientation of the amino acid side chains, essential for the biological activity of the peptide. This approach is particularly effective for shorter sequences [30]. Therefore, the combination of alanine-scanning and NMR studies may lead to better understand the structural requirements for any give peptide–receptor interaction and the design of shorter peptide. For example, the octapeptide CPQPRPLC, a VEGF receptor binding peptide, requires only the RPL motif for interaction. This information allowed for the design of a much shorter peptidomimetic compound, $_{\text{D}}$(LPR), using the retroinversion approach [8, 13]. The $_{\text{D}}$(LPR) peptidomimetic compound (and its cyclic version) both preserved its VEGF receptor binding activity and inhibited retinal angiogenesis in vitro and in vivo [8, 11]. In summary, phage display allowed for the identification of a novel endothelial cell binding peptide, which was converted by phage and NMR studies into a peptidomimetic compound with potential clinical applications. We envision that this pipeline could be successfully applied to any protein or biological system of interest for the identification of targeting peptides and the development of peptidomimetic compounds for preclinical development:

1. Based on the peptide primary sequence and previous understanding of amino acid residues important for peptide activity, design shorter version of retroinverted peptides.
2. Order small batches (1 mg) of each retroinverted peptide, and test them with a phage binding assay. The retroinverted peptide should be able to compete with the native peptide for receptor binding.
3. Retroinverted peptides that successfully inhibit phage binding can also be assessed by NMR to confirm binding to the receptor.
4. Once a suitable retroinverted peptide version is identified, order larger batches (10–100 mg) to be used in biological assays, such as cell migration/proliferation and in vivo models.

References

1. Santos R, Ursu O, Gaulton A et al (2017) A comprehensive map of molecular drug targets. Nat Rev Drug Discov 16(1):19–34
2. Wang L, Wang N, Zhang W et al (2022) Therapeutic peptides: current applications and future directions. Sig Transduct Target Ther 7(1):48
3. Muttenthaler M, King GF, Adams DJ, Alewood PF (2021) Trends in peptide drug discovery. Nat Rev Drug Discov 20(4):309–325
4. Moraes BC, Ribeiro-Filho HV, Roldão AP et al (2022) Structural analysis of TrkA mutations in patients with congenital insensitivity to pain reveals PLCγ as an analgesic drug target. Sci Signal 15(731):eabm6046
5. Gray BP, Brown KC (2013) Combinatorial peptide libraries: mining for cell-binding peptides. Chem Ver 114(2):1020–1081
6. Smith GP (2019) Phage display: simple evolution in a Petri Dish (Nobel Lecture). Angew Chem Int Ed Engl 58(41):14428–14437
7. Cardó-Vila M, Zurita AJ, Giordano RJ et al (2008) A ligand peptide motif selected from a cancer patient is a receptor-interacting site within human interleukin-11. PLoS One 3(10):e3452
8. Giordano RJ, Cardó-Vila M, Salameh A et al (2010) From combinatorial peptide selection to drug prototype (I): targeting the vascular endothelial growth factor receptor pathway. Proc Natl Acad Sci U S A 107(11):5112–5117
9. Cardó-Vila M, Giordano RJ, Sidman RL et al (2010) From combinatorial peptide selection to drug prototype (II): targeting the epidermal growth factor receptor pathway. Proc Natl Acad Sci U S A 107(11):5118–5123
10. Pasqualini R, Millikan RE, Christianson DR et al (2015) Targeting the interleukin-11 receptor α in metastatic prostate cancer: a first-in-man study. Cancer 121(14): 2411–2421
11. Sidman RL, Li J, Lawrence M et al (2015) The peptidomimetic Vasotide targets two retinal VEGF receptors and reduces pathological angiogenesis in murine and nonhuman primate models of retinal disease. Sci Transl Med 7(309):309ra165
12. Giordano RJ, Lahdenranta J, Zhen L et al (2008) Targeted induction of lung endothelial cell apoptosis causes emphysema-like changes in the mouse. J Biol Chem 283(43): 29447–29460

13. Giordano RJ, Anobom CD, Cardó-Vila M et al (2005) Structural basis for the interaction of a vascular endothelial growth factor mimic peptide motif and its corresponding receptors. Chem Biol 12(10):1075–1083
14. Chorev M (2005) The partial retro-inverso modification: a road traveled together. Biopolymers 80(2–3):67–84
15. Fischer PM (2003) The design, synthesis and application of stereochemical and directional peptide isomers: a critical review. Curr Protein Pept Sci 4(5):339–356
16. Seidman CE, Struhl K, Sheen J et al (2001) Introduction of plasmid DNA into cells. Curr Protoc Mol Biol Chapter 1:Unit1.8
17. Barbas CFIII, Burton DR, Scott JK, Silverman GJ (eds) (2001) Phage display: a laboratory manual. Cold Spring Harbor Laboratory, Cold Spring Harbor
18. Michaloski JS, Redondo AR, Leila S, Magalhães LS et al (2016) Discovery of pan-VEGF inhibitory peptides directed to the extracellular ligand-binding domains of the VEGF receptors. Sci Adv 2:e1600611
19. Dias-Neto E, Nunes DN, Giordano RJ et al (2009) Next-generation phage display: integrating and comparing available molecular tools to enable cost-effective high-throughput analysis. PLoS One 4(12):e8338
20. Wuthrich K (1986) NMR of proteins and nucleic acids. Wiley-Interscience Publication, Wiley, New York
21. Tinoco LW, Da Silva A, Leite A et al (2002) NMR structure of PW2 bound to SDS micelles. A tryptophan-rich anticoccidial peptide selected from phage display libraries. J Biol Chem 277(39):36351–36356
22. Tinoco LW, Gomes-Neto F, Valente AP, Almeida FCL (2007) Effect of micelle interface on the binding of anticoccidial PW2 peptide. J Biomol NMR 39(4):315–322
23. Vincenzi M, Mercurio FA, Leone M (2020) NMR spectroscopy in the conformational analysis of peptides: an overview. Curr Med Chem 28(14):2729–2782
24. Agarwal N, Nair MS, Mazumder A (2018) Characterization of nanomaterials using nuclear magnetic resonance spectroscopy. In: Characterization of nanomaterials. Woodhead Publishing, Sawston, Cambridge. pp 61–102
25. Simpson JH (2012) Through-bond effects. Organic structure determination using 2-D NMR spectroscopy: a problem-based approach. Academic Press, pp 123–168
26. Bax AD, Donald GD (1969) MLEV-17-based two-dimensional homonuclear magnetization transfer spectroscopy. J Magn Reson 65(2): 355–360
27. Kumar, A, Rani Grace, RC (eds) (2017) Nuclear overhauser effect. Encyclopedia of spectroscopy and spectrometry, 3rd edn, 1946–1955. Academic Press, San Diego
28. Ziegler P (2017) TopSpin guide book. Bruker Corporation. Germany, available at: https://www.lsu.edu/science/chemistry/research/nmr/files/basic_1d_2d.pdf
29. Perez JJ (2018) Designing peptidomimetics. Curr Top Med Chem 18(7):566–590
30. Doti N, Mardirossian M, Sandomenico A (2021) Applications of retro-inverso peptides. Int J Mol Sci 22(16):8677

Part II

Engineering Phages

Chapter 6

Contracting the Host Range of Bacteriophage T7 Using a Continuous Evolution System

Tzvi Holtzman, Ram Nechooshtan, Ido Yosef, and Udi Qimron

Abstract

Bacteriophage T7 is an intracellular virus that recognizes its host via tail and tail fiber proteins known as receptor-binding proteins (RBPs). The RBPs attach to a specific lipopolysaccharide (LPS) displayed on the host. While there are various reports of phage host range expansion resulting from mutations in the RBP encoding genes, there is little evidence for contraction of host range. Notably, most experimental systems have not monitored changes in host range in the presence of several hosts simultaneously. Here, we use a continuous evolution system to show that T7 phages grown in the presence of five restrictive strains and one permissive host, each with a different LPS, gradually cease to recognize the restrictive strains. Remarkably, this result was obtained in experiments with six different permissive hosts. The altered specificity is due to mutations in the RBPs as determined by gene sequencing. The results of using this system demonstrate a major role for RBPs in restricting the range of futile infections, and this process can be harnessed to reduce the host range in applications such as recognition and elimination of a specific bacterial serotype by bacteriophages.

Key words Continuous evolution system, Continuous fermentation, Aseptic techniques, Specialist phage, Permissive host, Restrictive strain

1 Introduction

Bacteriophages (phages) are ubiquitous biological entities that are found in habitats occupied by bacteria. As such, they influence the ecosystem of their environment. Phages and bacteria exert mutual selection pressures in a never-ending molecular arms race, which forces phages to evolve and adapt to changing conditions. One of the ways in which phages can adapt is through modifying their host range.

A bacteriophage's host range is defined as the list of hosts that it is capable of infecting [1]. This range is dependent on host factors including defense mechanisms such as CRISPR-Cas [2] and restriction–modification systems [3] and the presence of phage receptors [4]). It may also depend on environmental factors such

Huan Peng et al. (eds.), *Phage Engineering and Analysis: Methods and Protocols*,
Methods in Molecular Biology, vol. 2793, https://doi.org/10.1007/978-1-0716-3798-2_6,

as temperature and pH [5, 6]), and on phage-encoded features such as the receptor-binding proteins (RBPs) [7]). Specialist bacteriophages commonly display a narrow host range, i.e., they infect a limited number of bacterial strains of the same species. In contrast, generalist bacteriophages inherently display a broad host range [8, 9]. The host range can be artificially expanded [7] or contracted: for example, the bacteriophage lambda has been reported to evolve to improve binding to a specific receptor while losing the ability to bind a previously recognized receptor [10].

From an evolutionary point of view, the host range of a phage is reminiscent of the situation of animals that choose one food source over another or specialize in foraging and digesting a specific food source. Bull and colleagues [11] built a mathematical model of phage–bacterium interactions based on the optimal foraging theory. This uses the ability of phage T7 to discriminate between two bacterial strains in order to validate the mathematical predictions. Importantly, the phage can evolve to infect one strain of *Escherichia coli* while avoiding another strain that differs only in the surface molecules. Their results identified a single phage gene, 17, as responsible for the discrimination between the hosts.

Gene 17 encodes tail fiber proteins that are essential for absorption because they recognize the outer membrane lipopolysaccharide (LPS) of *E. coli* [12]. The abundant LPS molecules on the bacterial outer membrane are composed of lipid A, which is embedded in the outer membrane, and an appended core region containing keto-deoxy-d-manno-8-octanoic acid, heptose, galactose, and glucose. The biosynthetic pathway responsible for the core region is partly catalyzed by enzymes, which are encoded by the *E. coli* genes *waaC*, *waaF*, *waaG*, *waaO*, *waaR*, and *waaU*. It is possible to construct *E. coli* hosts lacking all these genes, with the exception of *waaU*, which is presumed to be an essential gene [13, 14]. Deletion of any one of the other genes produces mutant forms of the LPS on the outer membrane of the bacteria.

Our aim was to investigate whether the bacteriophage T7 can adapt and evolve to distinguish between altered forms of the natural LPS receptor. Specifically, we aimed to evolve T7 specialists with the ability to recognize hosts with an altered LPS while avoiding hosts with other LPS mutations, or even with wild-type (WT) LPS. Unlike Bull and colleagues [11] who propagated T7 phage on a mixture of two hosts at a time, we evolved T7 in a mixture of up to six hosts—each with a different type of LPS. Furthermore, we used a continuous evolution system that enables us to propagate ~100 phage generations in as little as 95 h. The use of a continuous culture [15] for phage evolution studies is well described. Some systems have been used to study the coevolution of bacteria and phage [16], where positive selection is the driving force for the evolutionary process. Other evolution systems have been used to provide improved biomolecules [17, 18]. Here, we used our continuous evolution system to study the development of phage avoidance [19]. To this end, only one of the six hosts in the mixture in a given experiment is permissive for the propagation of T7, while the other strains are restrictive due to deletion of *trxA*. This gene encodes thioredoxin, and is essential for T7 phage replication because the protein serves as a subunit of the phage DNA polymerase [20]. While the absence of thioredoxin halts phage replication and propagation, it does not affect adsorption and DNA injection. This protocol drives the evolution of phage T7 toward recognizing a specific LPS form while avoiding others. Using this method, we evolved six different specialist T7 phages with mutations in tail genes 11 and 12, and in the tail fiber gene 17. The evolved phages all exhibit poor recognition of LPS types other than the one they evolved to recognize. The results of these experiments suggest that a major role of the RBPs is to prevent futile injection of phage DNA by avoiding restrictive strains. This contrasts with the common intuitive perception that RBPs evolved to improve the recognition of the natural host.

2 Materials

System description: A schematic drawing of the continuous evolution system is shown in Fig. 1. The system is composed of two parts (*see* Subheadings 2.1 and 2.2) that are prepared and autoclaved separately (*see* Subheadings 3.1 and 3.2), and are then assembled under sterile conditions (*see* Subheading 3.4).

2.1 Continuous Evolution System: Section I "Wet Part"

The term refers to the mode of preparation and the autoclave program used for sterilization.

The "wet part" comprises two 10 L GL-45 glass media bottles that form the LB medium reservoir that feeds the chemostat. The first media bottle is connected with a silicon tube to the second media bottle, which in turn is connected by a silicon tube to the chemostat vessel (*see* Subheading 3.1 for "wet part" preparation).

2.2 Continuous Evolution System: Section II "Dry Part"

The term refers to the mode of preparation and the autoclave program used for sterilization.

The "dry part" of the system contains two spinner flasks, where the first is the chemostat and the second is the lagoon. The inlet of the chemostat receives fresh LB medium from the media reservoir and the seed host strain culture. A silicon tube connects the outlet of the chemostat to the inlet of the lagoon in order to feed the lagoon continuously with bacterial culture. This tube also contains branches that serve as a sampling port and provide a method to bypass the lagoon to the waste bottle when required. The outlet of the lagoon is connected by a silicon tube to the waste bottle and has a side tube that serves as a sampling port (*see* Subheading 3.2 for "dry part" preparation).

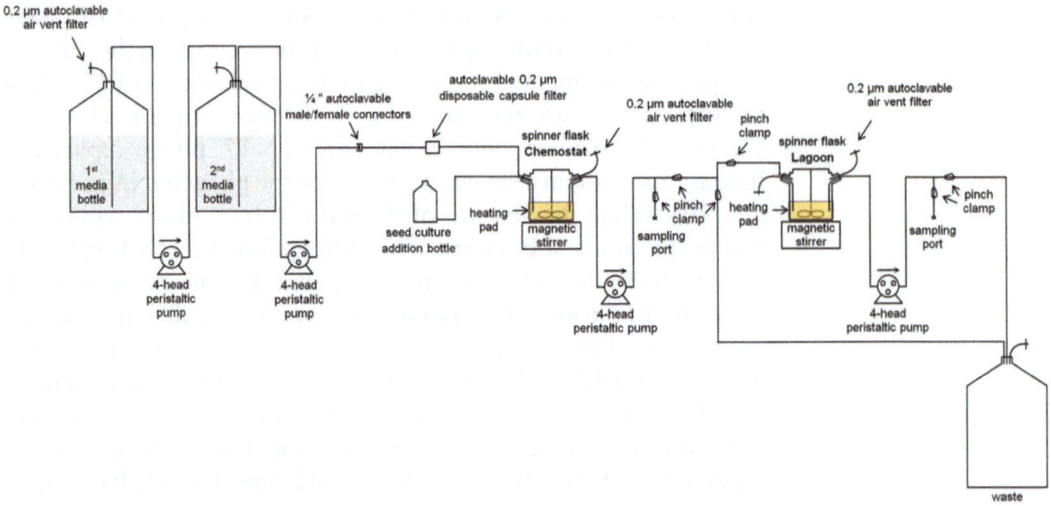

Fig. 1 A schematic drawing of the entire continuous culture system for phage evolution. The system contains two media bottles that serve as media reservoirs. Media are drawn through the two bottles to the chemostat by a peristaltic pump. In the same way, bacteria are pumped into the lagoon at a rate that maintains the logarithmic state of the culture. T7 bacteriophages are inoculated initially into the lagoon where they continuously infect the bacteria mixture. Both the chemostat and the lagoon are diluted at the same constant rate. Excess infected media are pumped into a waste container by the peristaltic pump

2.3 Strains

1. Glycerol stocks of *E. coli* JW5856, JW3596, JW3595, JW3602, JW5856, JW3606, JW1656, IYB5709, IYB5758, TH4, TH5, TH6, TH7, and TH8 (*see* **Note 1**).
2. The WT bacteriophage T7 used in these experiments has a Gp12 S694P substitution compared to the published sequence (GenBank accession no. CAA24430.1) but is otherwise similar to GenBank AY264774.

2.4 Solutions

All solutions are prepared using double-distilled water (DDW). All reagents are prepared and stored at room temperature (RT) unless indicated otherwise:

1. LB medium: Add 10 g Bacto-tryptone, 10 g yeast extract, and 5 g NaCl to 800 mL DDW. Stir until clear, adjust volume to 1 L with DDW, sterilize by autoclaving, and store at RT.
2. Kanamycin 50 mg/ml (Kan): Dissolve 0.5 g Kan in 10 mL DDW. Filter through a 0.22 µm syringe filter and store at −20 °C.
3. LB agar plates: Add 15 g agar to 1 L LB (*see* Subheading 2.4.1) and sterilize by autoclaving. Cool to about 45 °C, add Kan (*see* Subheading 2.4.2) to a final concentration of 25 µg/mL, and pour 20 mL into Petri plates. Allow to solidify and leave to dry overnight at RT. Store at 4 °C for up to 1 month.

4. Overlay or top agar: Prepare LB medium (*see* Subheading 2.4.1) with the addition of agar to a concentration of 5 g/L. Dispense the medium in 100 mL aliquots into screw-capped glass bottles. Sterilize by autoclaving. Store at 4 °C for up to 6 months. Before use, heat the bottle containing the top agar in a microwave to melt the agar, and keep the bottle on a 45–50 °C heating block until use.

2.5 Other Materials Used

These include 10 L glass vessels, 20 L glass vessel, spinner flasks, silicon tubing, autoclavable air vent filters, autoclavable 0.22 μm disposable capsule filter, pinch clamps, male/female connectors, magnetic stirrers, 4-channel peristaltic pump, temperature-controlled heating pads (for spinners) or heating room/large incubator, and aspirator bottles with a silicon tubing outlet.

3 Methods

3.1 Preparation of Continuous Evolution System "Wet Part" (Fig. 2)

1. Weigh LB medium ingredients (80 g Bacto-tryptone, 80 g yeast extract, 40 g NaCl), and add to each of the 10 L GL-45 glass bottles. Add DDW to 8 L.
2. Prepare a two-port cap for the first media bottle: one port is connected by a silicon tube to an autoclavable 0.2 μm air vent filter. Connect the second port on the inside to a dip tube long enough to reach the bottom of the bottle and on the outside to a silicon tube that is connected to the second media bottle (*see* **Note 2**).

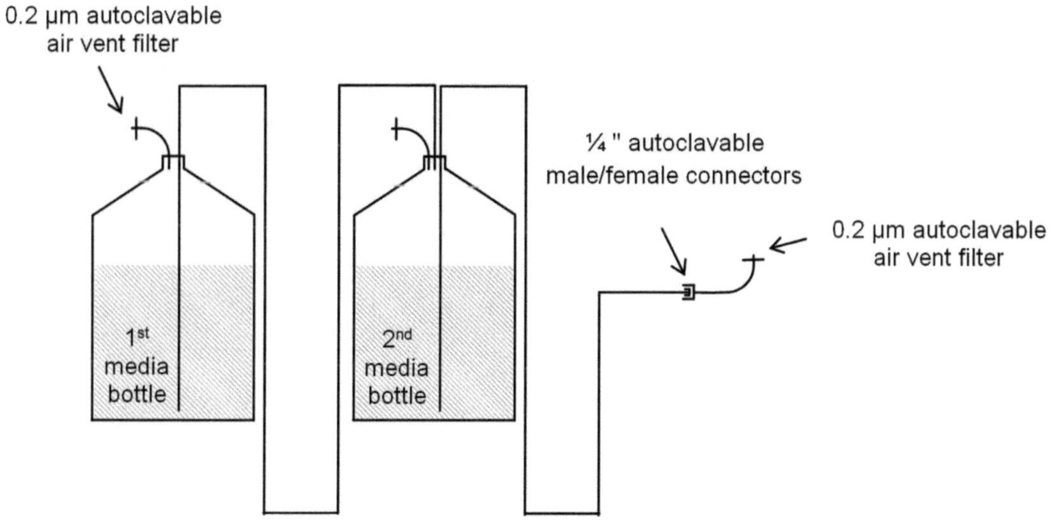

Fig. 2 A schematic drawing of the "wet part" of the continuous evolution system. See text for details

3. Prepare a three-port cap for the second media bottle: connect one port by a silicon tube to an autoclavable 0.2 μm air vent filter. Use a tube to connect the second port to the second port in the cap of the first media bottle. Connect the third port on the inside to a dip tube long enough to reach the bottom of the bottle, and on the outside, use a silicon tube to connect to a ¼" autoclavable male connector (*see* **Note 3**). Plug the male connector into a ¼" female connector connected to a 0.2 μm autoclavable air vent filter.
4. Sterilize the "wet part" of the continuous evolution system in an autoclave equipped with a liquids program. The temperature probe that controls the autoclave cycle should be dipped in a 10 L GL-45 glass bottle containing 8 L of DDW, similar to the volume of LB media in the media bottles. The sterilization cycle should be no less than 15 min at 121 °C in order to ensure full sterilization of the medium. After sterilization, allow to cool to RT and assemble the system by combining with the "dry part" (*see* Subheading 3.4).

3.2 Preparation of Continuous Evolution System "Dry Part" (Fig. 3)

1. Prepare a two-port entrance cap for the chemostat (*see* **Note 4**):

   Connect one end of a silicon tube to a port with a dip tube and the second end to a ¼" female connector, which is plugged into a ¼" male connector during the sterilization process and connected by a tube to a 0.2 μm autoclavable air vent filter. Connect one end of a silicon tube to a port in the cap with a short tube. Connect the second end of the silicon tube to the exit side of an autoclavable 0.2 μm disposable liquid capsule

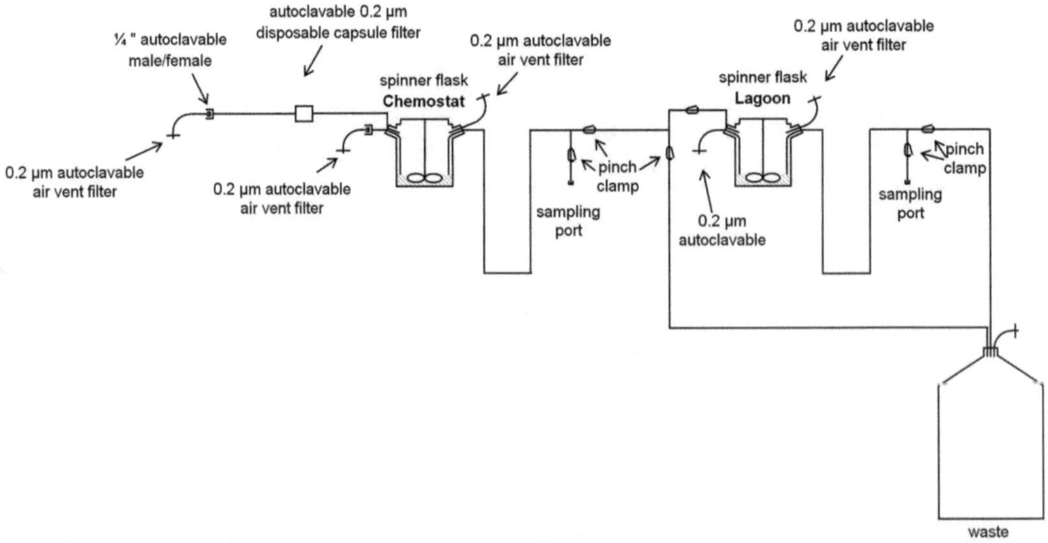

Fig. 3 A schematic drawing of the "dry part" of the continuous evolution system. See text for details

filter (*see* **Note 5**). Connect one end of a silicon tube to the entrance side of the filter and the second end to a ¼" female connector which is plugged into a ¼" male connector during the sterilization process and connected by a tube to a 0.2 μm autoclavable air vent filter.

2. Prepare two sampling ports: a silicon tube with a pinch clamp is connected to a ¼" male connector that is plugged into a ¼" female connector during the sterilization process and connected by a tube to a 0.2 μm autoclavable air vent filter.
3. Prepare a two-port exit cap for the chemostat: connect one end of a silicon tube to a port with a short tube, and the second end to an autoclavable 0.2 μm air vent filter. Connect one end of a silicon tube to the dip tube port of the cap. Connect the second end of this tube to the stem of a Y junction connector. Connect a sampling port (*see* Subheading 8) to one outlet. Pull a silicon tube through a pinch clamp, and connect one end of this silicon tube to the second outlet. Connect the other end of the tube to the stem of a second Y junction connector. Pull another silicon tube through a pinch clamp, and connect one end of this silicon tube to one outlet. The other end of this tube will be connected to the lagoon. Pull a silicon tube through a pinch clamp, and connect it to the second outlet of the second Y junction connector, where the other end will be connected to the waste bottle (*see* **Note 6**).
4. Prepare a two-port entrance cap for the lagoon: one port with a dip tube is connected by a silicon tube to an autoclavable 0.2 μm air vent filter. Another silicon tube is used to connect the second port to a branch of the second Y junction connector that is connected to the outlet of the chemostat (*see* Subheading 3.2.3).
5. Prepare a two-port exit cap for the lagoon: one port with a short tube is connected by a silicon tube to an autoclavable 0.2 μm air vent filter. Connect one end of a silicon tube to the second port with a dip tube (*see* **Note 7**) and the other end to the stem of a Y junction connector. Connect to one branch of the Y junction connector to a sampling port. Pull a silicon tube through a pinch clamp, and connect one end of this silicon tube to the second outlet, where this tube will be connected to the waste bottle.
6. Prepare the waste bottle: use a 20 L bottle with a three-port GL-45 cap. Connect one end of a silicon tube to a port in the cap, and the other end to an autoclavable 0.2 μm air vent filter. Connect the silicon tube from the exit cap of the chemostat (*see* Subheading 9) to a port in the cap. Connect a silicon tube from the exit of the lagoon (*see* Subheading 3.2.5) to a port in the cap.

7. Autoclave the dry system in an autoclave with a solids cycle, using a pre-vacuum cycle when available. Make sure to open all pinch clamps and to partially open the caps of the chemostat and lagoon spinner flasks and the vent of the liquid filter before sterilizing to ensure optimal air removal and steam penetration. Close all pinch clamps and the vent of the liquid filter, and tighten the caps of the spinner flasks after sterilization.

3.3 Bacterial Strain Starter Preparation

1. Grow each host strain separately (according to the experiment design; *see* **Note 8**) with shaking at 37 °C in a 125 mL flask containing 25 mL LB medium (*see* Subheading 2.4.1) including 25 μg/mL Kan (*see* Subheading 2.4.2) until the density is ~0.5 OD_{600}.
2. In a laminar flow hood, combine equal parts of host strains to a total volume of 125 mL. Add LB medium if needed.

3.4 Continuous Evolution System Assembly and Bacterial Seeding (Fig. 1)

1. All manipulations should be carried out in a laminar flow hood to avoid contamination. Be sure to wear gloves and disinfect all working surfaces and gloves with EtOH (*see* **Note 11**).
2. Place the two LB containing vessels in a laminar flow hood. Open the screw cap of each vessel carefully, and add Kan (*see* Subheading 2.4.2) aseptically to a final concentration of 25 μg/ mL. Immediately close each vessel and remove them from the hood.
3. Place the ends of the "wet part" and the "dry part" tubing in the hood (*see* Subheading 2.1 and 2.2). Disinfect the male/ female connectors with EtOH, and disconnect the female connector from the "wet part" and the male connector from the "dry part." Immediately reconnect both male and female connectors to avoid contamination. Remove the tubing from the hood.
4. In a laminar flow hood, connect a sterile aspirator bottle with a silicon tubing outlet to the chemostat vessel. Transfer the combined host strains (*see* Subheading 3.3) to the aspirator, and allow the cultures to enter the spinner vessel by gravitation. Remove the chemostat and the aspirator from the hood.

3.5 Continuous Evolution System Operation

1. Connect media vessels, the chemostat, and the lagoon spinner flask tubes to a four-channel peristaltic pump.
2. Put the heating pad jacket around the chemostat spinner flask and set the temperature to 37 °C.
3. Place the chemostat spinner flask on a magnetic stirrer and start stirring the culture.
4. Open the pinch clamps on the outlet tube of the chemostat connecting the spinner flask to the waste vessel, and close the

pinch clamp on the tube that connects the chemostat to the lagoon (*see* Figure 1). Verify that the pinch clamp on the chemostat sampling port is closed.

5. Start the peristaltic pump, and allow fresh media to flow from media bottle #1 to the chemostat spinner flask through media bottle #2 at a constant flow rate of 250 mL/h (2 volumes per hour (2V/h); *see* **Notes 9** and **10**).
6. Run the chemostat until the culture reaches a steady state of ~1 OD_{600} (*see* Subheading 3.7), and then transfer the culture to the lagoon for bacteriophage inoculation (*see* Subheading 3.6). This stage usually requires an overnight incubation.
7. Close the pinch clamp on the outlet tube of the chemostat connecting the spinner flask to the waste vessel, and open the pinch clamps on the tube that connects the chemostat to the lagoon (*see* Figure 1). Open the pinch clamp on the tube connecting the lagoon spinner flask to the waste vessel. Verify that the pinch clamp on the lagoon sampling port is closed. Allow the bacterial culture to flow from the chemostat to the lagoon and from the lagoon to the waste vessel at the same flow rate of 2 V/h. Verify that the volume of the culture in the lagoon is ~125 mL, at which point the lagoon is ready for bacteriophage inoculation (*see* Subheading 3.6)
8. Meanwhile, put heating pad jacket around the lagoon spinner flask, set the temperature to 37 °C, and place the lagoon spinner flask on a magnetic stirrer, and start stirring the culture.
9. The experiment can now continue until there is no material left in the reservoir.

3.6 Bacteriophage Inoculation

1. Prepare T7 bacteriophage to be injected to the lagoon (*see* **Note 11**): aseptically add T7 bacteriophage at a concentration equal to an MOI of 1 into a 15 mL sterile tube containing 3 mL sterile PBS.
2. Transfer the contents of the tube to a sterile syringe with a needle.
3. Place the lagoon vessel in a laminar flow hood (or disinfect the injection tube with EtOH), and use the needle to inject the bacteriophage solution into one of the inlet tubes of the lagoon. Immediately seal the tube with a clamp. Place the lagoon on a magnetic stirrer.
4. Close the tube that feeds the lagoon with host strains, and open the tube connecting the chemostat to the waste vessel, allowing the chemostat flow culture to bypass the lagoon. This allows the initial propagation of the bacteriophage in the lagoon without washout while maintaining a steady-state chemostat culture.

5. After 2–3 hours, open the tube that feeds the lagoon with host strains from the chemostat, and close the tube connecting the chemostat to the waste vessel, allowing the chemostat culture to flow into the lagoon.

3.7 Sampling and Bacteria Strain Ratio Maintenance

Sample the chemostat and the lagoon vessels aseptically via sampling ports twice a day (*see* **Note 12**). Maintain the bacterial ratio in the chemostat by adding a freshly prepared bacterial mixture to the chemostat:

1. Chemostat sampling: Place the chemostat sampling port in a laminar flow hood. Close the pinch clamp on the tube connecting the chemostat to the lagoon, and open the pinch clamp on the tube of the sampling port, allowing the culture to flow through the port for 5–10 min ("bleeding"). Collect 15 mL samples (measure OD_{600} of the chemostat culture; *see* Subheading 3.8). Close the pinch clamp on the tube of the chemostat sampling port, and open the pinch clamp on the tube connecting the chemostat to the lagoon, to allow the chemostat culture to flow into the lagoon.
2. Lagoon sampling: Place the lagoon sampling port in a laminar flow hood. Close the pinch clamp on the tube connecting the lagoon to the waste vessel, and open the pinch clamp on the tube of lagoon the sampling port, to allow the culture to flow through the port for 5–10 min ("bleeding"). Collect 15 mL samples for lysate preparation and measure OD_{600} (*see* Subheading 3.8). Close the pinch clamp on the tube of the sampling port, and open the pinch clamp on the tube connecting the lagoon to the waste vessel, to allow the lagoon culture to flow into the waste vessel.
3. Bacterial strain ratio maintenance (*see* **Note 13**): Grow the appropriate host strains (*see* **Note 8**) separately overnight at 37 °C in 20 mL LB medium (*see* Subheading 2.4.1) and 25 μg/mL Kan (*see* Subheading 2.4.2). Measure the OD_{600}, and mix equal parts of all the required bacterial strains and adjust to a final ~1 OD_{600} in 300 mL LB (*see* Subheading 2.4.1) and 25 μg/mL Kan (*see* Subheading 2.4.2). In a laminar flow hood, transfer the bacterial mixture to the aspirator with silicon tubing connected to the chemostat vessel. Open the pinch clamp on the tube connecting the aspirator to the chemostat vessel, and allow the culture to flow by gravity. Close the media supplement to the chemostat, and allow the volume in the vessel to reach the initial volume of 125 mL. Open the media supplement to the chemostat and allow fresh media to enter the vessel.

3.8 Sample Analysis

Use the collected samples to evaluate the state of the cultures in the chemostat and lagoon, and use them for lysate preparation and plaque assay (from lagoon samples).

1. Determine the turbidity of the sample by OD_{600} measurement of 1 mL of the sample. Dilute with LB if needed.
2. Use 5 mL of the lagoon samples to prepare lysates (*see* Subheading 3.9).
3. Use the lysates to quantify the number of infectious phage particles in each sample by spotting serial dilutions on LB agar plates containing a layer of soft agar including the appropriate host cells (*see* Subheading 3.10, **Note 8** and [21]). Use JW1656 strain (Δ*ydhQ*::*Kan*^{R}) instead of JW5856 (Δ*trxA*::*Kan*^{R}) strain to allow plaque formation. Count and normalize to the values obtained by plating on the permissive host.
4. After analyzing the plating efficiencies of the mutants evolved in the lagoon, if required, one can induce mutations in the evolved phages by addition of the chemical mutagen ethyl methanesulfonate (EMS). The treated phages can then be reintroduced to the hosts in the lagoon of the continuous evolution system for a second run.

3.9 Lysate Preparation

1. Add 200–300 μL chloroform to a 5 mL sample extracted from the lagoon.
2. Vortex, and aliquot the lysate into 1.7 mL sterile microcentrifuge tubes.
3. Centrifuge the lysate tubes for 1 min at 13,000 x *g* in a microcentrifuge at RT.
4. Store at 4 °C.

3.10 Plaque Assay

This assay is performed as previously described [21]:

1. Dry the required number of LB agar plates (*see* Subheading 2.4.3) by leaving them partially uncovered in a laminar flow hood for 15 min. Label the dry plates with the required dilution and host strain details. Leave one plate as a "control" (without phages).
2. Melt a bottle of top agar (*see* Subheading 2.4.4) in a microwave, and place in a heating block set at 50 °C.
3. Dispense 900 μL PBS to a row of sterile microcentrifuge tubes labeled with the appropriate tenfold dilutions.
4. Add 100 μL of phage lysate (*see* Subheading 3.9) to the first tube, vortex, and transfer 100 μL to the second tube in the series. Change the pipette tip before each transfer. Continue making all tenfold serial dilutions using a new pipette tip for each transfer.

5. Quickly transfer 3 mL of melted top agar, 100 μL dilution of phages, and 300 μL of the appropriate overnight culture of the restrictive and permissive host strains to a sterile 15 mL tube (*see* **Note 8**). Vortex and immediately pour the mixture over the surface of the dried and labeled LB agar plate. Repeat for all dilutions.
6. Allow the overlaid plates to set for 30–60 min; then, incubate the plates at 37 °C until plaques appear. Count the plaques and determine the titer of the original lysate. Normalize the plating efficiencies obtained to the plating efficiency on the permissive host.

3.11 EMS Treatment of Phages

This assay is performed as previously described [19, 22]:

1. Grow bacterial strain JW1656 (*see* **Note 1**) in LB medium (*see* Subheading 2.4.1) supplemented with 25 μg/mL Kan (*see* Subheading 2.4.2) overnight at 37 °C.
2. Next day, dilute the culture 1:50 in 5 mL of the same medium, and grow at 37 °C for several hours until the culture reaches mid log phase.
3. Infect the culture with 5 μL of the evolved T7 extracted from the end of the continuous evolution experiment (see Subheading 3.7 and **Note 8**).
4. Add EMS to a final concentration of 1 % to the culture infected with the evolved phage, and incubate at 37 °C until lysis appears.
5. Add 0.125 g solid NaCl to the 5 mL lysate, swirl to dissolve, and pour the lysate into a centrifuge bottle.
6. Centrifuge at 4600 *g* for 10 min at 4 °C.
7. Transfer the supernatant into another centrifuge bottle containing a stir bar. Add 0.5 g PEG 8000, swirl to dissolve, and store overnight at 4 °C.
8. Centrifuge the lysate–PEG mixture at 4600 *g* for 3 hours at 4 °C, and then decant the supernatant.
9. Recentrifuge the pellet at 4600 *g* for 10 min at 4 °C in order to move the material from the walls to the bottom of the tube. Decant the remaining supernatant, and allow the pellet to drain by inverting the bottle on paper towels for a few minutes. Wipe the inside of the rim of the bottle to remove as much liquid as possible.
10. Resuspend the PEG pellet in 50 μL of 1 M NaCl, 10 mM Tris–HCl, pH 8.0, 1 mM EDTA, and vortex vigorously for 30 s to extract the phage.
11. Centrifuge for 2 min at 14,000 *g*.

12. Transfer the supernatant into a microcentrifuge tube, and centrifuge for 2 min at 14,000 *g* to remove any debris.
13. Transfer the supernatant into a new microcentrifuge tube and store at 4 °C.
14. Perform a plaque assay to determine the titer (*see* Subheading 3.10).

4 Notes

1. The following table (Table 1) summarizes the genotype of the *E. coli* strains used in the experiments [19].

Single knockout mutants: JW5856, JW3596, JW3595, JW3602, JW3601, JW3606, and JW1656 were acquired from the Keio collection [13]. IYB5709 is a strain lacking *trxA* from the Keio collection [13] where the resistance cassette was flipped by pCP20 as previously described [23]. This strain was then used as an ancestral strain in homologous recombination to produce double mutants IYB5758, TH4, TH5, TH6, TH7, and TH8 as previously described [19].

Table 1
A summary of the genotype of the *E. coli* strains used

Strain name	Genotype	Source
JW5856	BW25113 Δ*trxA*::*Kan*^{R}	[13]
JW3596	BW25113 Δ*waaC*::*Kan*^{R}	[13]
JW3595	BW25113 Δ*waaF*::*Kan*^{R}	[13]
JW3602	BW25113 Δ*waaO*::*Kan*^{R}	[13]
JW3601	BW25113 Δ*waaR*::*Kan*^{R}	[13]
JW3606	BW25113 Δ*waaG*::*Kan*^{R}	[13]
JW1656	BW25113 Δ*ydhQ*::*Kan*^{R}	[13]
IYB5709	BW25113 Δ*trxA*	[19]
IYB5758	BW25113 Δ*trxA* Δ*waaC*::*Kan*^{R}	[7]
TH4	BW25113 Δ*trxA* Δ*waaF*::*Kan*^{R}	[19]
TH5	BW25113 Δ*trxA* Δ*waaO*::*Kan*^{R}	[19]
TH6	BW25113 Δ*trxA* Δ*waaR*::*Kan*^{R}	[19]
TH7	BW25113 Δ*trxA* Δ*waaG*::*Kan*^{R}	[19]
TH8	BW25113 Δ*trxA* Δ*ydhQ*::*Kan*^{R}	[19]

2. The silicon tube should be long enough to pass through the peristaltic pump that transfers the media from the first to the second media bottles.
3. The silicon tube should be long enough to allow the end to reach the laminar flow hood so that the connection to the chemostat can be made under sterile conditions. The tube should be the same as the tube connecting the first and the second media bottles. The same four-channel peristaltic pump can then be used for both tubes in order to ensure equal inlet and outlet flow.
4. The chemostat and lagoon vessels are both 500 mL spinner flasks with a magnetic impeller. Each spinner flask has two arms with GL45 caps. Each cap has two ports where one is equipped with a dip tube. By drawing the medium out of the spinner flask through the dip tube faster than the rate it is inserted, the volume of the culture in the flask can be adjusted to the length of the dip tube.
5. The liquid capsule has a flow direction. Make sure that the capsule is installed in the direction indicated by the manufacturer.
6. The length of the tube should be long enough for the chemostat to reach the laminar flow hood in order to provide a sterile working environment. The inner diameter of this tube should be slightly larger than the inner diameter of the tube that connects media bottle #2 to the chemostat.
7. The length of the tube should be long enough for the lagoon to reach the laminar flow hood in order to provide a sterile working environment. The inner diameter of this tube should be slightly larger than the inner diameter of the tube that connects media bottle #2 to the chemostat.
8. The following table (Table 2) summarizes the different bacterial strain combinations (permissive and restrictive) used in the experiments.

Table 2
Different bacterial strain combinations (permissive and restrictive) used

Experiment No.	Permissive LPS	Permissive host	Restrictive hosts
#1	WaaC	JW3596	JW5856, TH4, TH5, TH6, TH7
#2	WaaF	JW3595	JW5856, IYB5758, TH5, TH6, TH7
#3	WaaO	JW3602	JW5856, IYB5758, TH4, TH6, TH7
#4	WaaR	JW3601	JW5856, IYB5758, TH4, TH5, TH7
#5	WaaG	JW3606	JW5856, IYB5758, TH4, TH5, TH6
#6	WT LPS	JW1656	IYB5758, TH4, TH5, TH6, TH7

9. A preliminary experiment with the selected silicon tubing must be done in order to determine the setting for the four-channel peristaltic pump to be used. Since the inlet and outlet tubes of the chemostat and the outlet tube of the lagoon are all controlled by the same pump, the inlet flow of the medium to the chemostat vessel is the main factor that determines the dilution rate of the system. The same silicon tube connected to the pump is used in later experiments. One end of this tube is dipped in a beaker containing water with the other edge in a measuring cylinder, and the volume of liquid transferred per hour at various pump speeds is measured. A plot of transferred volume against pump set value can then be used to calculate the specific pump-tube flow rate equation and determine the setting that will provide the desired flow rate.
10. In order to exploit the advantages of the system and follow multiple generations of both bacteria and phages, one must consider the dilution rate of both the chemostat and lagoon cultures [15]. Lowering the dilution rate (in terms of volume per hours) will increase the chemostat OD_{600} and make the culture closer to a batch culture instead of a chemostat, thus lowering the generation turnover. In contrast, increasing the dilution rate will decrease the chemostat OD_{600} and increase the generation turnover. It is very important to make sure that the dilution rate is not too close to the critical dilution rate, at which the bacteria are washed out of the spinner flasks. The optimal dilution rate is close to the critical dilution rate but still in the "safe zone." The influence of the dilution rate on the chemostat culture can be determined by an experiment in which only a chemostat culture is grown (same assembly of the system, but without the lagoon spinner flask). After bacterial inoculation, the system runs at a low dilution rate for 24 h, after which the OD_{600} is measured. The dilution rate is then increased (by changing the pump rate), and the OD_{600} is measured after a further 24 h. The experiment continues until there is a sharp decrease in the culture OD_{600}. These results provide information about the desired dilution rate.
11. All manipulations involving connecting tubes, addition of antibiotic supplements, and bacterial inoculations must be carried out aseptically in a laminar flow hood. Any other interventions to the active system (phage injection, bacterial strain ratio maintenance) must also be made under aseptic conditions in order to prevent contaminations.
12. The chemostat and lagoon cultures are sampled by clamping the inlet silicon tubes of the lagoon and the waste vessel closed and opening the sampling port clamps. It is important to bleed the culture to a waste container for 5–10 min before collecting the samples and to close the sampling ports and open the inlet tubes after sampling, to restore culture flow.

13. This procedure is performed every 24 hours in order to maintain the initial ratio of all host strains and avoid a specific host strain takeover of the culture.

Acknowledgments

UQ is supported by the European Research Council—Horizon 2020 research and innovation program, grant no. 818878.

References

1. Hyman P, Abedon ST (2010) Bacteriophage host range and bacterial resistance. Adv Appl Microbiol 70:217–248
2. Barrangou R et al (2007) CRISPR provides acquired resistance against viruses in prokaryotes. Science 315(5819):1709–1712
3. Bertani G, Weigle JJ (1953) Host controlled variation in bacterial viruses. J Bacteriol 65(2): 113–121
4. Smith SG et al (2007) A molecular Swiss army knife: OmpA structure, function and expression. FEMS Microbiol Lett 273(1):1–11
5. Fister S et al (2016) Influence of environmental factors on phage-bacteria interaction and on the efficacy and infectivity of phage P100. Front Microbiol 7:1152
6. Kerby GP, Gowdy RA et al (1949) Purification pH stability and sedimentation properties of the T7 bacteriophage of Escherichia coli. J Immunol 63(1):93–107
7. Yosef I et al (2017) Extending the host range of bacteriophage particles for DNA transduction. Mol Cell 66(5):721–728.e3
8. Yu P et al (2016) Isolation of polyvalent bacteriophages by sequential multiple-host approaches. Appl Environ Microbiol 82(3): 808–815
9. Zborowsky S, Lindell D (2019) Resistance in marine cyanobacteria differs against specialist and generalist cyanophages. Proc Natl Acad Sci USA 116(34):16899–16908
10. Meyer JR et al (2016) Ecological speciation of bacteriophage lambda in allopatry and sympatry. Science 354(6317):1301–1304
11. Heineman RH, Springman R, Bull JJ (2008) Optimal foraging by bacteriophages through host avoidance. Am Nat 171(4):E149–E157
12. González-García VA et al (2015) Characterization of the initial steps in the T7 DNA ejection process. Bacteriophage 5(3):e1056904
13. Baba T et al (2006) Construction of Escherichia coli K-12 in-frame, single-gene knockout mutants: the Keio collection. Mol Syst Biol 2: 2006.0008
14. Qimron U et al (2006) Genomewide screens for Escherichia coli genes affecting growth of T7 bacteriophage. Proc Natl Acad Sci USA 103(50):19039–19044
15. Enfors S-O (2011) Continuous fermentation. In: Fermentation process engineering. Royal Institute of Technology, Stockholm, pp 55–70
16. Lenski RE, Levin BR (1985) Constraints on the coevolution of bacteria and virulent phage: a model, some experiments, and predictions for natural communities. Am Nat 125(4): 585–602
17. Esvelt KM, Carlson JC, Liu DR (2011) A system for the continuous directed evolution of biomolecules. Nature 472:499
18. Badran AH et al (2016) Continuous evolution of bacillus thuringiensis toxins overcomes insect resistance. Nature 533(7601):58–63
19. Holtzman T et al (2020) A continuous evolution system for contracting the host range of bacteriophage T7. Sci Rep 10(1):307
20. Mark DF, Richardson CC (1976) Escherichia coli thioredoxin: a subunit of bacteriophage T7 DNA polymerase. Proc Natl Acad Sci USA 73(3):780–784
21. Kropinski AM et al (2009) Enumeration of bacteriophages by double agar overlay plaque assay. Methods Mol Biol 501
22. Manor M, Qimron U (2017) Selection of genetically modified bacteriophages using the CRISPR-Cas system. Bio Protocol 7(15)
23. Datsenko KA, Wanner BL (2000) One-step inactivation of chromosomal genes in Escherichia coli K-12 using PCR products. Proc Natl Acad Sci USA 97(12):6640–6645

Chapter 7

Single-Cell Proteomics by Barcoded Phage-Displayed Screening via an Integrated Microfluidic Chip

Yujiao Wang, Jing Zhao, Zhenwei Jiang, Yuan Ma, and Rui Zhang

Abstract

Recent advancements in the profiling of proteomes at the single-cell level necessitate the development of quantitative and versatile platforms, particularly for analyzing rare cells like circulating tumor cells (CTCs). In this chapter, we present an integrated microfluidic chip that utilizes magnetic nanoparticles to capture single tumor cells with exceptional efficiency. This chip enables on-chip incubation and facilitates in situ analysis of cell-surface protein expression. By combining phage-based barcoding with next-generation sequencing technology, we successfully monitored changes in the expression of multiple surface markers induced by CTC adherence. This innovative platform holds significant potential for comprehensive screening of multiple surface antigens simultaneously in rare cells, offering single-cell resolution. Consequently, it will contribute valuable insights into biological heterogeneity and human disease.

Key words Phage-based profiling, Single cell, Protein expression, Barcoded phage-displayed antibody screening, Circulating tumor cells, Microfluidics

1 Introduction

Cell types that play a crucial role in the growth, development, and progression of diseases are often found in extremely low abundance. This rare cell type includes stem and progenitor cells [1], cancer stem cells [2], and circulating tumor cells [3]. While significant advancements have been made in characterizing distinct subpopulations using single-cell transcriptomic and proteomic approaches, such as mass spectrometry, genetic barcoding technologies, and other techniques [4–9], obtaining detailed spatial and temporal phenotypic information from rare single cells in the presence of numerous nontargeted cells remains a challenge. Specifically, the analysis of circulating tumor cells (CTCs) poses a particularly difficult problem due to their rarity in peripheral blood, leading to limited proteomic studies on CTCs [10].

Huan Peng et al. (eds.), *Phage Engineering and Analysis: Methods and Protocols*,
Methods in Molecular Biology, vol. 2793, https://doi.org/10.1007/978-1-0716-3798-2_7,

An emerging tool in the field of biomedical engineering, microfluidic chips, has proven highly valuable. These chips consist of interconnected microchannels or grooves engraved into various materials to achieve desired outcomes. Depending on their functions, microfluidic chips can be categorized as droplet microfluidic chips [11], organ-on-chip [12], microfluidic hydrogel chip [13], microfluidic chip integrated with 3D culture technology [14, 15], microfluidic chip merged with detection instruments, microfluidic chip for single-cell analysis, and microfluidic model organism. Microfluidic chips take advantage of precise fabrication techniques and the physical properties of fluid flow while operating at the microscale to provide enhanced precision and controllability in rare cell proteomic profiling [16–18].

Phage display is a very powerful tool capable of screening antibodies and peptides with targeted specific binding was invented in 1958 [19], and its versatility also enables the study of binding interactions of peptides with a variety of substrates, such as metals, minerals, small molecules, polymers, and nanomaterials [20]. Integrating microfluidic chips with phage display technology opens up possibilities for constructing high-throughput screening methods, which could significantly impact the discovery of disease markers and therapeutic targets.

In this chapter, we have developed an integrated platform that facilitates the capture of rare live cells and enables multiplexed profiling of cell-surface expression at single-cell resolution using barcoded phage-displayed antibodies [21].

2 Materials

2.1 Experimental Reagents

2-YT Broth:

1. Mix 31 g of 2-YT in 1 L of sterile water.
2. Autoclave (cycle: Liq 30).

2-YT Agar:

1. In 300 mL of sterile water, mix 9.3 g of 2-YT and 4.5 g of agar.
2. Autoclave (cycle: Liq 30).
3. If applicable, add antibiotics to desired concentration and swirl to mix.
4. Pour enough media that the bottom of the dish is covered.
5. Once the agar is solidified and the plates are labeled (with the type of media, antibiotics added, and date), store in the cold room.

PEG-NaCl (20% PEG-8000, 2.5 M NaCl):

1. Add 100 g of PEG and 75 g of NaCl to 400 mL of H_2O.
2. Dissolve, and add water up to 500 mL.

3. Pass through a vacuum filter.
4. Open the bottom of the filter in a sterile TC hood.
5. Store at RT.

PCR Reaction System:
1. 25 μL of 2 × KAPA HiFi HotStart ReadyMix (Roche).
2. 17 μL of dH_2O.
3. 6 μL of Primer mix (2.5 μM final).
4. 2 μL of phages (undiluted, 1/10 dilution, or 1/100 dilution).

Some Routine Reagents:
1. 0.1 M HCl in PBS.
2. 1 M Tris (pH 11).
3. 100 μg/mL carbenicillin.
4. 25 μg/mL kanamycin.
5. PBS with Ca^{2+}/Mg^{2+} and 0.5% BSA.

2.2 Experimental Equipment

1. PCR machine.
2. Clean bench.
3. High-speed centrifuge.
4. Electronic scale.
5. Autoclave.
6. Spectrophotometer.
7. Gel electrophoresis instrument.
8. Horizontal electrophoresis tank.
9. Thermostatic incubator.
10. Gel Imaging Systems.
11. Ultralow temperature freezers.
12. 0.22 μm syringe filter.

3 Methods

3.1 Preparation before Phage Display

3.1.1 Working with OmniMAX E. coli Cells (Invitrogen) Culture

5 mL 2YT broth inoculated with a single colony of OmniMAX *E. coli*.

Grow O/N at 37 °C, 200 rpm.

Inoculate cultures with 250 μL of O/N culture in 25 mL 2YT broth.

Grow until culture reaches OD600 of 0.4–0.8 for phage amplification and titer determination.

3.1.2 Working with Phage

Preparation of Phage from Library Stock

The M13 preselected barcoded phages displaying EGFR01, EGFR02, FZD1, FZD2, FZD7, and Luciferase were provided by Dr. Sachdev Sidhu (Terrence Donnelly Centre for Cellular and Biomolecular Research). And the phage pool was provided by Dr. Jim Wells (University of California at San Francisco). The phage library has been amplified. The stock concentration is about 10^{14}/mL:

1. Dilute library prep tenfold using PBS in a high-spin tolerant tube.
2. Add at least 1/5 volume of PEG-NaCl, invert the tube three times, and keep tubes on ice for 20 min, inverting multiple times during precipitation.
3. Spin at 17,600 × g for 20 min at 4 °C.
4. Dump supernatant into the tub, spin at 17,600 × g for 2 min at 4 °C, and use the filtered tip to remove the remaining supernatant.
5. Resuspend the phage pellet in 1 mL PBS with Ca^{2+}/Mg^{2+} and 0.5% BSA.
6. Spin at 17,600 × g for 20 min at 4 °C to pellet any insoluble matter, and transfer the supernatant to a 1.5 mL low-retention microcentrifuge tube.
7. Set aside 10 μL of the library prep for titer, and save at 4 °C.
8. Keep the phage on ice until the next step or save at 4 °C.

Preparation of Helper Phage

The M13KO7 helper phage (NEB) helper phage (M13KO7) has been amplified. The stock concentration is 10^{13}/mL:

1. Pick a single OmniMAX cell colony from a fresh 2Y plate (less than 1 week old) using a pipette tip and inoculate 2 ml of 2YT broth.
2. Incubate with shaking at 200 rpm at 37 °C for 6–8 h.
3. Add M13KO7 helper phage to a final concentration of 10^{10}/ml.
4. Incubate with shaking at 200 rpm at 37 °C for 30 min.
5. Transfer the culture to 1 liter of 2YT/kan media in a 4 liter baffled flask.
6. Incubate with shaking at 200 rpm at 37 °C overnight.
7. Centrifuge the culture for 10 min at 17,600 × g at 4 °C.
8. Precipitate phage particles by transferring the supernatant to new tubes containing 1/5 volume of PEG/NaCl, and incubate for 20 min on ice.
9. Centrifuge for 25 min at 17,600 × g at 4 °C.

10. Decant the supernatant; centrifuge for 2 min at 17,600 × g and aspirate the remaining supernatant.
11. Resuspend the phage in PBS and centrifuge for 10 min at 16,000 × g to remove insoluble matter. The protocol yields approximately 100 ml of M13KO7 helper phage stock at a concentration of 10^{13}/ml.

Titration of Phage

1. Add 5 μL of phage to 45 μL of PBS, mix, and create a series of ten-fold dilutions in the same row in a 96-well plate.
2. Add 5 μL of each phage dilution to 45 μL of OmniMAX cell culture in the wells of the next row on the plate.
3. Incubate at 37 °C at 200 rpm for 30 min.
4. Plate 5 μL of the infected cells from each of the wells on to a pre-warmed 2-YT agar plate with antibiotics, and grow O/N at 37 °C.
5. Calculate the total phage/ml as follows: Phage/ml = number of colonies on a carbenicillin plate in the most dilute sample × 200 × 10^i (where i = the maximum number of dilutions in which colonies are apparent).

3.2 Phage Display of Cells Captured on the Chip

3.2.1 Bind the Phages to the Cell on the Chip

1. Binding between single cells and 10^{12}–10^{13} phage was carried out with phage infusion at 200 μL/h for 1 h on the chip.

3.2.2 Wash Phages that Are Not Bound to Cells on the Chip

1. PBS washing the phages that are not binding to the single cells at 200 μL/h for 1 h.

3.2.3 Phage Elution

1. Punch out the capture zone containing single cells and punch into a microcentrifuge tube.
2. Add 1000 μL of 0.1 M HCl to the tube to submerge the capture zone, vortex to resuspend the cells and transfer the cell suspension to a 1.5 mL tube, and incubate at RT for 10 minutes.
3. Spin at 15,000 × g for 5 min at 4 °C to pellet the broken cells.
4. During the centrifugation, add 125 μL 1 M Tris (pH 11) to a new 1.5 mL tube.
5. Transfer the supernatant with 0.1 M HCl to the tube with 125 μL 1 M Tris (PH 11). Mix to adjust to neutral PH. The ratio of 0.1 M HCl to 1 M Tris (PH 11) is 8:1 in volume.
6. Keep the eluted phage on ice until bacterial cells are at an OD600 of 0.4–0.8. Take 10 μL for phage titration.

3.2.4 Infect OmniMAX Cells to Phage Propagation

1. The neutralized solution containing phage was then added to 5 mL of OmniMAX *E. coli* cells (Invitrogen) at an OD600 of 0.6–0.8 in 2-YT broth.
2. Incubate at 37 °C for 30 min at 200 rpm.
3. Using filter tips add helper phage (M13K07) to a final concentration of 10^{10}/mL to each 5 mL infection.
4. Incubate at 37 °C for 45 min.
5. Transfer the 5 mL cultures to 125 mL flasks with 45 mL 2YT + 1‰ carbenicillin [100 ug/mL] + 1‰ kanamycin [25 ug/mL] for a final volume of 50 mL.
6. Incubate cultures O/N at 37 °C at 200 rpm.

3.2.5 Phage Precipitation from O/N Culture

1. Transfer the bacteria/phage O/N culture to high-spin tolerant Nalgene tubes.
2. Spin at 17,600 × g for 10 min and collect supernatant.
3. Add the supernatant and at least 1/5 volume of PEG-NaCl to fresh high-spin tolerant Nalgene tubes. Invert the tube three times, and keep tubes on ice for 20 min, inverting multiple times during precipitation.
4. Spin at 17,600 × g for 2 min at 4 °C, and remove the remaining supernatant using filtered pipette tips.
5. Resuspend phage pellet in 1 mL PBS with Ca^{2+}/Mg^{2+} and 0.5% BSA.
6. Spin at 17,600 × g for 20 minutes at 4 °C to pellet the insoluble matter, and transfer the supernatant to a 1.5 mL low-retention microcentrifuge tube.
7. Store at 4 °C until PCR amplification of the H3 barcoded regions.

3.3 NGS Primer-Adaptors PCR Prep and Sequencing Analysis

Mix used for each 50 μL reaction: 25 μL of 2 × KAPA HiFi HotStart ReadyMix (Roche), 17 μL of dH_2O, 6 μL of Primer mix (2.5 μM final), and 2 μL of phages (undiluted, 1/10 dilution, or 1/100 dilution).

1. Make stocks of combined primer sets at 2.5 μM final.
2. Make three different phage dilutions for each sample: no dilution, 1/10 dilution, and 1/100 dilution.
3. Perform a PCR reaction for each dilution. Make PCR master mix and perform PCR cycles.
4. PCR products were run on a 1% agarose gel (expected size: ~200 bp):
   - If the reactions were successful with gel bands for all three dilutions for each sample, combine the PCR reactions of dilutions 1/10 and 1/100 into one tube.

- If there were no gel bands in the 1/100 dilution sample, then combine the No dilution and 1/10 dilution PCR reactions into one tube for each sample.
- If there were no bands for all, or if only the No dilution PCR reaction gave bands, then repeat PCR cycle for all dilutions with an increased input volume for phage.

5. Run the entire volume from the combined PCR reactions on a 1.0% agarose standard gel at 100 volts for 40 min.
6. Cut out the PCR bands and gel purify (for the same sample loaded into two wells, combine both bands into one sterile 2 mL tube).
7. Use the Qiagen kit protocol without heating the gels (let it dissolve at RT). After the last wash step (before elution), open the tube caps and let all remaining ethanol air-dry for 5 min at RT. For elution, add 25 μL of water to each tube, and incubate at RT for 5 min before spinning down. Then, repeat again with another 25 μL of water.
8. Use the nanodrop spectrophotometer to determine approximate DNA concentrations. Blank wells with water and measure samples in duplicates. Calculate the average final concentrations (make note of standard deviation and if high re-spec the sample).
9. Normalize all samples to the sample with the lowest concentration (ng/μL). Then, combine 10 μL of each sample (concentrations normalized) into a single tube containing all samples.
10. For Deep-Sequencing we want a 30–50 nM range in >40 μL volume per sample for submission.
11. Samples are now ready for Illumina Deep-Sequencing submission. Keep samples at 4 °C if sending them out immediately (1–2 days); otherwise (if keeping them for more than a day), freeze at −20 °C.
12. Bring the samples to Donnelly Sequencing Centre (University of Toronto) for next-generation sequencing (NGS) on a HiSeq4000 (Illumina).

3.4 Rare Cell Capture Using Antibody-Mediated Magnetic Sorting

A single-cell proteomics (SCP) chip is needed to separate rare cells labeled with magnetic nanoparticles. The SCP chip as Fig. 1 is recommended. The capture efficiency can be maintained at ~90% with no capture of nontarget cells. A gradient centrifugation process should be conducted to restore the high capture efficiencies when the rare cells are in whole blood. Cells can be incubated for up to 4 days in this SCP chip if it is fibronectin coated. Fresh cell culture medium need be supplemented every 2 days through the flow channels.

Fig. 1 Assembled SCP chip with single cells trapped per each capture zone

3.4.1 Simulations Guiding Device Design

The SCP chip should be well simulated first and built a simplified 2D model to simulate the interaction between nanobead–conjugated tumor cells and metglas-enhanced magnetic field in the flow channel. The magnetic force a tumor cell experiences can be expressed as

$$F_m = nV_{bead}\frac{\Delta\chi_{bead}}{\mu_0}(B\nabla)B \qquad (1)$$

Here n is the number of beads bound to tumor cell, V_{bead} is the volume of a nanobead, $\Delta\chi_{bead}$ is the relative magnetic susceptibility of nanobeads, μ_0 is the magnetic permeability of the free space, and B is the magnetic field.

Term $(B\nabla)B$ is determined by external magnetic field and metglas magnetic guide. The magnetic force is proportional to $(B\nabla)B$. Flowing tumor cell will be attracted laterally toward the magnetic guide due to this magnetic force.

The magnetic force of the tumor cell which is moving toward the magnetic guide is balanced by the hydrodynamic drag force produced by the ambient fluid. The drag force is expressed as

$$F_d = 6\pi\mu RU \qquad (2)$$

Here η is the viscosity of ambient fluid, R is the radius of tumor cell, and U is the tumor cell's velocity in the X-direction (perpendicular to metglas magnetic guide).

When $F_{mx} = F_d$, the tumor cell will under balanced conditions

The velocity of the tumor cell in the lateral direction is

$$U = \frac{nV_{bead}\frac{\Delta x_{bead}}{\mu_0}(B\nabla B)_x}{6\pi\mu R} \tag{3}$$

To ensure the cell be trapped by the trap structure near the magnetic guide, the cell should reach the sidewall before encountering the trap structure. In other words, the time for a cell to reach the sidewall (T) should be shorter than the time for the cell to encounter the trap structure (T_0). The movement of the tumor cell in the X-direction can be simplified as

$$dt = \frac{dx}{U} \tag{4}$$

If the initial position of the tumor cell is x_0, time required for the cell to reach the sidewall will be

$$T = \int_{x_0}^{W} \frac{dx}{U} \tag{5}$$

Here W is the width of the flow channel. Formula can be discretized as

$$T = \sum \frac{\Delta x_i}{U_i} = \sum \frac{6\pi\mu R\mu_0 \Delta x_i}{nV_{bead}\Delta x_{bead}|B\nabla B|_{x_i}} \tag{6}$$

In the above formula, Δx_i and $|BB\nabla|x_i$ can be extracted from COMSOL simulation.

3.4.2 Device Design and Fabrication

Use AutoCAD (Autodesk) to design the SCP chip. The chip includes three main components (Fig. 2):

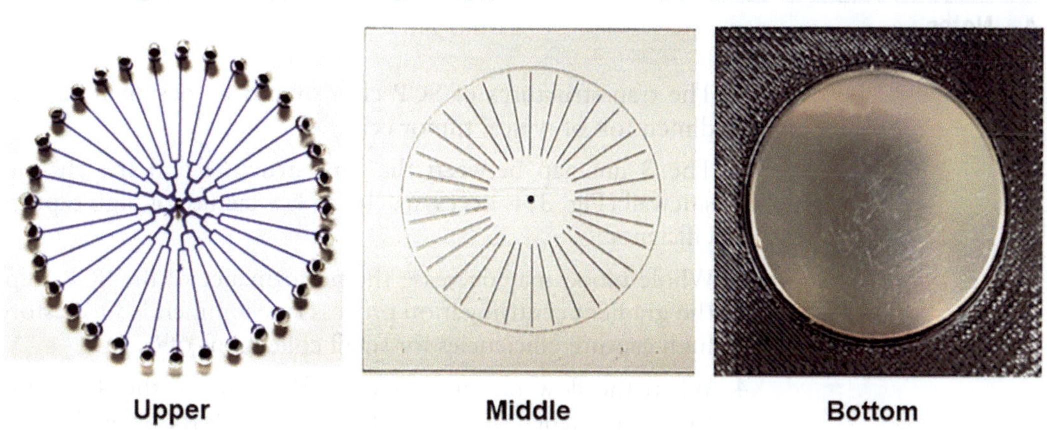

Fig. 2 Snapshot of the SCP-chip upper, middle, and bottom layer

1. The upper flow channel contains 32 constriction channels (*see* **Note 1**), and each channel contains 4 quadrant shaped trap structures (*see* **Note 2**).
2. The middle portion contains Metglas 2714A strips as magnetic guide.
3. The bottom component is a cylinder neodymium magnet encapsulated into the base by 3D printing. All components are independently changeable to simplify fabrication and can be precisely aligned before use.

3.4.3 Target Cell Capture

Target cells are labeled with antibody-conjugated magnetic particles in buffer or processed blood samples (*see* **Note 3**). Calculated magnetic force leads the cell to the trap structure when magnetically labeled cells flow through the constriction channel (**Notes 4 and 5**).

3.4.4 Culture of Target Cell on Chip

The SCP chip is designed to mimic in vivo conditions; CTCs captured can survive in the chip for days. This makes the study of different phenotypic phases of the cells in situ and dynamic surface protein expression profiling using barcoded phage antibodies possible. Cells can be incubated in the fibronectin-coated SCP chip for up to 4 days if fresh cell culture medium is supplemented every other day through the flow channels.

3.4.5 Collecting of Cell-Bound Phage

There are three methods to collect cell-bound phage: (i) acidic elution; (ii) release of cells and overnight incubation; and (iii) physical excision of the capture zone and overnight incubation. The physical excision method is recommended for it will introduce minimal nonspecific phage.

4 Notes

1. The trap structures of SCP chip should be designed per the dimension of typical tumor cells.
2. The 3 μm gap between the trap structure and the channel sidewall (Fig. 3) is necessary. It is a key facture of high capture efficiency.
3. Whole blood may decrease the performance of the SCP chip; the gradient centrifugation process is recommended to restore high capture efficiencies for small collections of cells.
4. When the flow rate increased to 2000 μL/h, the fluid drag force approached the level of the magnetic force while the drag force dominates at a flow rate of 4000 μL/h.
5. High cell concentrations should be avoided as they promoted nontargeted cell capture.

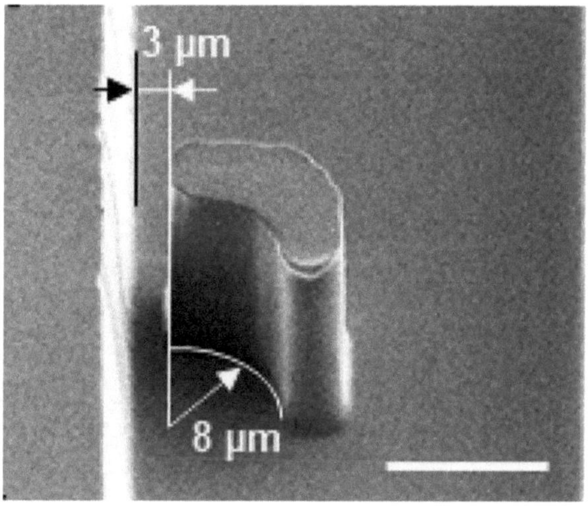

Fig. 3 Representative SEM image of the capture area

References

1. Orkin SH, Zon LI (2008) Hematopoiesis: an evolving paradigm for stem cell biology. Cell 132(4):631–644
2. Kreso A, Dick John E (2014) Evolution of the cancer stem cell model. Cell Stem Cell 14(3): 275–291
3. Plaks V, Koopman CD, Werb Z (2013) Circulating tumor cells. Science 341(6151): 1186–1188
4. Labib M, Kelley SO (2020) Single-cell analysis targeting the proteome. Nat Rev Chem 4(3): 143–158
5. Li X, Fan B, Cao S, Chen D, Zhao X, Men D, Yue W, Wang J, Chen J (2017) A microfluidic flow cytometer enabling absolute quantification of single-cell intracellular proteins. Lab Chip 17(18):3129–3137
6. Palii CG, Cheng Q, Gillespie MA, Shannon P, Mazurczyk M, Napolitani G, Price ND, Ranish JA, Morrissey E, Higgs DR, Brand M (2019) Single-cell proteomics reveal that quantitative changes in co-expressed lineage-specific transcription factors determine cell fate. Cell Stem Cell 24(5):812–820.e815
7. Forbes TP, Forry SP (2012) Microfluidic magnetophoretic separations of immunomagnetically labeled rare mammalian cells. Lab Chip 12(8):1471–1479
8. Kang JH, Krause S, Tobin H, Mammoto A, Kanapathipillai M, Ingber DE (2012) A combined micromagnetic-microfluidic device for rapid capture and culture of rare circulating tumor cells. Lab Chip 12(12):2175–2181
9. Armbrecht L, Müller RS, Nikoloff J, Dittrich PS (2019) Single-cell protein profiling in microchambers with barcoded beads. Microsyst Nanoeng 5(1):55
10. Poudineh M, Sargent EH, Pantel K, Kelley SO (2018) Profiling circulating tumour cells and other biomarkers of invasive cancers. Nat Biomed Eng 2(2):72–84
11. Seemann R, Brinkmann M, Pfohl T, Herminghaus S (2012) Droplet based microfluidics. Rep Prog Phys 75(1):016601
12. Wu Q, Liu J, Wang X, Feng L, Wu J, Zhu X, Wen W, Gong X (2020) Organ-on-a-chip: recent breakthroughs and future prospects. Biomed Eng Online 19(1):9
13. Nie J, Gao Q, Wang Y, Zeng J, Zhao H, Sun Y, Shen J, Ramezani H, Fu Z, Liu Z, Xiang M, Fu J, Zhao P, Chen W, He Y (2018) Vessel-on-a-chip with hydrogel-based microfluidics. Small 14(45):1802368
14. Carvalho MR, Lima D, Reis RL, Correlo VM, Oliveira JM (2015) Evaluating biomaterial- and microfluidic-based 3D tumor models. Trends Biotechnol 33(11):667–678
15. van Duinen V, Trietsch SJ, Joore J, Vulto P, Hankemeier T (2015) Microfluidic 3D cell culture: from tools to tissue models. Curr Opin Biotechnol 35:118–126

16. Zhang W, Kai K, Choi DS, Iwamoto T, Nguyen YH, Wong H, Landis MD, Ueno NT, Chang J, Qin L (2012) Microfluidics separation reveals the stem-cell–like deformability of tumor-initiating cells. Proc Natl Acad Sci 109(46):18707–18712
17. Chen K, Dopico P, Varillas J, Zhang J, George TJ, Fan ZH (2019) Integration of lateral filter arrays with immunoaffinity for circulating-tumor-cell isolation. Angew Chem Int Ed 58(23):7606–7610
18. Cohen L, Walt DR (2019) Highly sensitive and multiplexed protein measurements. Chem Rev 119(1):293–321
19. Ledsgaard L, Ljungars A, Rimbault C, Sørensen CV, Tulika T, Wade J, Wouters Y, McCafferty J, Laustsen AH (2022) Advances in antibody phage display technology. Drug Discov Today 27(8):2151–2169
20. Cung K, Slater RL, Cui Y, Jones SE, Ahmad H, Naik RR, McAlpine MC (2012) Rapid, multiplexed microfluidic phage display. Lab Chip 12(3):562–565
21. Ma Y, Chen K, Xia F, Atwal R, Wang H, Ahmed SU, Cardarelli L, Lui I, Duong B, Wang Z, Wells JA, Sidhu SS, Kelley SO (2021) Phage-based profiling of rare single cells using nanoparticle-directed capture. ACS Nano 15(12):19202–19210

Chapter 8

Targeted Genome Editing of Virulent *Pseudomonas* Phages Using CRISPR-Cas3

Kaat Schroven, Marleen Voet, Rob Lavigne, and Hanne Hendrix

Abstract

The vast number of unknown phage-encoded ORFan genes and limited insights into the genome organization of phages illustrate the need for efficient genome engineering tools to study bacteriophage genes in their natural context. In addition, there is an application-driven desire to alter phage properties, which is hampered by time constraints for phage genome engineering in the bacterial host. We here describe an optimized CRISPR-Cas3 system in *Pseudomonas* for straightforward editing of the genome of virulent bacteriophages. The two-vector system combines a broad host range CRISPR-Cas3 targeting plasmid with a SEVA plasmid for homologous directed repair, which enables the creation of clean deletions, insertions, or substitutions in the phage genome within a week. After creating the two plasmids separately, a co-transformation to *P. aeruginosa* cells is performed. A subsequent infection with the targeted phage allows the CRISPR-Cas3 system to cut the DNA specifically and facilitate or select for homologous recombination. This system has also been successfully applied for *P. aeruginosa* and *Pseudomonas putida* genome engineering. The method is straightforward, efficient, and universal, enabling to extrapolate the system to other phage–host pairs.

Key words Bacteriophage, Genome engineering, CRISPR-Cas, *Pseudomonas aeruginosa*, SEVA, Cas3

1 Introduction

The majority of the genes derived from bacteriophages are annotated as hypothetical, and due to the generation of big data, this knowledge gap is only increasing with time [1, 2]. These genes are referred to as "ORFans" or open reading frame orphans. The number of phage-derived ORFans is significantly higher than those encountered in bacteria [3]. Despite the efforts that have been made lately to elucidate the function of those hypothetical proteins, for instance via next-generation interaction proteomics and Colabfold, the majority still remain uncharacterized [4, 5]. As such, there is a need for the development of additional tools, for example, approaches to efficiently edit the viral genome [6].

Huan Peng et al. (eds.), *Phage Engineering and Analysis: Methods and Protocols*,
Methods in Molecular Biology, vol. 2793, https://doi.org/10.1007/978-1-0716-3798-2_8,

In addition, phage engineering holds promises to create engineered phages as well as designer phages, for therapeutic applications. As such, phage genes can be deleted, and novel genes and regulatory elements can be introduced in such a way that a phage is created with the desired properties [7].

Since virulent phages can only be manipulated during the short infection cycle within the host cell as their genomes never integrate into the bacterial chromosome, genome editing is less straightforward compared to temperate phages. In vivo approaches that have been used for the editing of virulent phages are random mutagenesis, homologous recombination, and recombineering. For random mutagenesis, phages are exposed to ultraviolet light or chemical agents that will result in a pool of random mutant phages that emerge [8, 9]. The homologous recombination strategy relies on either the co-infection of a host cell with two related phages or the infection of a cell containing a specific donor plasmid with the desired mutation. During phage infection, homologous recombination will take place between the phage genome and the desired mutated region, which is flanked by homologous phage sequences. However, recombination efficiencies are low and require extensive screening [10]. Engineering efficiencies can be improved by using recombinases, including the Bacteriophage Recombineering of Electroporated DNA (BRED) method, which consists of linear phage DNA and synthetic DNA with the desired mutation that are simultaneously introduced into the bacterial host cell expressing phage recombination genes [11].

Recently, more advanced approaches are reported that use homologous recombination combined with CRISPR-Cas systems (clustered regulatory interspaced short palindromic repeats—CRISPR-associated proteins) for virulent phage genome editing, which are characterized by their low costs, high efficiency, and simple design [12]. The CRISPR-Cas systems originate from bacteria and serve as adaptive immune systems that protects them from predation by bacteriophages. When foreign DNA enters the cell, Cas nucleases will cleave it, and the created fragments will be built into a CRISPR array as novel spacers. Upon entrance of the same DNA sequence, the spacers will be transcribed into CRISPR RNA (crRNA) and used to form the single-guide RNA (sgRNA) which in turn will guide the Cas proteins to cleave the DNA that is being recognized [13]. The type II CRISPR-Cas9 system is the most studied system among the different Cas systems, which create double-stranded breaks in the genome. By providing a DNA repair template, homologous recombination will occur, resulting in a precise genome editing approach [6, 14]. Alternatively, the CRISPR-Cas system can be used to counter-select for the wild-type phage [15, 16]. In this manner, the CRISPR-Cas9 has been used to edit the genome of phages infecting *Vibrio natriegens* [14], *Vibrio cholerae*, *Mycobacterium smegmatis* [16], *Lactococcus lactis*

[6], *Klebsiella pneumoniae* [17], *Bacillus subtilis* [18], *Escherichia coli* [19], *Listeria ivanovii* [20], and *Streptococcus thermophilus* [21].

Despite efforts made, only a few studies report a successful modification of the genome of virulent *Pseudomonas aeruginosa*-infecting phages. Via the sequential PCR amplification and assembly of different genomic fragments in a yeast artificial chromosome (YAC), a reduced phage genome was created. This genome was subsequently electroporated into *P. aeruginosa*, and the plaques resulting from the engineered phage were retrieved [22]. Furthermore, the virulent phage phiKZ, which develops a "phage nucleus" structure during replication, was modified via homologous recombination combined with the type V CRISPR-Cas13a system. This CRISPR-Cas system specifically targets mRNAs and was designed in this purpose to target an essential transcript, to select for phages that successfully underwent the homologous repair event and thereby acquired the anti-CRISPR *acrVIA1* gene in their genome [15].

We here describe a method to modify the genome of virulent *P. aeruginosa*-infecting phages in a precise and efficient way, using the optimized CRISPR-Cas3 vector system developed by Csörgő et al. (2020) [23], combined with the Standard European Vector Architecture (SEVA) vector system for homologous directed repair (Fig. 1). The type I CRISPR-Cas3 system is known to target virulent *P. aeruginosa* phages, and combines nuclease and helicase activity, degrading DNA processively [23, 24]. First, we describe how the two vector systems are constructed (Subheading 3.1) and introduced in *P. aeruginosa* (Subheading 3.2). Next, it is explained how the CRISPR-Cas system can be used to target virulent *P. aeruginosa* phages to introduce specific deletions, insertions, or substitutions (Subheading 3.3).

2 Materials

2.1 Creation of the Targeting and HDR Plasmids

1. Chemically competent *Escherichia coli* cells, e.g., One Shot TOP10 chemically competent *E. coli* (Thermo Fisher Scientific).
2. Autoclaved LB medium: 1% Tryptone, 0.5% yeast extract, 1% NaCl (supplemented with 1.5% or 0.5% bacteriological agar to prepare solid or soft medium, respectively) in ultrapure water.
3. Autoclaved SOC medium: 2% Tryptone, 0.5% yeast extract, 10 mM NaCl, 2.5 mM KCl, 10 mM $MagCl_2$, 10 mM $MgSO_4$, 20 mM glucose in ultrapure water.
4. Annealing buffer: 7.8 mM Tris–HCl pH 8.5, 50 mM NaCl, 1 mM EDTA in ultrapure water.

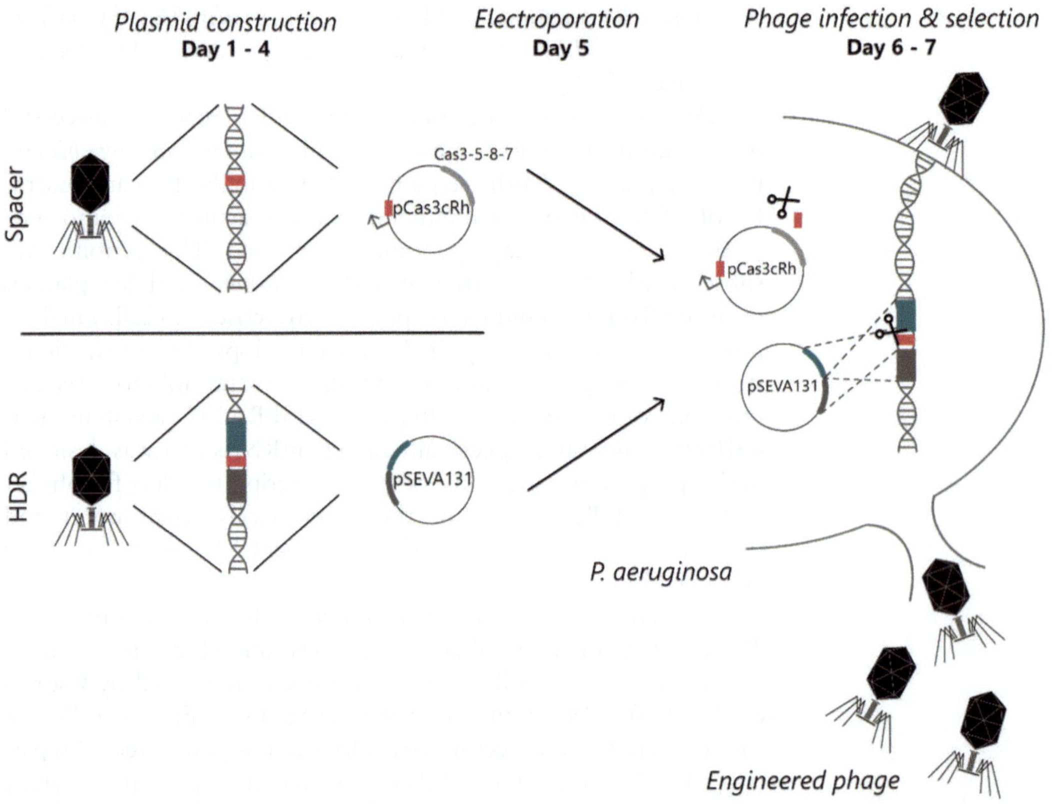

Fig. 1 General overview of the CRISPR-Cas3-based engineering method for *P. aeruginosa*-infecting phages. Days 1–4: A 34 bp spacer (marked in red) is cloned into the modified repeat sequence of the pCas3cRh vector (targeting plasmid) by Golden Gate cloning with type IIs restriction enzyme BsaI. In parallel, a homologous repair template containing two >500 bp arms (marked in gray and cyan) is inserted into the pSEVA131 vector (HDR plasmid) using Golden Gate cloning with type IIs restriction enzyme BsaI (Subheading 3.1). Day 5: Both plasmids are introduced in *P. aeruginosa* by means of electroporation (Subheading 3.2). Days 6–7: On day 6, phage infection is performed in the recombinant *P. aeruginosa* cells expressing the Cas3 system using the double agar method. The Cas3 enzyme cleaves the genomic DNA at the target location, and triggers the homologous recombination event using the repair template or exerts a selective pressure toward recombinant phages. After overnight incubation, multiple single plaques are subjected to PCR and Sanger sequencing, to verify the presence of the desired genomic modification (Subheading 3.3)

5. Antibiotics: 1000x stock of ampicillin sodium salt (100 mg/mL; final concentration, 100 μg/mL; Ap100), 1000x stock of gentamicin sulfate (15 mg/mL; final concentration, 15 μg/mL; Gm15).
6. Sterile ultrapure water.
7. pCas3cRh plasmid (Addgene: Plasmid #133773).
8. pSEVA131 plasmid (Standard European Vector Architecture 4.0, SEVA database).
9. Template phage DNA.

10. Primers:
    (a) Spacer oligos.
    (b) pSEVA131 primers.
    (c) HDR primers.
    (d) pCas3cRh_F: CAGGAAATGCGGTGAGC
    (e) pCas3cRh_R: GAGCAGCTAATTCACCGC
    (f) pSEVA131_F: AGGGCGGCGGATTTGTCC
    (g) pSEVA131_R: GCGGCAACCGAGCGTTC
11. Agarose.
12. 50 μg/ml ethidium bromide ().
13. 6X loading buffer: 40% (w/v) sucrose, 0.1% (w/v) bromophenol blue.
14. High-fidelity DNA polymerase with corresponding PCR buffer.
15. Kapa HiFi polymerase (Roche Sequencing Store).
16. DreamTaq DNA polymerase with corresponding PCR buffer.
17. 10 mM dATP.
18. 10 mM dNTP solution.
19. T4 polynucleotide kinase (PNK) with corresponding buffer.
20. Thermosensitive Alkaline Phosphatase (FastAP) with corresponding buffer.
21. Eco31I (BsaI) with corresponding buffer.
22. T4 DNA ligase with corresponding buffer.
23. Glycerol.
24. DNA marker.
25. PCR purification kit (e.g., GeneJet PCR Purification kit, Thermo Fisher Scientific).
26. Plasmid miniprep kit (e.g., GeneJet Plasmid Miniprep kit, Thermo Fisher Scientific).
27. 0.5 mL PCR tubes.
28. Glass tubes.
29. 10 mm petri dishes.
30. Sterile toothpicks.
31. Eight-well PCR tube strips or 96-well PCR plate.
32. Micropipettes (single and multichannel).
33. PCR thermocycler (single tubes and 96-well plates).
34. 37 °C incubator.
35. Gel electrophoresis device.
36. DNA spectrophotometer.
37. Temperature-controlled water bath.

2.2 Electroporation of the Plasmids into the P. aeruginosa Host

1. *Pseudomonas aeruginosa* propagation host (e.g., strain PAO1 [25]).
2. Autoclaved LB medium (liquid and solid).
3. Antibiotics: 1000x stock of carbenicillin Dinatriumsalz (200 mg/mL; final concentration, 200 μg/ mL; Cb200), 1000x stock of gentamicin sulfate (50 mg/mL; final concentration, 50 μg/mL; Gm50).
4. Filter-sterilized 10% sucrose solution: 300 mM final concentration sucrose in ultrapure water.
5. Glass tubes.
6. 1.5 mL and 2 mL Eppendorf tubes.
7. 10 mm petri dishes.
8. Single-channel micropipettes.
9. Electroporator (e.g., Gene Pulser Xcell (Bio-Rad, Hercules, California, USA)).
10. Electroporation cuvettes (2 mm gap).
11. Microcentrifuge.

2.3 Phage Infection and Selection for the Engineered Phage

1. Autoclaved LB medium (liquid and solid).
2. Antibiotics: 1000x stock of carbenicillin Dinatriumsalz (200 mg/mL; final concentration, 200 μg/ mL), 1000x stock of gentamicin sulfate (50 mg/mL; final concentration, 50 μg/mL).
3. Phage stock of the wild-type phage (stock solution of approximately 10^{11} PFU/mL).
4. Autoclaved phage buffer: 10 mM Tris–HCl pH 7.5, 100 mM $MgSO_4$, 150 mM NaCl.
5. Sterile ultrapure water.
6. DreamTaq DNA polymerase with corresponding PCR buffer.
7. 10 mM dNTP mix.
8. Control primers.
9. Ultrapure water.
10. DNA marker.
11. Glass tubes.
12. Eight-well PCR tube strip or 96-well PCR plate.
13. 10 mm petri dishes.
14. 37 °C incubator.
15. PCR thermocycler.
16. Gel electrophoresis device.
17. DNA spectrophotometer.

3 Methods

3.1 Creation of the Targeting and HDR Plasmids

Two different vectors are used in this genome editing protocol. A spacer sequence is introduced into the pCas3cRh vector, to specifically target the phage gene/region of interest. On the other hand, the pSEVA131 vector (*see* **Note 1**) is used to carry the homologous directed repair (HDR) arms, to ensure a clean modification of the gene.

3.1.1 Spacer Cloning in pCas3cRh

The pCas3cRh vector contains the Cas genes (*cas3, cas5, cas7, and cas8*), the crRNA with spacer sequence, and the gentamicin resistance gene. The vector can be purchased from Addgene and is sent as a stab culture of transformed *E. coli* NEB5α strain. To purify the plasmid from the strain, a plasmid miniprep or maxiprep kit can be used.

Two oligos are designed which consist of the spacer sequence, corresponding to both strands, and an additional sequence which ensures a successful ligation in the restricted pCas3cRh vector. For spacer selection, a 34 bp sequence (protospacer) downstream of a 5′-TTC-3′ sequence (PAM sequence) is chosen (*see* **Notes 2** and **3**). Template oligos:

Forward: 5' - GAAAC xxx (34 bp) G - 3'.

Reverse: 5' - GCGAC xxx (reverse complement 34 bp) G - 3'.

After phosphorylation and pairwise annealing of the two oligos, the template is ligated into the BsaI-restricted pCas3cRh vector.

Spacer Annealing

1. For each oligo, mix 4 μL of 100 μM oligo, 2 μL 10x buffer, 2 μL 10 mM dATP, and 4 μL T4 PNK in 8 μL mQ.
2. Incubate the mixture for 1 h at 37 °C, followed by a 10-min enzyme inactivation step at 75 °C.
3. Anneal the oligos by adding 2 μL of each phosphorylated oligo to 46 μL annealing buffer.
4. Incubate the mixture for 5 min at 95 °C, and cool down to 25 °C gradually (0.1 °C/sec).
5. Perform a PCR purification step to collect the annealed spacer (*see* **Note 4**).
6. Measure the DNA concentration.

Restriction of pCas3cRh with BsaI

1. Prepare a reaction mixture containing 5 μg pCas3cRh vector (*see* **Note 5**), 1.5 μL BsaI (10/μL), 1× buffer, in a total volume of 35 μL.
2. Incubate the mixture for 1 h at 37 °C, followed by a 20 min enzyme inactivation step at 65 °C.

3. Purify the restricted DNA with a PCR purification kit in 30 μL elution buffer.
4. Dephosphorylate the purified DNA by adding 1x buffer FastAP and 1.5 μL FastAP (1 U/ μL) to 30 μL pCas3cRh/BsaI.
5. Incubate for 15 min at 37 °C, and inactivate the enzyme for 5 min at 75 °C.
6. Purify the DNA with a PCR purification kit.
7. Determine the DNA concentration.

Ligation of the Spacer in pCas3cRh/BsaI

1. In a total volume of 20 μL, mix 50 ng pCas3cRh/BsaI/FastAP with 0.87 ng annealed spacer (*see* **Note 6**), 1x T4 DNA ligase buffer, and 1 μL T4 DNA ligase (1 U/μL) (*see* **Note 7**).
2. Incubate at 16 °C, overnight (*see* **Note 8**).
3. Inactivate the enzyme via an incubation step at 65 °C for 10 min.
4. Transform 5 μL of the ligation mixture to chemical-competent *E. coli* cells. Plate the complete cell volume on three LB/Gm15 plates (50 μL–100 μL—rest volume).
5. To verify the constructed spacer, inoculate two colonies separately in 4 mL LB/Gm15.
6. Extract the plasmids using a plasmid DNA extraction kit, and verify the sequence via sequencing (e.g., Sanger sequencing) by making use of the pCas3cRh primer pair (pCas3cRh_F and pCas3cRh_R). The expected sequence from forward to reverse primer is:

   5′ – TCGCGCCCCGCACGGGCGCGTGGATTGAAA – SPACER – TCGCCCGGCAAAACCGGGCGTGGATTGAAA – 3′.

3.1.2 Construction of Homologous Repair Template in pSEVA131

Although any canonical SEVA vector can be used to provide the HDR template (*see* **Note 1**), the pSEVA131 is preferred due to its medium-copy number (pBBR1 origin of replication) and appropriate resistance gene for selection in both *E. coli* (ampicillin) and *P. aeruginosa* (carbenicillin). Different primer sets are needed for the construction of the pSEVA131 vector with HDR arms (Fig. 2).

Two primer pairs, one for each homology arm, are needed to construct the HDR template (*see* **Notes 9** and **10**). For deletions, the HDR template consists of two joined homology arms (≥ 500 bp each), identical to the sequences directly up- and downstream of the region to be deleted. For integrations and substitutions, the repair template consists of the desired insertion or substitution, flanked by the up- and downstream homology arms (*see* **Note 11**). Importantly, the Cas3 target site should be removed from the HDR template, either by deletion, PAM mutation, or

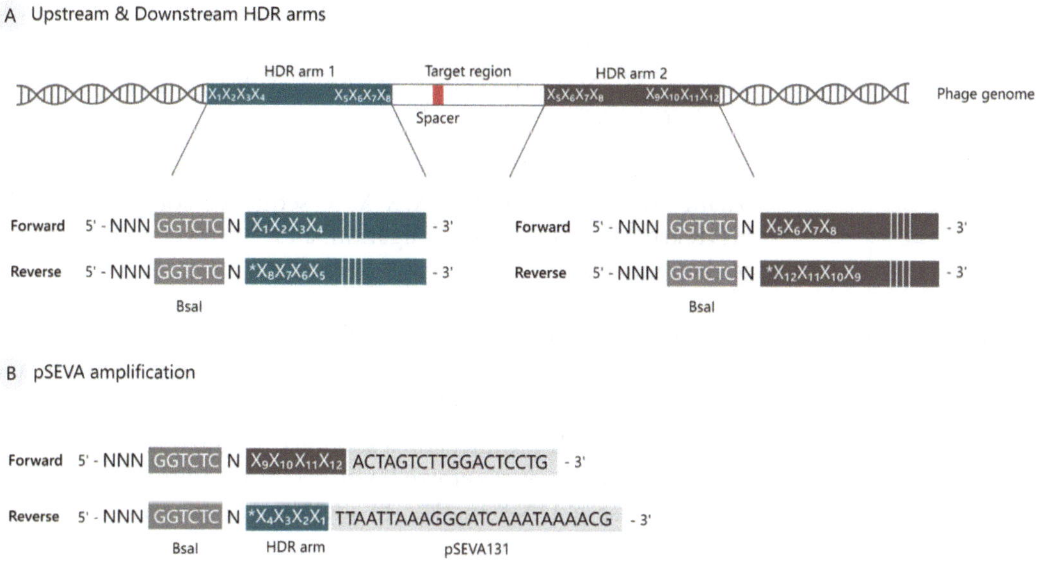

Fig. 2 Graphical representation of the primer design for the pSEVA131-HDR system. (**a**) Primers for the amplification of the upstream and downstream homologous repair region of the targeted sequence. A tail is introduced containing a BsaI recognition site fused to the outer 4 bp sequence of the HDR region, to ensure a scarless ligation. In this graphical summary, the HDR arms for a genomic deletion are visualized. (**b**) An additional primer pair ensures the linearization of the pSEVA131 vector while introducing a tail at both ends. This tail contains the BsaI recognition site as well as the 4 bp sequence that corresponds to that of the HDR arms. This will ensure a scarless ligation of the fragments into the vector. With GGTCTCN = the BsaI recognition site, N = adjustable to balance the GC content, $X_xX_xX_xX_x$ = sequence compatible to the HDR arms to ensure scarless ligation, and * = reverse complement of the sequence

protospacer/PAM interruption (*see* **Note 12**). Make sure that the annealing part of the primers covers at least 16 bp. Template primers:

HDR arm 1.

Forward: 5′ – NNN GGTCTC N $X_1X_2X_3X_4$ xxx (> 12 bp) - 3'.

Reverse: 5′ – NNN GGTCTC N *$X_8X_7X_6X_5$ xxx (> *12 bp) - 3'.

HDR arm 2.

Forward: 5′ – NNN GGTCTC N $X_5X_6X_7X_8$ xxx (> 12 bp) - 3'.

Reverse: 5′ – NNN GGTCTC N *$X_{12}X_{11}X_{10}X_9$ xxx (> *12 bp) - 3′.

With GGTCTC = the BsaI recognition site, N = adjustable to balance the GC content, $X_xX_xX_xX_x$ = sequence compatible to the HDR arms to ensure scarless ligation, xxx (> 12 bp) = sequence compatible to the HDR arms to ensure primer annealing, and * = reverse complement of the sequence.

Additionally, a primer pair is designed to amplify the pSEVA131 vector while introducing BsaI restriction sites. Template primers:

Forward: 5′- NNN GGTCTC N $X_9X_{10}X_{11}X_{12}$ ACTAGTCTTGG ACTCCTG – 3'.

Reverse: 5′- NNN GGTCTC N *$X_4X_3X_2X_1$ TTAATTAAAGGC ATCAAATAAAACG – 3'.

With GGTCTC = the BsaI recognition site, N = adjustable to balance the GC content, $X_xX_xX_xX_x$ = sequence compatible to the HDR arms to ensure scarless ligation, and * = reverse complement of the sequence.

Construction of the HDR Fragments

1. For each fragment, PCR amplify the inserts with a high-fidelity DNA polymerase using the phage genome as a template and following the manufacturer's instructions.
2. Run the PCR products on a solidified 1% agarose gel. For this, each PCR product is mixed with a DNA loading dye (*see* **Note 13**). Use a DNA marker to identify the length of the constructed fragments.
3. Purify the mixtures with a PCR purification kit.

Amplification of pSEVA131

1. Amplify the pSEVA131 plasmid by mixing 1 ng pSEVA131 with 12.5 μL ReadyMix Kapa, 0.37 μL of each primer (20 μM) in a total volume of 25 μL.
2. Program: 4 min at 95 °C; 30 cycles of 20 s at 98 °C, 15 s at 65 °C and 2 min 25 s at 72 °C; 10 min at 72 °C; hold at 12 °C.
3. Purify the PCR mixture with a PCR purification kit.
4. Determine the DNA concentration.

Construction of the HDR Fragments in the pSEVA131 Vector

1. Ligate the HDR fragments into the amplified pSEVA131 vector via a restriction–ligation cycling program. Prepare the reaction mixture with 100 ng amplified pSEVA131, 50 ng of each fragment, 1 μL BsaI (1 U/μL), 3 μL T4 DNA ligase (1 U/μL), and 1x DNA ligation buffer, in a total volume of 20 μL.
2. Perform a cyclic program on the mixture: 50 cycles of 2 min at 37 °C and 3 min at 16 °C; 5 min at 50 °C; 5 min at 80 °C; hold at 12 °C.
3. Transform the mixture (5–8 μL) to *E. coli* chemical-competent cells, and plate the complete cell volume on three LB/Ap100 plates (50 μL–100 μL—rest volume). Incubate overnight at 37 °C.
4. To analyze the resulting transformants, pick eight random clones and subject them to a colony PCR analysis (*see* **Note 14**). For this, inoculate each clone in 5 μL sterile water in an eight-well PCR tube strip, and immediately streak on a LB/Ap100 plate (*see* **Note 15**). The agar plate is incubated overnight at 37 °C. The plate will be later used to inoculate liquid cultures of positive clones.

5. To each well, add the following components: 0.05 μL DreamTaq DNA polymerase, 0.5 μL 20 μM of each primer (pSEVA131_F&R), 2.5 μL 10x DreamTaq DNA polymerase Green buffer, and 0.5 μL 10 mM dNTP; mix in a total volume of 25 μL (*see* **Note 16**). Mix the solution by pipetting multiple times with a multichannel pipette.
6. Run a PCR program in a PCR heat block: 5 min at 95 °C; 30 cycles of 30 s at 95 °C, 30 s at 52 °C and 2 min 30 s at 72 °C; 5 min at 72 °C; hold at 12 °C.
7. To verify the length of the amplified PCR product, load and run on a 1% agarose gel alongside a DNA marker. The expected length of the construct is 195 bp plus the sum of the length of the fragments.
8. Select three clones with the correct fragment length and inoculate in 4 mL LB/Ap100.
9. The next day, extract the plasmids via a DNA plasmid extraction kit, and verify the sequence by sequencing (e.g., Sanger sequencing) using the pSEVA131_F and R primer pair (*see* **Note 17**). The expected sequence from forward to reverse primer is:

5′ – CCTTTCGTTTTATTTGATGCCTTTAATTAA – HDR fragment - ACTAGTCTTGGACTCCTGTTGATAGATCCA – 3′.

3.2 Electroporation of the Plasmids into the P. aeruginosa Host

Following the successful creation of the targeting and HDR plasmids, they both have to be transformed into the *P. aeruginosa* propagation host. After freshly preparing electrocompetent *P. aeruginosa* cells according to the method described by Choi and colleagues [26], the plasmids are co-electroporated to the cells, and transformants are verified by colony PCR:

1. Prepare an overnight culture of the propagation host in LB (*see* **Note 18**).
2. Begin by making the cells electrocompetent. For this, bring 1.5 mL of the overnight culture into a 2 mL Eppendorf tube in duplicate, and spin down for 4 min at 6000 *g*. Discard the supernatant and wash the cells with 1 mL of a 10% sucrose solution and centrifuge again (*see* **Note 19**). Repeat the washing step three times in total. Finally, pool the two pellets in 100 μL 10% sucrose solution.
3. Add 100 ng pCas3cRh and 100 ng pSEVA plasmid (*see* **Notes 20** and **21**).
4. Incubate for 5 min at room temperature.
5. Transfer the cell solution to a 0.2 mm cuvette. Set up the parameters of the electroporator (25 μF, 200 Ω and 1.8 kV), and place the cuvette in the device.

6. Quickly add 450 μL of liquid LB medium to the cuvette, and transfer the content to a fresh 1.5 mL Eppendorf tube (*see* **Note 22**).
7. Incubate the tubes for 1 h 20 min at 37 °C.
8. Plate the complete cell volume on three LB/Gm50/Cb200 plates (50 μL–100 μL—rest volume), and incubate for 1–2 days (*see* **Note 23**).

3.3 Phage Infection and Selection for the Engineered Phage

Once a correct host strain carrying both plasmids is obtained, phage infection using the double agar overlay method is performed. The resulting plaques are subjected to PCR using control primers that bind the phage genome just outside the homologous arms.

A primer set is constructed to be able to assess the genomic modification occurred in the phage genome. For this, a primer pair is designed that binds outside of the homology arms. In this way, the entire modified region is covered (*see* **Note 24**):

1. Prepare an overnight culture in 4 mL LB/Cb200/Gm50 with a clone from the electroporation plate (*P. aeruginosa* harboring the targeting and HDR plasmids).
2. The next day, make a dilution series of the wild-type phage stock in sterile water. Use dilutions 10^{-6}, 10^{-8}, and 10^{-10} for the infection assay (*see* **Note 25**).
3. Perform the double agar overlay method for phage infection. For this, mix 200 μL of the overnight culture with 100 μL of the phage dilution in 4 mL soft agar, and pour on top of a LB/Gm50/Cb200 plate (*see* **Note 26**).
4. Incubate overnight at 37 °C.
5. Single plaques that appear on the plates are checked for successful genome modification, by performing a PCR reaction with control primers. For this, pick several plaques with a toothpick and resuspend separately in 20 μL 1x phage buffer (*see* **Note 27**).
6. For each phage plaque solution, bring 5 μL in a well of a 8-well strip or 96-well microtiter plate, and mix with 0.05 μL DreamTaq DNA polymerase, 0.5 μL 20 μM of each primer (control primer set), 2.5 μL 10x DreamTaq Green buffer, and 0.5 μL 10 mM dNTP in a total volume of 25 μL (*see* **Note 16**). Mix the solution by pipetting up and down multiple times with a multichannel pipette.
7. Subject the plate to a thermal program: 5 min at 95 °C; 30 cycles of 30 s at 95 °C, 30 s at x °C and 2 min 30 s at 72 °C; 5 min at 72 °C; hold at 12 °C, with x the lowest melting temperature of the primer −1 °C.

8. Run the PCR products together with a DNA marker on a solidified 1% agarose gel to detect the presence of the genomic modification (*see* **Notes 28** and **29**).
9. Sequence the PCR fragments that show a product with the correct length (e.g., Sanger sequencing) (*see* **Note 30**).

4 Notes

1. Depending on the host strain, another SEVA vector can be used as well (e.g., if a different antibiotic resistance marker is preferred).
2. To avoid off-target activity, check originality of the protospacer in the genome of the phage, cloning strain, and host strain, especially for the first 15 bp.
3. The PAM sequence is chosen within the sequence that is to be deleted or substituted, or in case of an integration within 15 bp of the splice site. If no suitable PAM site is available in this region, a site in close proximity can be used as well. Nevertheless, the PAM site or protospacer should be removed from the phage DNA after homologous recombination.
4. As the constructed fragment is a primer–dimer of 40 bp long, no isopropanol is added to extract the DNA.
5. If desired, the amount of plasmid DNA can be reduced.
6. The molecular weight of the vector is approximately 11,502 bp × 649 g/mol = 7,464,798 g/mol. In this cloning reaction, 50 ng corresponds to 0.0067 pmol. Since a fragment to vector ratio of 5:1 is preferred, 0.0335 pmol of the fragment is needed. The weight of the fragment is approximately 40 bp × 649 g/ mol = 25,960 g/mol. Consequently, 0.87 ng of the fragment is used in this ligation reaction.
7. The restriction and ligation steps can be combined in one step. For this, mix 100 ng pCas3cRh, 50 ng annealed primer pair, 3 μL T4 DNA ligase (1 U/μL), 1 μL BsaI (1 U/μL), and 1x DNA ligation buffer, in a total volume of 20 μL. Subject the mixture to 30 restriction–ligation cycles (37 °C for 2 min; 16 °C for 3 min), followed by a ligase inactivation step for 50 min at 50°c and a BsaI inactivation step for 5 min at 80 °C.
8. This incubation time can be shortened to 1 h at 22 °C if desired.
9. Instead of the creation of the HDR arms by PCR, a gene block can be synthesized synthetically. Especially in case of an insertion or substitution, this might be advantageous, as the PAM or protospacer sequence need to be changed.

10. Check for the presence of the BsaI restriction site in the selected region. While ≥500 bp homologous arms are preferred, shorter arms are generally sufficient for recombination.
11. Apart from the two fragments with the homologous arms, a third fragment with the desired insertion can be needed. Also for substitution, a third fragment can be used to obtain all mutations, including the PAM substitution.
12. In case the PAM/protospacer sequence is located in a coding region, design the mutation(s) based on codon usage pattern.
13. To use the minimal amount of PCR mixture, 3 μL of the mix can be diluted with 2 μL water before the loading dye is added.
14. For difficult constructs, e.g., high GC content, secondary structures, and long constructs, more colonies can be screened.
15. A raster can be drawn on the solid agar plate, which corresponds to the wells of the 8-well PCR tube strips. With the same toothpick, inoculate the strain on the LB plate as well as in the corresponding well of the strip.
16. If desired, a master mix can be prepared for the total number of colonies that are screened. Prepare the mix on ice and add the enzyme at the last step.
17. If the concentration is too low for sequencing (< 40 ng/μL), send the PCR product for sequencing, whether or not after PCR purification.
18. *P. aeruginosa* is notorious for its biofilm production. To circumvent this, inoculation cultures can be made in small Erlenmeyer's containing 15 mL LB medium. Moreover, if possible, try to do the electroporation in the morning.
19. Remove the sucrose solution by discarding the supernatant on a tissue to minimize cell loss. In the last step, a pipette is used to remove all the supernatants and the residing biofilm.
20. Two controls can be included to address the effect of the spacer in phage infection. For this, electroporate 50 ng of pCas3cRh (empty) and 50 ng pCas3cRh(spacer) separately to electrocompetent cells without the addition of the pSEVA131 vector.
21. If the electroporation was unsuccessful, the plasmids can be introduced one by one into *P. aeruginosa*. In our experience, the order of the introduction of the different plasmids does not matter.
22. Prewarming your LB medium may help for a successful transformation.
23. The same protocol can be used for bacterial genome editing, which can be followed by a plasmid-curing step [27].
24. If the modified region is too large to amplify by PCR (for instance when a genomic insertion occurred), one can opt for

a primer pair that is located in the modified region. For this, design one primer located on the unmodified genomic DNA and another primer that binds to the homology region. This ensures that the produced product covers the target region.

25. Other dilutions can also be used to obtain single plaques.
26. Although the Cas genes on the targeting plasmid, pCas3cRh, are under control of the RhaRS/pRhaBAD-inducible system, no IPTG induction is needed in *P. aeruginosa* strain PAO1. Nevertheless, in other host strains, the addition of 0.1% rhamnose to the agar plates and soft agar may be needed.
27. The phage plaque in buffer can be filtered prior to PCR analysis, to minimize the amount of bacterial DNA present.
28. In case of substitution, immediately sequence the PCR products, as no difference in length can be observed.
29. Two or more DNA bands present in one lane on the DNA gel may indicate an impure phage plaque. In that case, one may streak the plaque on a bacterial lawn in soft agar and check multiple new plaques with PCR.
30. A PCR purification step is not necessary prior to sequencing. Mix 0.5 μL of the PCR product with 4.5 μL water.

Acknowledgments

The pCas3cRh vector was kindly provided by prof. J. Bondy-Denomy (UCSF). This project received funding from the European Research Council (ERC) under the European Union's Horizon 2020 Research and Innovation Programme (Grant Agreements 819800), and from the KU Leuven (C1 project 'ACES', C16/20/001).

References

1. Edwards RA, Rohwer F (2005) Viral metagenomics. Nat Rev Microbiol 3:504–510
2. Ru X, Li L, Wang C (2019) Identification of phage viral proteins with hybrid sequence features. Front Microbiol 10: 507
3. Yin Y, Fischer D (2008) Identification and investigation of ORFans in the viral world. BMC Genomics 9:24
4. Fossati A, Mozumdar D, Kokontis C et al (2023) Next-generation interaction proteomics for quantitative Jumbophage-bacteria interaction mapping. Nat Commun 14:5156
5. Mirdita M, Schütze K, Moriwaki Y et al (2022) ColabFold: making protein folding accessible to all. Nat Methods 196(19):679–682
6. Lemay M-L, Tremblay DM, Moineau S (2017) Genome engineering of virulent Lactococcal phages using CRISPR-Cas9. ACS Synth Biol 6:1351–1358
7. Chen Y, Batra H, Dong J et al (2019) Genetic engineering of bacteriophages against infectious diseases. Front Microbiol 10:1–12
8. Favor AH, Llanos CD, Youngblut MD et al (2020) Optimizing bacteriophage engineering through an accelerated evolution platform. Sci Rep 10:13981
9. Drake JW (1966) Ultraviolet mutagenesis in bacteriophage T-4. I. Irradiation of extracellular phage particles. J Bacteriol 91:1775–1780

10. Mahler M, Rita Costa A, van Beljouw SPB et al (2023) Approaches for bacteriophage genome engineering. Trends Biotechnol 41(5):669–685
11. Marinelli LJ, Piuri M, Swigoň Ová Z et al (2008) BRED: a simple and powerful tool for constructing mutant and recombinant bacteriophage genomes BRED: a simple and powerful tool for constructing mutant and recombinant bacteriophage genomes. PLoS One 3(12): e3957
12. Chang P (2018) ADP-ribosylation and NAD+ utilizing enzymes. Methods in Molecular Biology 1813
13. Xu Y, Li Z (2020) CRISPR-Cas systems: overview, innovations and applications in human disease research and gene therapy. Comput Struct Biotechnol J 18:2401–2415
14. Zhang X, Zhang C, Liang C et al (2022) CRISPR-Cas9 based bacteriophage genome editing. Microbiol Spectr 10(4): e00820–22
15. Guan J, Oromí-bosch A, Mendoza SD et al (2022) Bacteriophage genome engineering with CRISPR – Cas13a. Nat Microbiol 7(12): 1956–1966
16. Wetzel KS, Guerrero-Bustamante CA, Dedrick RM et al (2021) CRISPY-BRED and CRISPY-BRIP: efficient bacteriophage engineering. Sci Rep 11:6796
17. Shen J, Zhou J, Chen G-Q et al (2018) Efficient genome engineering of a virulent Klebsiella bacteriophage using CRISPR-Cas9. J Virol 92(17):e00534–18
18. Schilling T, Dietrich S, Hoppert M et al (2018) A CRISPR-Cas9-based toolkit for fast and precise in vivo genetic engineering of Bacillus subtilis phages. Viruses 10(5):241
19. Tao P, Wu X, Tang W-C et al (2017) Engineering of bacteriophage T4 genome using CRISPR-Cas9. ACS Synth Biol 6:1952–1961
20. Hupfeld M, Trasanidou D, Ramazzini L et al (2018) A functional type II-A CRISPR-Cas system from listeria enables efficient genome editing of large non-integrating bacteriophage. Nucleic Acids Res 46:6920–6933
21. Martel B, Moineau S (2014) CRISPR-Cas: an efficient tool for genome engineering of virulent bacteriophages. Nucleic Acids Res 42: 9504–9513
22. Pires DP, Monteiro R, Mil-Homens D et al (2021) Designing P. aeruginosa synthetic phages with reduced genomes. Sci Rep 11:1–11
23. Csörgő B, León LM, Chau-Ly IJ et al (2020) A compact Cascade–Cas3 system for targeted genome engineering. Nat Methods 17:1183–1190
24. Mendoza SD, Nieweglowska ES, Govindarajan S et al (2020) A bacteriophage nucleus-like compartment shields DNA from CRISPR nucleases. Nature 577:244–248
25. Stover CK, Pham XQ, Erwin AL et al (2000) Complete genome sequence of *Pseudomonas aeruginosa* PAO1, an opportunistic pathogen. Nature 406:959–964
26. Choi K-H, Kumar A, Schweizer HP (2006) A 10-min method for preparation of highly electrocompetent Pseudomonas aeruginosa cells: application for DNA fragment transfer between chromosomes and plasmid transformation. J Microbiol Methods 64:391–397
27. Lammens E-M, Volke DC, Schroven K et al (2023) A SEVA-based, CRISPR-Cas3-assisted genome engineering approach for Pseudomonas with efficient vector curing. Microbiol Spectr 11(6):e0270723

Part III

Decorating Phages by Conjugation

Chapter 9

Preparation of Bioconjugates of Chimeric M13 Phage and Gold Nanorods

Huan Peng and Irene A. Chen

Abstract

Phage-nanomaterial conjugates are functional bio-nanofibers with various applications. While phage display can select for phages with desired genetically encoded functions and properties, nanomaterials can endow the phages with additional features at nanoscale dimensions. Therefore, combining phages with nanotechnology can construct bioconjugates with unique characteristics. One strategy for filamentous phages is to adsorb nanoparticles onto the side wall, composed of pVIII subunits, through electrostatic interactions. However, a noncovalent approach may cause offloading if the environment changes, potentially causing side effects especially for in vivo applications. Therefore, building stable phage-bioconjugates is an important need. We previously reported the construction of chimeric M13 phage conjugated with gold nanorods, named "phanorods," without weakening the binding affinity to the bacterial host cells. Herein, we give a detailed protocol for preparing the chimeric M13 phage and covalently conjugating gold nanorods to the phage.

Key words M13, Phage, Gold nanorods, Bioconjugates, pIII, Phage display, pVIII chemical modification, Phanorod

1 Introduction

Phage display technology was first invented in 1985 by George Smith, who shared the Nobel Prize in Chemistry in 2018 for this development [1, 2]. Heterologous fusion proteins can be displayed on the outer surface coat proteins of phages by inserting foreign DNA into the phage genome, often without destroying the structural characteristics of the original phage. The genetically modified phages can be produced by *E. coli* cells [3], allowing creation of libraries of mutant phages that display a large number of variant peptides or proteins. Phage display selection of these variants, such as by biopanning against an immobilized target, has proven to be a versatile and powerful method to identify peptides and proteins with desired functions [4].

Huan Peng et al. (eds.), *Phage Engineering and Analysis: Methods and Protocols*,
Methods in Molecular Biology, vol. 2793, https://doi.org/10.1007/978-1-0716-3708-2_9,

One of the most well-documented phage display systems is based on M13 [5–9]. M13 phage is a filamentous virus with nanoscale dimensions that infects *E. coli* expressing the F pilus. The circular single-stranded DNA genome of M13 encodes 11 proteins, and five of them are known as structural proteins (pVIII, pIII, pVI, pVII, and pIX; Fig. 1) [10]. The ~2700 copies of the major coat protein, pVIII, assemble to form the side wall of the phage, while several copies of each of the four minor structural proteins contribute to the capping on both tips of the phage particle. The other six proteins of the M13 genome are nonstructural proteins involved in infection and replication. Although all of the coat proteins are reported to be engineerable to display functional proteins or peptides, the pIII and pVIII proteins are the two most widely used proteins for phage display. The pIII protein contains the receptor-binding domain, which contacts the bacterial receptor and thus confers host specificity. While the pIII protein can display large proteins with up to five copies per viral particle, the pVIII protein can only tolerate short peptides but with many more copies (up to thousands of copies) [11]. Besides single display, dual- or triple-displayed phages are reported to display peptides with different properties on a single phage to construct multifunctional bio-nanofibers [12, 13]. Thus, the surface chemistry and functions of the phages can be precisely manipulated by genetic engineering of the phage proteins.

Phage display is generally limited to the canonical set of 20 L-amino acids used in biological protein synthesis. Therefore, chemical modification of the capsid proteins has been widely used in various applications [14–16]. Moreover, the side wall of the M13 phage is around 900 nm long, with ~2700 copies of pVIII proteins, offering abundant targets for chemical modification. The pVIII protein contains solvent-accessible amino acid residues with functional groups such as amines and carboxylic acids for modification [17, 18]. Although the pVIII protein lacks thiol residues, functionalization of the amine or carboxylic acid groups can introduce reactive sulfhydryl groups for coupling small molecules or proteins. Bioconjugation of the functional groups is largely dependent on the microenvironment, such as solvent accessibility, steric hindrance, and ionization. While the N-terminus of the pVIII protein is a widely used labeling target due to its high solvent accessibility [19], the ε-NH_2 group of Lys8 is also reported to be conjugated with fluorescent dyes and drugs [15], indicating a robust bioconjugation approach when using the pVIII protein. Alternatively, the carboxylate groups from aspartic and glutamic acids are used for coupling with DNA, enzymes, and small molecules due to solvent accessibility [20, 21].

Combining phage display and chemical modification, different types of functional bio-nanofibers such as hydrogels [22, 23], sensors [24, 25], and imaging agents [26, 27] have been constructed.

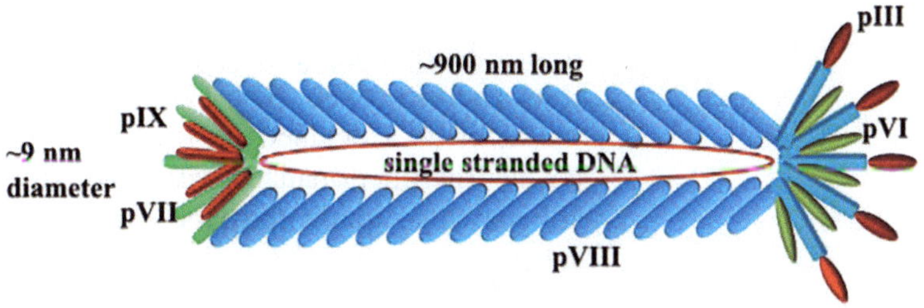

Fig. 1 Illustration of M13 phage

In particular, architectures combining phages and nanoparticles can demonstrate outstanding properties compared with conventional design. Nanoparticles possess unique physical, chemical, and optical properties due to their high surface area and nanoscale size (10–1000 nm) [28–31]. The Belcher group is a pioneer in the fabrication of phage-nanomaterials using M13. In one case, tetraglutamic acid (E4) was displayed at the N-terminus of the pVIII protein to increase the virus' negative charge, for combination with positively charged nanoparticles [32], while a SPARC (secreted protein acidic and rich in cysteine)-binding peptide was displayed on the pIII protein. The resulting phage could be used for in vivo recognition of prostate cancer. In another case, the Belcher group displayed E3 on the pVIII protein and the modified M13 phage was co-assembled with Cy3 and silver nanoparticles (AgNPs) for in vitro bacterial detection [33]. More recently, Mao et al. utilized dual-phage display technology to develop an ultrasensitive method to detect a biomarker of *Candida albicans* infection [34]. Two functional peptides were displayed on a single filamentous phage, one for recognizing anti-secreted aspartyl proteinase 2 IgG antibody in the sera of cancer patients and another for binding magnetic nanoparticles. The limit of detection was two orders of magnitude lower than that of the traditional antigen-based method.

The most widely used strategies for conjugation of nanoparticles on pVIII protein display functional peptides that can capture the nanoparticles on the capsid protein. However, usually the mechanism behind the capture is noncovalent, such as electrostatic interaction between the nanoparticles and the phage proteins, which may lead to the risk of uncontrolled or undesired offloading depending on the chemical environment. Covalent conjugation would be more stable, motivating the development of protocols to conjugate nanoparticles covalently to phage (e.g., to pVIII protein). We previously reported the rational design of chimeric phages that can bind infectious pathogens such as *Pseudomonas aeruginosa* in different environments. The receptor-binding protein from phages that naturally infect the host cell (e.g., *P. aeruginosa*) was displayed as a genetic fusion with the pIII protein [35]. For photothermal therapy, we developed a technique to covalently conjugate gold nanorods onto the pVIII protein via thiol chemistry [36, 37]. Herein, based on our previous research, we describe a detailed protocol to prepare chimeric M13 phage covalently functionalized with gold nanorods (Fig. 2). This protocol may also guide the labelling of pVIII with other functional species.

2 Materials

2.1 Preparing Gold Nanorods

1. Gold (III) chloride trihydrate ($HAuCl_3$, 99.9%).
2. Sodium borohydride ($NaBH_4$, 98%).
3. Milli-Q water.
4. 5-Bromosalicylic acid (5-BAA, >98.0%).
5. Hexadecyltrimethylammonium bromide (CTAB, >98.0%).
6. L-Ascorbic acid (≥99.5%).
7. Silver nitrate ($AgNO_3$, >99%).
8. Nitric acid (HNO_3).
9. Hydrochloric acid (HCl).
10. 100-mL round neck bottle.
11. Stir bar.
12. Magnetic stirrer.

2.2 Preparing Chimeric Phage

1. M13-NotI-Kan construct [38].
2. Sodium chloride (NaCl, 99%).
3. Tryptone (99%).
4. Yeast extract (99%).
5. Agar.
6. Isopropylthio-β-galactoside (IPTG, 99%).
7. Poly(ethylene glycol) (PEG-8000, MW: 8000 Da).
8. Chloramphenicol.
9. Kanamycin sulfate.
10. Mix and Go competent cells.
11. QIAprep Spin Miniprep Kit.
12. QIAquick Gel Extraction Kit.
13. KpnI-HF/NotI-HF restriction enzyme.
14. Cutsmart® digestion buffer.
15. T4 DNA ligase.
16. DNA ligation buffer.

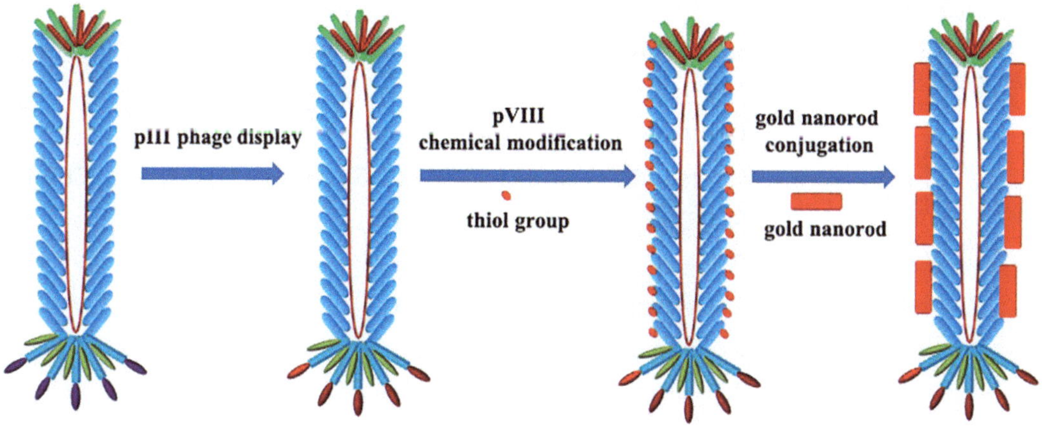

Fig. 2 Overall scheme for preparing the chimeric M13 phage–gold nanorod conjugates

17. Nalgene™ membrane and prefilter disks (0.45 μm).
18. Luria-Bertani (LB) media: Prepare by mixing 10 g tryptone, 10 g sodium chloride (NaCl), and 5 g yeast extract in 950 mL MilliQ H_2O and adjusting the pH to ~7 and final volume to 1000 mL. Autoclave the media and store at 4 °C. Add antibiotics when necessary after the media cools down.
19. LB agar plate: Prepare by mixing 10 g tryptone, 10 g sodium chloride (NaCl), 15 g agar, and 5 g yeast extract in 950 mL MilliQ H_2O, adjusting the pH to ~7 and final volume to 1000 mL. Autoclave the media and pour the liquid into plates when it cools down. Let plates cool and dry with the lids ajar and then store at 4 °C. Add antibiotics and IPTG when necessary after the liquid cools down before pouring the plates.
20. Forward primer, 5′- AAACTGGCAGATGCACGGTT -3′; reverse primer, 5′-AACCCGTCGGATTCTCCG-3′.
21. SsoAdvanced Universal SYBR Green Supermix.
22. PureLink™ Viral RNA/DNA Mini Kit.
23. M13mp18 single-stranded DNA.
24. PEG/NaCl solution: Dissolve 100 g PEG-8000 and 75 g NaCl in 400 ml water and bring to a final volume of 500 ml by stirring at room temperature overnight. Sterilize with autoclave and filter with 0.45 μm Nalgene™ membrane and prefilter disks.
25. TBS buffer (pH 7.5).
26. Agarose.
27. Tris–acetate–EDTA buffer.

2.3 Preparing Phage–Gold Nanorod Conjugates

1. N-Succinimidyl-S-acetylthiopropionate (SATP).
2. Hydroxylamine hydrochloride (99%).
3. Tris buffer (50 mM, pH ~ 3).
4. PBS buffer (pH from 7 to 8).
5. Thiol–PEG–acid (HOOC-PEG-SH; PEG average M_n 5000).
6. Dialysis kit (MWCO 3500 Da).
7. N_2 cylinder.
8. Dimethyl sulfoxide (DMSO).
9. Cysteine.
10. Ellman's reagent.

3 Methods

3.1 Preparing Gold Nanorods

The gold nanorods were prepared with modified procedures according to a reported protocol [39].

1. Clean all glassware with aqua regia (mixture of nitric acid and hydrochloric acid, molar ratio of 1:3), rinse them extensively with water, and dry before use.
2. Prepare the seed solution. Mix 10 mL of 0.2 M CTAB solution with 10 mL of 1 mM $HAuCl_4$ solution with gentle stirring (800 rpm). Prepare 0.6 mL ice-cold $NaBH_4$ solution (0.01 M) and diluted to 1 mL with ice-cold water. Add the $NaBH_4$ solution to the $HAuCl_4$/CTAB mixture solution under vigorous stirring (1500 rpm). Stop stirring 2 min later and let the seed solution age at room temperature for 15 min.
3. Prepare the growth solution. Completely dissolve 18 g CTAB in 500 mL warm water with 2.5 g 5-BAA under vigorous stirring in a 2-L flask. Let the chemicals completely dissolve in warm water and avoid excessive foaming from CTAB. Let the solution slowly cool down to 30 °C and then add 25 mL of 4 mM $AgNO_3$ solution under gentle stirring (800 rpm). Put the mixture in a water bath of 30 °C for 30 min without stirring. Add 500 mL $HAuCl_4$ solution (1 mM) and stir for 15 min gently (500 rpm). Add 5 mL of ascorbic acid (0.064 M) and vigorously stir until the red color disappears.
4. Gold nanorods growth. Add 1.6 mL seed solution to the growth solution with gentle stirring (600 rpm). Stop stirring and let the nanorods grow overnight at 30 °C.
5. Purification of the gold nanorods. Collect gold nanorods by centrifugation (10,000 rpm) for 30 min. Resuspend the nanorods in 200 mL water using a sonication bath for 20 min.

Repeat the centrifugation–suspension cycle for three times and resuspend the nanorods in 50 mL water.

6. Measure the concentration of the gold nanorods by inductively coupled plasma mass spectrometry (ICP-MS).

3.2 Preparing Chimeric Phages

The following procedure requires an infected colony (or culture) of the cloned M13 construct.

1. Inoculate a 5 mL starter culture (LB media +50 μg/mL kanamycin +0.1 mM IPTG). Grow to saturation. Note: Antibiotic and IPTG may or may not be required for specific constructs.
2. Inoculate 100 mL of media with 1 mL of the starter culture. Grow overnight.
3. Spin down at 4000 g for 15 min, 4 °C. Pipette clear supernatant out and save the supernatant.
4. Repeat spin. Pipette supernatant out, leaving behind 5 mL per tube to avoid pipetting up bacteria.
5. To the supernatant, add 20 mL of 2.5 M NaCl/20% w/w PEG-8000. Incubate on ice for 1 h to precipitate phages.
6. Spin down at 12,000 rpm at 4 °C for 20 min to pellet phages. Discard supernatant. Resuspend pellet in 4 mL of TBS; aliquot to four microfuge tubes.
7. Spin for 5 min to pellet cell debris.
8. Transfer the supernatant to fresh centrifuge tubes. Add 200 μL NaCl/PEG solution per tube. Incubate on ice for 15 min to reprecipitate phages.
9. Centrifuge for 10 min at 4 °C. Discard supernatant. Spin again for 1 min and remove as much supernatant as possible.
10. Resuspend phage pellet in TBS (200 uL per tube); vortex as needed.

3.3 Quantification of the Phages by Quantitative PCR

There are multiple possible methods for quantifying phages. These include UV absorbance at 270 nm (based on total DNA and protein absorbance), ELISA with anti-pVIII antibody, plaque titer (if particles are infectious), and qPCR (described here). All methods have advantages and disadvantages, so the best method (s) should be selected based on the experimental situation.

1. Extract the DNA of the chimeric phage with PureLink™ Viral RNA/DNA Mini Kit following the manufacturer's instructions.
2. Run real-time PCR with SsoAdvanced Universal SYBR Green Supermix and Bio-Rad C1000 PCR machine to quantify the DNA amount from the phages. Suggested forward primer, 5′-AAACTGGCAGATGCACGGTT-3′; reverse primer,

5′-AACCCGTCGGATTCTCCG-3′; PCR conditions, 95 °C for 10 min and then 45 cycles of 95 °C for 15 s and 60 °C for 60 s. Use M13mp18 single-stranded DNA as standard.

3. Compare with the standard curve to determine the concentration of the chimeric phages.

3.4 Thiolation of Phages

The phage thiolation was performed with a modified procedure based on a reported protocol [40].

1. Completely remove the oxygen of the phage solution (2 mL, 10^{12} pfu/mL, in PBS buffer, pH 8) in a 20-mL round bottle flask by purging with dry nitrogen for 30 min.
2. Add 1 mM SATP (0.5 mL, 10% v/v DMSO/PBS buffer) to the phage solution. Gently stir (500 rpm) at room temperature overnight.
3. Remove the DMSO by dialysis (MWCO 3500 Da) against PBS buffer at 4 °C for 12 h. Change the buffer every 2 h. Add 1/5 volume of PEG/NaCl solution to the phage solution and incubate on ice for 30 min.
4. Separate the phages by centrifugation (10,000 rpm, 15 min). Resuspend the phage in 2 mL PBS buffer (pH 7.4 with 25 mM EDTA and 0.5 M hydroxylamine). Gently stir for 3 h (500 rpm) at room temperature.
5. Transfer the phage solution into a dialysis tube (MWCO 3500 Da) and dialyze against PBS buffer (pH 7.5) for 24 h in 4 °C. Change the buffer every 2 h.
6. Concentrate the phage solution by ultracentrifugation (MWCO 50 KDa).
7. Quantify the thiol groups with Ellman's reagent according to the manufacturer's instructions. Use cysteine as standard. Quantify the phage by real-time PCR following procedures mentioned above. Determine the thiol groups per phage by dividing the number of thiols by the amount of the phages.

3.5 Preparation of Phage–Gold Nanorod Bioconjugates

1. Suspend the gold nanorods in 1 mL Tris buffer (50 mM, pH ~ 3), and adjust the concentration to 6.8 nM.
2. Add 200 uL of thiolated phage (10^{11} pfu/mL) to the gold nanorod solution drop by drop with an Eppendorf pipette.
3. Gently rotate the solution at room temperature for 2 h.
4. Separate the conjugates by centrifugation at 8000 rpm for 30 min. Resuspend the pellet in 1 mL water.
5. Repeat **step 4** twice.
6. Resuspend the conjugates in 0.5 mL PBS buffer. Add HOOC-PEG-SH (final concentration 2.0 μM) dropwise with an Eppendorf pipette.

7. Rotate the solution gently at room temperature for 24 h.
8. Separate the bioconjugates as described in **steps 4** and **5**.
9. Resuspend the bioconjugates in 200 uL of water.
10. Confirm the morphology of the phage–gold nanorod conjugates with transmission electron microscopy (TEM).

4 Notes

1. Aqua regia is very corrosive; make sure wear to personal protective equipment when handling it!
2. If the CTAB precipitates during the solution cooling, dissolve it completely by heating.
3. Avoid high temperature when using the sonication bath.
4. You may repeat **step 7** in Subheading 3.2 if there seems to be a lot of bacterial debris.
5. It is important to keep the temperature stable and the solution undisturbed in the gold nanorod growth process.

Acknowledgements

This study was supported by the National Natural Science Foundation of China (Grant No. 32201100) (H.P.) and the U.S. National Institutes of Health (DP2GM123457 to IAC).

References

1. Smith GP, Petrenko VA (1997) Phage display. Chem Rev 97:391–410
2. Smith GP (2019) Phage display: simple evolution in a petri dish (Nobel lecture). Angew Chem Int Ed 58:14428–14437
3. Wang R, Li H-D, Cao Y, Wang Z-Y, Yang T, Wang J-H (2023) M13 phage: a versatile building block for a highly specific analysis platform. Anal Bioanal Chem 415:3927
4. Molek P, Bratkovič T (2015) Bacteriophages as scaffolds for bipartite display: designing Swiss Army knives on a nanoscale. Bioconjug Chem 26:367–378
5. Wang X, Zhu X, Wang D, Li X, Wang J, Yin G, Huang Z, Pu X (2023) Identification of a specific phage as growth factor alternative promoting the recruitment and differentiation of MSCs in bone tissue regeneration. ACS Biomater Sci Eng 9:2426–2437
6. Owens AE, Iannuzzelli JA, Gu Y, Fasan R (2020) MOrPH-PhD: an integrated phage display platform for the discovery of functional genetically encoded peptide macrocycles. ACS Central Sci 6:368–381
7. Chen M, Samuelson JC (2016) A DsbA-deficient periplasm enables functional display of a protein with redox-sensitive folding on M13 phage. Biochemistry 55:3175–3179
8. Orner BP, Liu L, Murphy RM, Kiessling LL (2006) Phage display affords peptides that modulate β-amyloid aggregation. J Am Chem Soc 128:11882–11889
9. Wu C-H, Liu IJ, Lu R-M, Wu H-C (2016) Advancement and applications of peptide phage display technology in biomedical science. J Biomed Sci 23:8
10. Smeal SW, Schmitt MA, Pereira RR, Prasad A, Fisk JD (2017) Simulation of the M13 life cycle I: assembly of a genetically-structured deterministic chemical kinetic simulation. Virology 500:259–274

11. Jaroszewicz W, Morcinek-Orłowska J, Pierzynowska K, Gaffke L, Węgrzyn G (2021) Phage display and other peptide display technologies. FEMS Microbiol Rev 46
12. Mohan K, Weiss GA (2014) Dual genetically encoded phage-displayed ligands. Anal Biochem 453:1–3
13. Løset GÅ, Bogen B, Sandlie I (2011) Expanding the versatility of phage display I: efficient display of peptide-tags on protein VII of the filamentous phage. PLoS One 6:e14702
14. Bernard JML, Francis MB (2014) Chemical strategies for the covalent modification of filamentous phage. Front Microbiol 5
15. Carmody CM, Goddard JM, Nugen SR (2021) Bacteriophage capsid modification by genetic and chemical methods. Bioconjug Chem 32: 466–481
16. Davenport BJ, Catala A, Weston SM, Johnson RM, Ardanuy J, Hammond HL, Dillen C, Frieman MB, Catalano CE, Morrison TE (2022) Phage-like particle vaccines are highly immunogenic and protect against pathogenic coronavirus infection and disease. npj Vaccines 7:57
17. Hess GT, Cragnolini JJ, Popp MW, Allen MA, Dougan SK, Spooner E, Ploegh HL, Belcher AM, Guimaraes CP (2012) M13 bacteriophage display framework that allows Sortase-mediated modification of surface-accessible phage proteins. Bioconjug Chem 23:1478–1487
18. Mohan K, Weiss GA (2016) Chemically modifying viruses for diverse applications. ACS Chem Biol 11:1167–1179
19. Li K, Chen Y, Li S, Nguyen HG, Niu Z, You S, Mello CM, Lu X, Wang Q (2010) Chemical modification of M13 bacteriophage and its application in cancer cell imaging. Bioconjug Chem 21:1369–1377
20. Peng H, Borg RE, Nguyen ABN, Chen IA (2020) Chimeric phage nanoparticles for rapid characterization of bacterial pathogens: detection in complex biological samples and determination of antibiotic sensitivity. Acs Sensors 5:1491–1499
21. Hess GT, Guimaraes CP, Spooner E, Ploegh HL, Belcher AM (2013) Orthogonal labeling of M13 minor capsid proteins with DNA to self-assemble end-to-end multiphage structures. ACS Synth Biol 2:490–496
22. Peivandi A, Tian L, Mahabir R, Hosseinidoust Z (2019) Hierarchically structured, self-healing, fluorescent, bioactive hydrogels with self-organizing bundles of phage Nanofilaments. Chem Mater 31:5442–5449
23. Souza GR, Christianson DR, Staquicini FI, Ozawa MG, Snyder EY, Sidman RL, Miller JH, Arap W, Pasqualini R (2006) Networks of gold nanoparticles and bacteriophage as biological sensors and cell-targeting agents. Proc Natl Acad Sci 103:1215–1220
24. Kim K-H, Nguyen TM, Ha S-H, Choi EJ, Kim Y, Kim W-G, Oh J-W, Kim J-M (2020) M13 bacteriophage-assisted morphological engineering of crack-based sensors for highly sensitive and wide linear range strain sensing. ACS Appl Mater Int 12:45590–45601
25. Lee JH, Fan B, Samdin TD, Monteiro DA, Desai MS, Scheideler O, Jin H-E, Kim S, Lee S-W (2017) Phage-based structural color sensors and their pattern recognition sensing system. ACS Nano 11:3632–3641
26. Tsedev U, Lin C-W, Hess GT, Sarkaria JN, Lam FC, Belcher AM (2022) Phage particles of controlled length and genome for in vivo targeted glioblastoma imaging and therapeutic delivery. ACS Nano 16:11676–11691
27. Deutscher SL (2010) Phage display in molecular imaging and diagnosis of cancer. Chem Rev 110:3196–3211
28. Monahan M, Homer M, Zhang S, Zheng R, Chen C-L, De Yoreo J, Cossairt BM (2022) Impact of nanoparticle size and surface chemistry on Peptoid self-assembly. ACS Nano 16: 8095–8106
29. Jiang Y, McNeill J (2017) Light-harvesting and amplified energy transfer in conjugated polymer nanoparticles. Chem Rev 117:838–859
30. Chakraborty A, Boer JC, Selomulya C, Plebanski M (2018) Amino acid functionalized inorganic nanoparticles as cutting-edge therapeutic and diagnostic agents. Bioconjug Chem 29:657–671
31. Ha M, Kim J-H, You M, Li Q, Fan C, Nam J-M (2019) Multicomponent Plasmonic nanoparticles: from Heterostructured nanoparticles to colloidal composite nanostructures. Chem Rev 119:12208–12278
32. Nam KT, Kim D-W, Yoo PJ, Chiang C-Y, Meethong N, Hammond PT, Chiang Y-M, Belcher AM (2006) Virus-enabled synthesis and assembly of nanowires for lithium ion battery electrodes. Science 312:885–888
33. Huang S, Qi J, deQuilettes DW, Huang M, Lin C-W, Bardhan NM, Dang X, Bulović V, Belcher AM (2019) M13 virus-based framework for high fluorescence enhancement. Small 15: 1901233
34. Wang Y, Ju Z, Cao B, Gao X, Zhu Y, Qiu P, Xu H, Pan P, Bao H, Wang L, Mao C (2015) Ultrasensitive rapid detection of human serum antibody biomarkers by biomarker-capturing viral nanofibers. ACS Nano 9:4475–4483

35. Peng H, Chen IA (2019) Rapid colorimetric detection of bacterial species through the capture of gold nanoparticles by chimeric phages. ACS Nano 13:1244–1252
36. Peng H, Rossetto D, Mansy SS, Jordan MC, Roos KP, Chen IA (2022) Treatment of wound infections in a mouse model using Zn2+-releasing phage bound to gold Nanorods. ACS Nano 16:4756–4774
37. Peng H, Borg RE, Dow LP, Pruitt BL, Chen IA (2020) Controlled phage therapy by photothermal ablation of specific bacterial species using gold nanorods targeted by chimeric phages. Proc Natl Acad Sci 117:1951–1961
38. Lin A, Jimenez J, Derr J, Vera P, Manapat ML, Esvelt KM, Villanueva L, Liu DR, Chen IA (2011) Inhibition of bacterial conjugation by phage M13 and its protein g3p: quantitative analysis and model. PLoS One 6:e19991
39. Ye X, Jin L, Caglayan H, Chen J, Xing G, Zheng C, Doan-Nguyen V, Kang Y, Engheta N, Kagan CR, Murray CB (2012) Improved size-tunable synthesis of monodisperse gold Nanorods through the use of aromatic additives. ACS Nano 6:2804–2817
40. Kernan DL, Wen AM, Pitek AS, Steinmetz NF (2017) Featured article: delivery of chemotherapeutic vcMMAE using tobacco mosaic virus nanoparticles. Exp Biol Med 242:1405–1411

Chapter 10

Liquid Glycan Array

Chih-Lan Lin, Eric J. Carpenter, Taoran Li, Tareq Ahmed, and Ratmir Derda

Abstract

The M13 phage platform is a stable and monodisperse nanoscale carrier, which can be modified with different molecules by chemical conjugation strategies. Here, we describe M13 phage acylated on pVIII protein with a dibenzocyclooctyne reacting with azido glycan to yield 30–1500 copy numbers of glycan per phage and monitored by MALDI-TOF spectrometry to generate multivalent glycoconjugates that contain desired densities of glycans. We prepared the liquid glycan arrays (LiGA) such that both the structure and density of glycans were encoded in the DNA of the bacteriophage. The LiGA can be used to validate the binding properties of glycans to purified lectins and explore the effect of glycan density on such binding. From a mixture of multivalent glycan probes, LiGAs can also identify the glycoconjugates with optimal avidity necessary for binding to lectins on living cells in vitro and live animals in vivo.

Key words M13 phage, Liquid glycan arrays, Phage-displayed glycans, Multivalent, Lectin binding

1 Introduction

The most frequent posttranslational modification of proteins is glycosylation [1]. More than 50% of human proteins are glycosylated. Oligosaccharides on the protein surface play essential roles in physicochemical and chemical behaviors such as resistance to proteolysis and clearance of glycoproteins [2]. For example, sialic acids regulate immunity through recognition by Siglecs-displayed on immune cells [3]. And polylactosamine, a compound of repeating galactose and glucosamine, plays a role in immune response and cancer metastasis [4]. One of the most important tools to study the biological role of glycans is a glycan microarray. But currently there are some limitations and challenges of traditional glass glycan microarrays including limited ability to study dynamic competition between multiple glycans interacting with a GBP, fluorescent-labeled lectins may introduce additional interference with the true binding process, and the size and the orientation of the glycan

Huan Peng et al. (eds.), *Phage Engineering and Analysis: Methods and Protocols*,
Methods in Molecular Biology, vol. 2793, https://doi.org/10.1007/978-1-0716-3798-2_10,

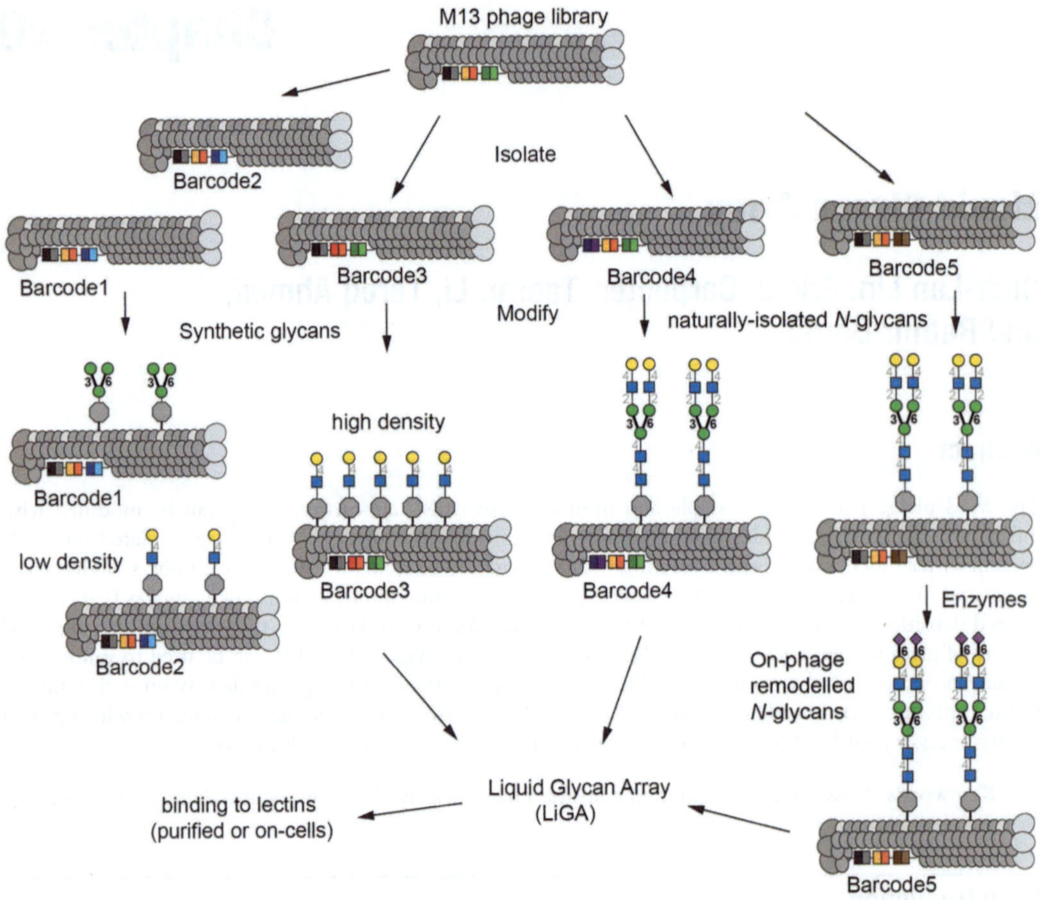

Fig. 1 Illustration of a liquid glycan arrays

may impact binding on solid surfaces. Binding of live cells to glycan arrays displayed on glass surfaces has been reported but, in general it is challenging; furthermore, using glycan array to study protein–glycan interaction inside live animals, in vivo, is impossible. To overcome these limitations, display of glycans on scaffolds different from the glass surface has also been explored.

Previously, our group published genetically encoded, multivalent liquid glycan array (LiGA) on M13 phage [5, 6]. DNA-encoded glass-based glycan arrays have been reported by Chevolot and co-workers [7] and DNA-encoded libraries of glycans on DNA strands have been published by Flitsch and co-workers [8]. Unlike the other DNA-encoded strategies described above, LiGA is built on M13 phage with DNA barcodes inside the phage genome and the multivalent display of glycans on phage can be chemically conjugated to a subset of the ~2700 copies of the major coat protein pVIII. The multivalent presentation of glycans on the surface of the phage and the protected DNA barcodes of LiGA make it a tool to study the interactions of glycans with GBPs in vitro and in vivo (Fig. 1).

To assemble a glycan library of phage containing distinct DNA-encoded barcodes of each different glycan, we used strain-promoted azido alkyne cycloaddition (SPAAC) as a ligation strategy to click alkyl-azido linked glycans with DBCO-modified M13 phage. Glycans installed on the surface of phage can be remodelled by enzymes. In situ analysis of the ligation reaction or products of chemoenzymatic remodeling by MALDI-TOF makes it convenient to improve reaction conditions and trace the progress of reactions.

2 Materials

Chemical reagents were purchased from either Sigma-Aldrich or Thermo Fisher Scientific unless noted otherwise. MALDI-TOF MS spectra were recorded on an ABI Sciex Voyager Elite MALDI MS equipped with a MALDI-TOF pulsed nitrogen laser (337 nm) (3 ns pulse up to 300 μJ/pulse) operating in full-scan MS in positive ionization mode. Nanodrop™ (Thermo Fisher) was used to measure the absorbance of protein and DNA solutions.

HEPES buffer contained 20 mM HEPES, 150 mM NaCl, 2 mM $CaCl_2$, pH 7.4. PBS buffer contains 137 mM NaCl, 10 mM Na_2HPO_4, 2.7 mM KCl, pH 7.4. Solutions used for phage work were sterilized by filtration through 0.22-μm filters.

Concanavalin-A (ConA) was purchased from Sigma-Aldrich.

Sanger sequencing and deep sequencing were performed at the Molecular Biology Service Unit (University of Alberta) using an Illumina NextSeq 500 system. All DNA primers were ordered from Integrated DNA Technologies.

3 Methods

3.1 Construction of Silent Distal Barcoded (SDB) M13 Phage Library

1. Take the phage genome and use the Gibson Assembly cloning kit (NEB#E5510) from New England Biolabs to create a library of silent double-barcode codons (SDB) near the pIII cloning site.
2. Introduce the SDB regions into the M13KE vector using the NEBuilder HiFi DNA Assembly (NEB# E2621S). Begin with the double-stranded DNA of the M13KE vector containing the stuffer sequence CAGTTTACGTAGCTGCATCAGGGTG GAGGT, which corresponds to the peptide QFT*LHQ, with * indicating a stop codon.
3. Perform PCR amplification of the insert fragment with primers P1 and P2, and PCR amplify the vector fragment with primers P3 and P4:

Name Sequence (5->3):

P1 GAGATTTTCAACGTGAAAAACT **N**CT**N**TT**Y**GC-**N**AT**H**CC**N**CT**N**GTGGTACCTTTCTATTCTCA

P2 TTAAGACTCCTTATTACGCAGTA

P3 TTGCTAACATACTGCGTAATAAG

P4 TTTTTCACGTTGAAAATCTC

P5 GTGGTACCTTTCTATTCTCACTCGAGYGTNGARAARAAYGAYCARAARACNTAYCAYGCNGGNGGNGGNT-CGGCCGAAACTGTT-GAAAG

P6 CGAGTGAGAATAGAAAGGTAC

Take 50 ng of phage dsDNA and add 1 mM dNTPs, 0.5 μM primers, and 0.5 μL of Phusion High Fidelity DNA polymerase in 1x PCR buffer (NEB #B0518S) to make a total volume of 50 μL. Perform PCR with the following temperature cycling protocol:

(a) 98 °C 3 min.

(b) 98 °C 30 s.

(c) 60 °C 30 s.

(d) 72 °C 4 min.

(e) Repeat b–d for 35 cycles.

(f) 72 °C 10 min.

(g) 4 °C hold.

4. Take the PCR-amplified fragments and treat them with the restriction enzyme DPN 1 (NEB #R0176S). Then, perform gel purification.
5. For NEBuilder Hifi DNA assembly, mix 100 ng of vector with 4 ng of the insert. Add 10 μL of NEBuilder Hifi DNA assembly master mix and deionized H2O to reach a total volume of 20 μL, following the manufacturer's protocols.
6. Take the resulting ligated DNA and transform it into *E. coli* K12 ER2738. Propagate the transformed cells overnight at 37 °C. Afterward, centrifuge the overnight culture to separate the bacteriophage from the host cells.
7. Deep sequencing of the resulting SB1 QFT*LHQ cloning vector which now contains 6144 theoretical sequence combinations in the leader region is available at https://48hd.cloud/file/20161105-68OOooIC-NB.
8. Take primers P5 and P2, anneal them, and then amplify them using PCR with the 35-cycle protocol mentioned above to create a dsDNA insert for the SVEKNDQKTYHAGGG peptide cloning.
9. PCR amplify the vector SB1 QFT*LHQ using primers P4 and P6. Process the PCR fragments using the NEBuilder Hifi DNA assembly kit as described earlier.

10. Transform the ligated DNA into electrocompetent cells of *E. coli* SS320 (Lucigen). After transformation, centrifuge the overnight culture to remove host cells. Incubate the culture with 5% PEG-8000 and 0.5 M NaCl for 8 hours at 4 °C. Following the incubation, perform a 15-minute centrifugation at 13000 × *g* to concentrate the released phage. Resuspend the PEG-precipitated phages in PBS–glycerol 50% and store them at −20 °C.
11. The sequence of the vector containing the M13-SDB-SVEKY library is available on Gen Bank (#MN865131). Deep sequencing of the resulting library is available at https://48hd.cloud/file/20161215-67OOooOO-NB.

3.2 SDB Clone Isolation and Amplification

1. Take the M13-SDB-SVEKY library, as described earlier, and use it to isolate each monoclonal silently encoded phage.
2. Dilute a 10 μL aliquot of phage and plate it with a density of 100 plaques per plate. Pick single colonies manually and transfer them into clean 1.7-mL plastic tubes containing 0.5 mL of PBS–glycerol 50%. Incubate the tubes at room temperature for 30 minutes, and then heat them at 55 °C for 10 minutes to inactivate any remaining bacterial cells.
3. To a 125-mL baffled Erlenmeyer flask, add 20 μL solution of each colony suspension in 25 mL of LB medium with 250 μL of log phase *E. coli* K12 ER2738 and incubate for 7 hours at 37 °C.
4. After amplification, collect the phage clones from the culture supernatant by centrifugation at 4000 × *g* for 15 minutes.
5. Incubate the supernatant with 5% PEG-8000 and 0.5 M NaCl for 8 hours at 4 °C, and then perform a centrifugation at 14000 × *g* for 30 minutes to precipitate the phage particles.
6. Resuspend the phage in 1 mL of PBS and transfer the solution to a 1.5-mL Eppendorf tube and centrifuge at 14000 × g for 20 min.
7. Transfer supernatant solution to a new 1.5-mL Eppendorf tube and add 110 μL Triton X-100 solution, vortex briefly, and incubate for 1 hour at room temperature.
8. Then add 200 μL 5% PEG-8000 and 0.5 M NaCl and incubate for 1 hour on ice. After 1 hour, centrifuge the tube at 14000 × *g* for 20 min, and decant all supernatant.
9. Resuspend the phage pellet in 1 mL of PBS–glycerol 50%, titer the suspension, and store the phages at −20 °C until further use.
10. For SDB identification, submit a sample of 400 ng of the phage DNA for Sanger sequencing.

11. Select the clones that contain three or more base pair substitutions from one another (Hamming distance [H] ≥3). This allows correction of any point mutations that may have arisen during the analysis by deep sequencing. A list of isolated clones is available in Table 1.
12. Take a 2 μL solution of clonal phage and perform PCR amplification using barcoded sequencing primers. Follow the protocol in PCR Protocol Subheading 3.9. After amplification, analyze the products by Illumina sequencing.
13. The results of the Illumina sequencing of the isolated clones are available at https://48hd.cloud. Example of "SDB10" can be found at https://48hd.cloud/file/11846.

3.3 Chemical Modification of Phage Clones by DBCO-NHS

1. Take a solution of phage clone (50 μL, 10^{12}–10^{13} PFU/mL in PBS pH 7.4) and add DBCO-NHS (20 mM in DMF) to 1.7 mL Eppendorf™ tubes. Create three different concentrations of DBCO-NHS in the reaction mixture: 0.2 mM, 1.0 mM, and 2.0 mM. These concentrations typically yield 5–50% of pVIII modification after 45 min of incubation. (Prepare the three specific densities as mentioned.)
2. After conjugation of DBCO-NHS, individually purify each clone using a Zeba™ Spin Desalting column (7 K MWCO, 0.5 mL, Thermo Fisher). Add the reaction to the equilibrated Zeba™ column and centrifuge it at 1500 × g for 2 min. Store the filtered solution on ice or at 4 °C.
3. Use 2 μL of the filtrate from **step 2** to check the reaction progress immediately after Zeba™ column purification using MALDI-TOF MS (Fig. 2).

3.4 Chemical Modification of Phage Clones with Glycans to Build LiGA Components

1. After DBCO conjugation, take the filtrate and add solutions of lactose-specific azido glycans (10 mM stock in nuclease-free H_2O) to achieve a 2 mM concentration of azido-glycan in each solution. Incubate the solutions overnight at 4 °C.
2. Verify and quantify all chemical reactions using MALDI-TOF MS (Fig. 3). If any reactions are incomplete and a residual pVIII-DBCO peak is detected, address this by adding more azido-glycan and extending the incubation time.
3. Once the reactions are complete, purify the conjugates using a Zeba™ Spin Desalting column. Supplement the purified conjugates with glycerol and store them in a 50% glycerol stock at −20 °C. Also, analyze the conjugates using a plaque-forming assay to calculate the titer (PFU/mL).

Table 1
Examples of silent double barcodes (SDB) used to build Liquid glycan arrays

SDB_id	SB-1 region	SB-2 region
SDB1	CTGCTGTTCGCAATACCACTC	AGTGTGGAGAAGAATGATCAGAAGACTTATCATGCGGGTGGAGGT
SDB2	CTTCTATTCGCAATTCCGCTC	AGTGTGGAGAAGAATGATCAGAAGACTTATCATGCGGGTGGAGGT
SDB3	CTGCTTTTCGCAATTCCGCTT	AGTGTGGAGAAGAATGATCAGAAGACTTATCATGCGGGTGGAGGT
SDB6	CTGCTGTTTGCGATTCCACTG	AGTGTGGAGAAGAATGATCAGAAGACTTATCATGCGGGTGGAGGT
SDB9	CTTCTTTTTGCAATTCCTCTA	AGTGTGGAGAAGAATGATCAGAAGACTTATCATGCGGGTGGAGGT
SDB10	CTACTGTTTGCTATACCGCTG	AGTGTGGAGAAGAATGATCAGAAGACTTATCATGCGGGTGGAGGT
SDB12	CTTCTGTTCGCGATACCTCTA	AGTGTGGAGAAGAATGATCAGAAGACTTATCATGCGGGTGGAGGT
SDB13	CTACTTTTCGCAATTCCTCTG	AGTGTGGAGAAGAATGATCAGAAGACTTATCATGCGGGTGGAGGT
SDB15	CTGCTGTTCGCCATACCCCTT	AGTGTGGAGAAGAATGATCAGAAGACTTATCATGCGGGTGGAGGT
SDB17	CTACTCTTCGCGATTCCGCTT	AGTGTGGAGAAGAATGATCAGAAGACTTATCATGCGGGTGGAGGT
SDB18	CTGCTGTTTGCTATCCCTCTG	AGTGTGGAGAAGAATGATCAGAAGACTTATCATGCGGGTGGAGGT
SDB20	CTGCTCTTTGCCATCCCGCTT	AGTGTGGAGAAGAATGATCAGAAGACTTATCATGCGGGTGGAGGT
SDB21	CTACTCTTTGCAATTCCCCTT	AGTGTGGAGAAGAATGATCAGAAGACTTATCATGCGGGTGGAGGT
SDB22	CTACTGTTTGCTATCCCACTT	AGTGTGGAGAAGAATGATCAGAAGACTTATCATGCGGGTGGAGGT
SDB23	CTGCTCTTTGCAATACCTCTT	AGTGTGGAGAAGAATGATCAGAAGACTTATCATGCGGGTGGAGGT
SDB24	CTACTATTCGCGATCCCGCTC	AGTGTGGAGAAGAATGATCAGAAGACTTATCATGCGGGTGGAGGT
SDB26	CTGCTATTCGCTATCCCACTC	AGTGTGGAGAAGAATGATCAGAAGACTTATCATGCGGGTGGAGGT
SDB29	CTGCTATTTGCGATCCCGCTG	AGTGTGGAGAAGAATGATCAGAAGACTTATCATGCGGGTGGAGGT
SDB34	CTTCTTTTTGCGATTCCGCTG	AGTGTGGAGAAGAATGATCAGAAGACTTATCATGCGGGTGGAGGT
SDB35	CTGCTCTTCGCTATTCCACTT	AGTGTGGAGAAGAATGATCAGAAGACTTATCATGCGGGTGGAGGT

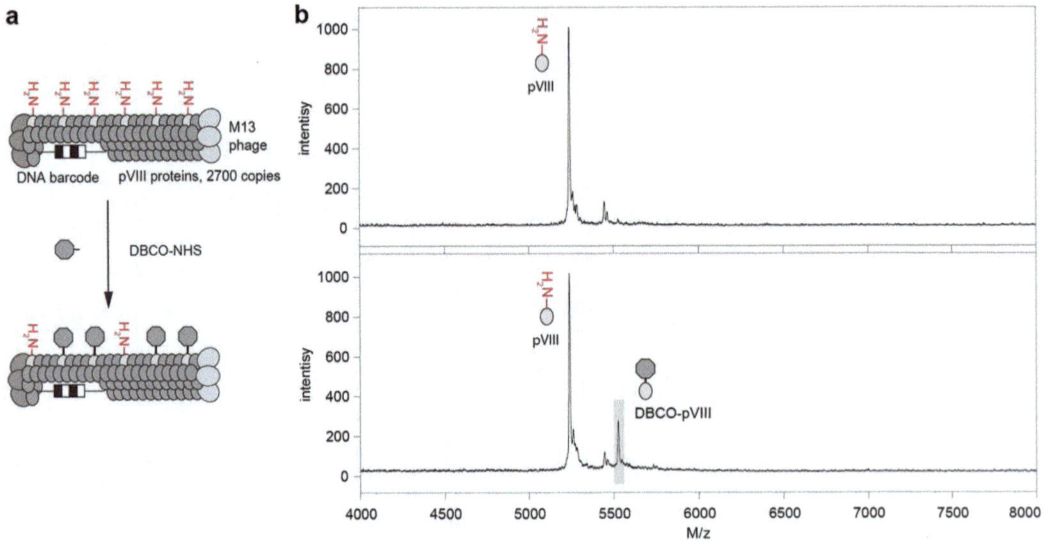

Fig. 2 MALDI-TOF MS spectra of DBCO-pVIII peak

3.5 Enzymatic Modification on Phage-Displayed Glycans

1. Procedures for phage surface glycosidase treatment: Cleavage of sialic acid on phage-displayed N-glycans by neuraminidase—after modification of phage clones with 7 to afford the conjugated phage (60 μL, ~10^{12}–10^{13} PFU/mL in PBS, pH 7.4), 5% PEG-8000, 0.5 M NaCl (12 μL) was added and the solution was kept for 1.5 h at 0°C. The solution was then centrifuged at 21000 × *g* for 10 min at room temperature. The supernatant was removed, and the pellet was resuspended in sodium acetate buffer (40 μL, 50 mM, pH 5.5 containing 5 mM $CaCl_2$) followed by the addition of *Clostridium perfringens* neuraminidase (1 μL, 0.02 units). The reaction was incubated at 37 °C and monitored by MALDI-TOF MS. The reaction was complete in 1 hr. to afford galactose-terminating *N*-glycans on the phage surface.
2. Procedures for phage surface glycosylation treatment: Phage clones with LacNAc (β-Gal-(1→4)-β-GlcNAc-$OCH_2CH_2N_3$) (60 μL, ~10^{12}–10^{13} PFU/mL in PBS, pH 7.4), 5% PEG-8000, 0.5 M NaCl (12 μ L) was added and the solution was kept for 1.5 h at 0°C. After 1.5 h, the solution was centrifuged at 21000 × *g* for 10 min. at room temperature. and supernatant was decanted. The pellet was resuspended in Tris–HCl buffer (25 μL, 100 mM with 20 mM $MnCl_2$, pH 8.5) containing CMP-Neu5Ac (300 μg), recombinant shrimp alkaline phosphatase (rSAP, 0.5 μL) and Pd26ST (10 μL, 1.29 mg/mL). The reaction mixture was incubated at 37 °C and monitored by MALDI-TOF MS. Pd26ST (5 μL, 1.5 mg/mL) was added twice, at 3 hours and at 6 hours reaction time.
3. Example MALDI-TOF spectra for Pd26ST modification of LacNAc displayed on phage is in Fig. 4.

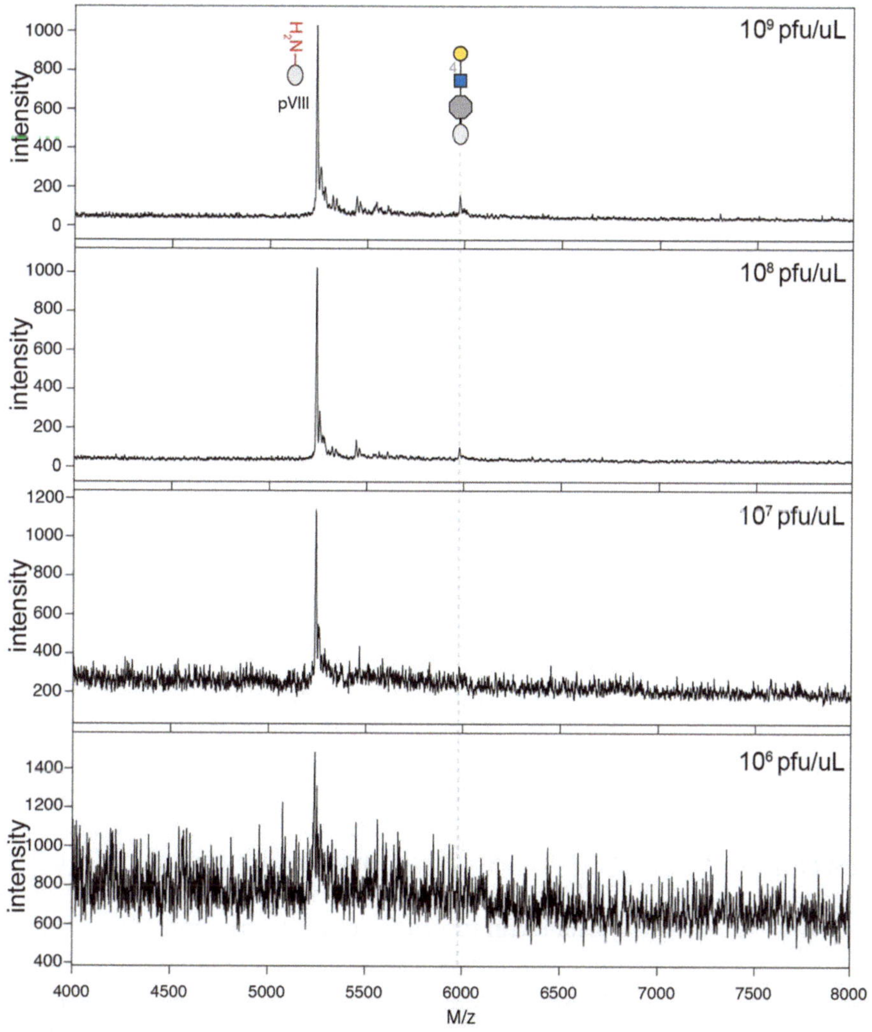

Fig. 3 MALDI-TOF MS analysis of glycan conjugate on phage with different concentrations

3.6 Analysis of Glycosylation of Phage Samples by MALDI-TOF MS

1. Take sinapinic acid (Sigma, 10 mg/mL) and prepare layer 1 by dissolving it in acetone–methanol (4:1). Prepare layer 2 by dissolving sinapinic acid (10 mg/mL) in acetonitrile–water (1:1) with 0.1% TFA. *See* **Note 1** for details of layer 1 and layer 2 compositions.
2. For sample preparation, combine 2 μL of the phage solution (~10^9–10^{13} PFU/mL) in PBS with 4 μL of layer 2. Then, deposit 0.7 μL of layer 1 on top. Once the solution is dry, add 1.5 μL of the layer 2 + phage mixture. To remove salts from the PBS, wash the spots three times with 0.1% TFA in water (10 μL).

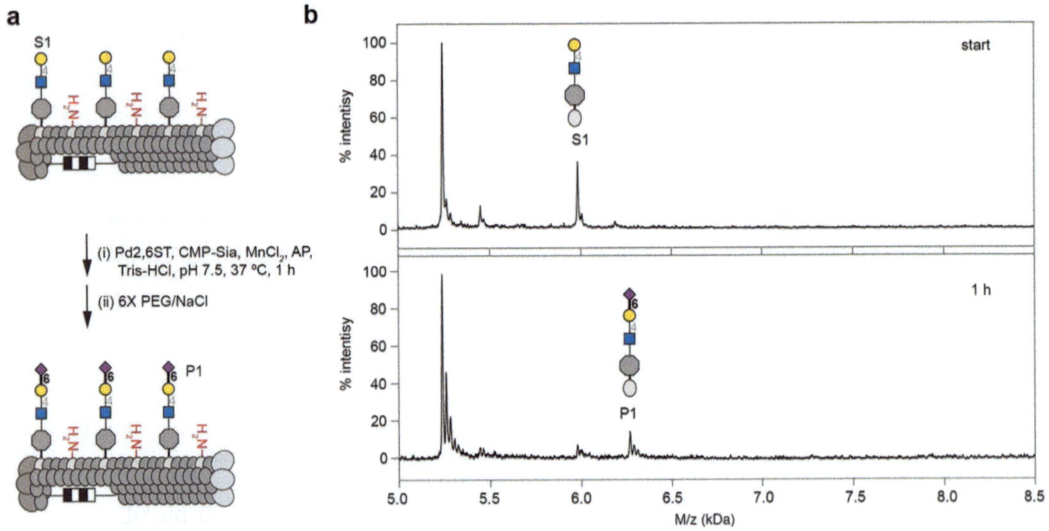

Fig. 4 MALDI-TOF MS results of Pd26ST-catalyzed sialylation

3. Analyze the prepared samples using MALDI-TOF MS. Estimate the ratio of modified to unmodified pVIII by fitting and plotting the data using MatLab.

3.7 Preparation of LiGA from Glycosylated Clones

1. Take 10^8 PFU of the desired glycan–phage conjugates and mix them in a single tube to prepare a specific LiGA SC. Characterize the mixture by titering, and use N × 10^6 PFU (N = the number of glycan–phage conjugates) for a typical lectin or cell-binding experiment.
2. Assign a two-letter identifier (e.g., "SC") to each unique LiGA mixture and create a "dictionary" named SC.xlsx. This table describes the correspondence between the DNA barcode sequences and the glycans in the LiGA mixture (refer to Table 2).
3. Use these dictionaries to translate the nucleotide sequences in the deep-sequencing files to the corresponding glycans, along with their density information. Examples of the dictionaries used can be found in the Supporting Information Table 2.

3.8 Binding of LiGA to Lectins Immobilized on Plate

1. Take lyophilized concanavalin A and dissolve it in HEPES to make a 1 mg/mL stock solution. Dilute this stock solution with HEPES to get a final concentration of 20 μg/mL. Add 75 μL of this diluted solution to four wells of a 96-well plate (Corning® #CLS3369) (These four wells will be used as four replicates of "test") and add solution of 2% BSA in HEPES to another four wells of the same plate (these four wells will be

Table 2
Example of LiGA dictionary

SDB	Alphanum	Order	Common name	Axis name (glycan-[density]
SDB217	PZ-10048	1	aMan	aMan-[840]
SDB121	PZ-8015	2	(Man)3	(Man)3-[1300]
SDB122	PZ-8015	3	(Man)3	(Man)3-[1730]
SDB137	Man3	4	(Man)3	(Man)3-[410]
SDB148	M5	5	GN	GN-[590]
SDB154	M5	6	GN	GN-[1050]
SDB123	PZ-5080	7	Lac-peg4	Lac-peg4-[1080]
SDB128	D9	8	Lac	Lac-[1030]
SDB105	D9	9	Lac	Lac-[1240]
SDB79	D10	10	LacNAc, LN	LacNAc, LN-[970]
SDB129	D10	11	LacNAc, LN	LacNAc, LN-[1110]
SDB52	D8	12	Lec	Lec-[680]
SDB113	Tr59	13	Galili-tri	Galili-tri-[1000]
SDB152	Tr61	14	Pk	Pk-[860]
SDB168	Tr260	15	Gala3-type1	Gala3-type1-[350]
SDB143	Tr60	16	B2 tri	B2 tri-[350]
SDB55	Tr62	17	P1 tri	P1 tri-[620]
SDB58	D21	18	LacDiNAc	LacDiNAc-[50]
SDB115	Tr54	19	LNT-2	LNT-2-[430]
SDB153	Tr55	20	GNLN	GNLN-[810]
SDB169	Tr307	21	3'GN type1	3'GN type1-[860]

The dictionary was subsequently used to translate from nucleotide sequences of each SDB in the deep-sequencing files to the corresponding glycans

used as four replicates of "control"). Cover the plate with sealing tape to prevent contamination and incubate it overnight at 4 °C. *See* **Note 2** and **3** for other utilities of panning on plate.

2. The next day, wash the wells three times by adding washing buffer (200 μL, 0.1% Tween-20 in HEPES) to each well and discarding the solution by inverting the plate on top of a paper towel.
3. After washing, add blocking solution (100 μL, 20 μg/μL BSA in HEPES) to each well and incubate for 1 hour at room temperature.

4. Discard the blocking solution by inverting the plate, and then wash the wells three times with washing buffer (200 μL, 0.1% Tween-20 in HEPES).
5. After blocking and triple washing, add 75 μL of LiGA (8×10^8 PFU/mL in HEPES) to each well of "test" and "control," and incubate for 2 hours at room temperature.
6. Discard the LiGA solution from the wells by inverting the plate. Then, wash the wells two times with washing buffer (200 μL, 0.1% Tween-20 in HEPES) and one time with HEPES (200 μL).
7. To elute the bound phage, add 75 μL of HCl (pH 2.0) to each well and incubate for only 9 minutes at room temperature. Transfer the contents of each well to an Eppendorf tube containing 37.5 μL of 5x Phusion HF buffer (NEB, #M0530S).

The neutralized solution from the four "test" wells and the four "control" wells are PCR-amplified (Subheading 3.9) separately giving rise to 8 DNA samples for Illumina sequencing (Subheading 3.10) and 8 separate Illumina sequencing datasets for DE analysis (Subheading 3.11).

3.9 Two-Step PCR

1. Take 5 μL of the DNA template solution in nuclease-free water and add it to a total volume of 50 μL. Prepare the reaction mixture for the first step with 1x Phusion® buffer, 50 μM of each dNTP, 500 μM $MgCl_2$, 1 μM NF10 primer, 1 μM −96 primer, and one unit of Phusion® High-Fidelity DNA Polymerase (NEB, #M0530S). The typical 50 μL reaction mixture for the first step is as follows:

1st step PCR:

(a) DNA template (phage solution).	5 μL
(b) DMSO	2 μL
(c) Phusion HF buffer (5×)	10 μL
(d) Phusion® polymerase (1 unit)	0.5 μL
(e) 10 mM dNTPs	1 μL
(f) NF10 primer (10 μM)	2.5 μL
(g) -96 primer (10 μM)	2.5 μL
(h) Nuclease-free water	26.5 μL (for a total volume of 50 μL)

2. Perform the first step PCR cycles using the following thermocycler settings:
   (a) 98 °C 3 min.
   (b) 98 °C 10 s.

(c) 50 °C 20 s.
(d) 72 °C 20 s.
(e) Repeat b–d for 30 cycles.
(f) 12 °C 1 min.
(g) 4 °C hold.

3. For the second step PCR, take 5 μL of the first step PCR product as the DNA template. Perform the amplification in a total volume of 50 μL, including 1x Phusion® buffer, 50 μM of each dNTP, 500 μM $MgCl_2$, 1 μM forward barcoded primer, 1 μM reverse barcoded primer, and one unit of Phusion® High-Fidelity DNA Polymerase (NEB, #M0530S). A typical 50 μL reaction mixture for the 2nd step contains:

2nd step PCR:

(a) PCR product from 1st step.	5 μL
(b) Phusion HF buffer (5x)	10 μL
(c) Phusion® polymerase (1 unit)	0.5 μL
(d) 10 mM dNTPs	1 μL
(e) Forward barcoded primer (10 μM)	2.5 μL
(f) Reverse barcoded primer (10 μM)	2.5 μL
(g) Nuclease-free water	28.5 μL (for a total volume of 50 μL)

4. Perform 2nd step PCR cycles using the following thermocycler settings:
   (a) 98°C 3 min.
   (b) 98°C 10 s.
   (c) 58°C 30 s.
   (d) 72°C 20 s.
   (e) Repeat b–d for 20 cycles.
   (f) 12°C 1 min.
   (g) 4°C hold.

5. Take the PCR products and run them on a 2% (w/v) agarose gel in Tris–borate–EDTA buffer at 100 volts for about 35 minutes. Use a low-molecular-weight DNA ladder as a standard (NEB, #N3233S) to quantify the PCR products.

3.10 Illumina Sequencing

1. The PCR products that contain different indexing barcodes were pooled, allowing 10 ng of each product in the mixture.

2. The mixture was purified by eGel, quantified by QuBit, and sequenced using the Illumina NextSeq paired-end 500/550 High Output Kit v2.5 (2x75 cycles).
3. Data was automatically uploaded to BaseSpace™ Sequence Hub. Processing of the data is described in Subheadings **3.11** and **3.12**.

3.11 General Data Processing Methods

1. Data analysis was performed in Python, MatLab, or R. Core scripts for analysis are available at GitHub. Here is the URL to access the scripts: https://github.com/derdalab/nglycans-synthesis-paper/tree/v1.0.
2. Comparison and testing differences for significance in LiGA data was performed essentially as differential enrichment analyses of phage-displayed libraries described in previous reports [9, 10] and based on differential expression (DE) analysis implemented in edgeR [9, 10].
3. The DE analysis has three major factors considered: (i) The abundance of each phage/glycan is estimated from the observed counts using a negative binomial model; (ii) the abundances between two sets of samples can be tested for significant differences, in order to deal with the many such tests, Benjamini–Hochberg (BH) adjustment is used to control the false discovery rate (FDR) at $\alpha = 0.05$ [10]; and (iii) normalization of data across multiple samples assumes that a group of the phages (azidoethanol-labeled phages for in vitro) is invariant between the test and control data.
4. To assess the significance of a glycan binding in a specific experiment, the differential enrichment of the levels of the DNA barcode associated with that glycan in "test" sets of DNA read was compared to the levels of the same barcode in "control" sets.
5. In binding to lectins, the "control" dataset was from association of the same LiGA mixture with blank carriers (BSA-coated wells).
6. Examples of "test" and "control" datasets can be retrieved from the https://48hd.cloud/ server as tables of DNA sequences and raw sequencing counts. *See* **Note 4** for examples of the data file's URL. In these datasets, DNA reads that could not be mapped to any entries in the LiGA dictionary were discarded.

3.12 Process of Illumina Sequencing Data

1. From the GitHub repository (https://github.com/derdalab/nglycans-synthesis-paper/tree/v1.0), download the files in the differential-testing directory. To do this conveniently, click on the green "Code" button at the top of the page and choose "Download ZIP" from the popup menu.

2. Unpack the Zip archive and navigate to the "differential-testing" subdirectory.
3. Prepare a dictionary file describing the mixture of sequences in your LiGA samples. At minimum it must contain an "SDB" column, a label column, and a "Alphanum" column. The "SDB" column gives the name of each phage sequence, the label column a glycan-density description of the glycophage used to label plots, etc., and the "Alphanum" column contains a second identifier for the glycan.
4. In preparing the dictionary and other files, it may be useful to use the example files in the repository as templates.
5. Prepare an order table. This table must contain an "Alphanum" column and an "Order" column. Each "Alphanum" entry in the dictionary should have a matching "Alphanum" entry here. The glycophages in the dictionary will be sorted by the corresponding order value.
6. Prepare the variable columns of a campaign file. This describes which data is compared by the script. It requires three columns which describe the data, "Filename" which holds the filename without any extension, "Type" which assigns each file to either the test or control side of the comparison, and "Columns" which is a comma separated list of numbers specifying which data columns are included. Note that an entry of "1,2,4" omits the third column.
7. Prepare the fixed columns of a campaign file. These must include "Dictionary," "Labels," and "OrderTable." These columns must have the same values (fixed) for each row of the table. The Labels column contains the name of the label column in your dictionary, while the "Dictionary" and "OrderTable" contain the names of your dictionary and order table, respectively.
8. Download each of the desired datasets from the https://48hd.cloud website. In each case, click on its name to view additional details and then on the Download button to copy a zip format archive containing the data to your computer.
9. Open the "differential-test.R" file in any text editor. If installed R Studio is a good choice.
10. At the top of the "differential-test.R," a number of variables are defined specifying the locations for various input files.
11. Ensure the LiGA.data, dict_dir, order_table_dir, and campaign_dir variables match the directories containing the LiGA data (from 48HD), dictionary, order table, and campaign files (your creations), respectively. This can be done by editing the script, moving/renaming files/directories, or both.

12. Save the "differential-test.R" file to ensure that any edits are used.
13. If you have not already installed all the prerequisite packages used by the script, you can do this now by entering `source ("00-install-prerequisites.R")` on the R command line. Alternatively, if you open this file with R Studio, click the "Source" button in the upper right-hand corner of the window. Once this step has been done once, it does not need to be repeated.
14. To use the script type `source("differential-test.R")` into the R command line. Alternatively, if the file is open in R Studio, this can be done by clicking the "Source" button in the upper right-hand corner of the window. As the commands are evaluated, some output is written to the R console, and output files are written.
15. The files with raw DNA reads, raw counts, and mapped glycans were uploaded to a server, https://48hd.cloud/. All LiGA sequencing data is publicly available at this location. Each experimental dataset has a unique alphanumeric name (e.g., **20220219-87SCcaBI-CT**) and unique static URL: https://48hd.cloud/file/10331.

4 Notes

1. Layer 1: 7.5 mg sinapinic acid +300 μL acetone +75 μL methanol. Layer 2: 10 mg sinapinic acid +500 μL acetonitrile +500 μL water. Store the layers at 4°C with tin foil wrapped around the layer tubes.
2. Many other lectins can be coated on the plate in identical fashion.
3. Biotinylated lectins can be coated on streptavidin-coated plate. Fc-lectins such as Siglecs[ref] can be coated on a protein-G-coated plate. In the latter case, BSA-blocked streptavidin-coated well or protein-G-coated well must be used as a "control."
4. Binding of LiGA to ConA. URL of set 1 (test): https://48hd.cloud/file/10331. URL of set 2 (control): https://48hd.cloud/file/10329.

References

1. An HJ, Froehlich JW, Lebrilla CB (2009) Determination of glycosylation sites and site-specific heterogeneity in glycoproteins. Curr Opin Chem Biol 13:421–426
2. Shirakawa A, Manabe Y, Fukase K (2021) Recent advances in the chemical biology of N-Glycans. Molecules 26:1040

3. Crocker PR, Paulson JC, Varki A (2007) Siglecs and their roles in the immune system. Nat Rev Immunol 7:255–266
4. Dennis JW, Granovsky M, Warren CE (1999) Glycoprotein glycosylation and cancer progression. Biochim Biophys Acta Gen Subj 1473: 21–34
5. Lin C-L et al (2023) Chemoenzymatic synthesis of genetically-encoded multivalent liquid N-glycan arrays, vol 14. derdalab/nglycans-synthesis-paper, p 5237. https://doi.org/10.5281/zenodo.8188084
6. Lin C-L et al (2023) Chemoenzymatic synthesis of genetically-encoded multivalent liquid N-glycan arrays. Nat Commun 14:5237
7. Chevolot Y et al (2007) DNA-based carbohydrate biochips: a platform for surface glyco-engineering. Angew Chem Int Ed 119:2450–2454
8. Thomas B et al (2017) Application of biocatalysis to on-DNA carbohydrate library synthesis. Chembiochem 18:858–863
9. Cheng K, Zhou Y, Neelamegham S (2017) DrawGlycan-SNFG: a robust tool to render glycans and glycopeptides with fragmentation information. Glycobiology 27:200–205
10. Tsuchiya S, Yamada I, Aoki-Kinoshita KF (2019) GlycanFormatConverter: a conversion tool for translating the complexities of glycans. Bioinformatics 35:2434–2440

Part IV

Structural Biology of Phages

Chapter 11

Electron Microscopy Methods for Phage-Based Study

Damayanti Bagchi

Abstract

Electron microscopy (EM) techniques play a vital role in virology research including phage discovery and their identification. The use of different staining protocols based on the concept of negative staining is one of the most important steps in the EM processing. This chapter will summarize the widely used EM protocols in phage research, their advantages, and limitations. Phage-based therapy, especially recently developed nanoparticle–phage conjugates, are expected to find clinical significance in the antimicrobial resistance (AMR) epidemic. EM techniques are important to characterize these conjugates and we will also discuss the methods here.

Key words Negative staining, Nanoparticle-phage conjugates, Synthesis strategy, Sample preparation, Uranyl acetate, Metal nanoparticle-phage conjugate

1 Introduction

Electron microscopy (EM), namely, transmission electron microscope (TEM), is a useful tool for examining phage morphology. Negative staining is a widely used contrast-enhancing method that uses heavy metals to increase the contrast between the background and the sample of interest (phage particle). Brenner and Horne first demonstrated the concept of negative staining for examining a virus structure almost 60 years ago [1] and this remains a useful technique for the EM community. As the heavy metal ions have a better ability to disperse electrons compared to less electron-dense biological molecules, it can generate difference in contrast that helps to identify phage clearly.

However, the success of EM imaging of phage depends on many factors. The purity of phage plays a critical factor in obtaining a clear and concise image with a clear background. The buffer used for washing and subsequent sample preparation step also contribute to image quality [2]. The incubation time and concentration of staining material need to be optimized for getting crisp images.

Huan Peng et al. (eds.), *Phage Engineering and Analysis: Methods and Protocols*,
Methods in Molecular Biology, vol. 2793, https://doi.org/10.1007/978-1-0716-3798-2_11,

Many protocols are available in the literature and it might be unclear for a new researcher. Here, we elaborate a step-by-step method with important points to be considered.

With the advent of phage-based therapeutics [3] in antibiotic resistant bacterial infections, it is also important to establish a more general EM protocol to visualize the conjugate materials. The heavy-metal staining methods become more challenging for nano-conjugates as the nanoparticle and phage particle itself have distinct contrast difference. In some cases, negative staining is not needed for seeing conjugates; as an example metal nanoparticles have greater contrast and phages can be visualized as low-contrast material. In some cases, alternative negative staining agents for phages can produce better clarity in imaging. The staining material, its composition and concentration, time of incubation, surface charge of conjugates, washing buffer, drying condition, and imaging parameters are all important factors for optimizing EM imaging [4]. We will discuss different conditions from some recent literature and try to establish a more generalized and easy-to-follow approach.

2 Materials

2.1 Phage Purification

1. 250-mL and 2-L Erlenmeyer flask.
2. A medium-size cold centrifuge able to reach 25,000 g, equipped with a fixed-angle rotor and can work at 4 °C.
3. Centrifuge tubes or jars depending upon the total amount of solution with conical bottoms and caps to prevent evaporation.
4. A well-mixed solution of 2.5 M NaCl/20% PEG-8000, prepared freshly and thoroughly homogeneous.
5. Tris-buffered saline (TBS).
6. Glycerol.

2.2 Staining Materials

1. Uranyl acetate, 1–2% (w/v).
2. Uranyl formate.
3. Phosphotungstate (potassium or sodium), 2%.
4. Lanthanide acetate stains, 1–2% (w/v).
5. Ammonium molybdate 1–2% (w/v).
6. Pointed tweezers.
7. EM grids, 200–400 square mesh carbon-coated copper grids.
8. Petri dishes laid out with filter paper or grid storage–transport box.

2.3 Preparation of Different Staining Materials

2.3.1 Preparation of Uranyl Acetate

1. Uranyl acetate (UA,1–2% [w/v]) is dissolved in boiled degassed ultrapure water.
2. The dissolution is slow, and pH should be maintained at 4–4.5 using 1 M KOH or NaOH.
3. After the dissolution and cooling of solution, filter through a 0.2-μm syringe filter.
4. Once prepared, the UA solution should be protected from light and stable at 4 °C up to 1 year.
5. This chemical is toxic and radioactive and should be handled with proper care. All preparations should be done under a fume hood with appropriate personal protective equipment.

2.3.2 Preparation of Uranyl Formate

6. 20 mg of powder uranyl formate (UF) is dissolved in 2 mL of degassed distilled water with continuous stirring.
7. 4 μL of 10 M NaOH is then added to the solution while stirring.
8. Change in pH should change the color of the solution to a darker yellow color, without any formation of precipitates.
9. Once thoroughly dissolved and cold, the solution is filtered through a 0.2-μm syringe filter.
10. Once prepared, it should be kept at 4 °C in the dark and only stable for 1–2 days.

2.3.3 Preparation of Phosphotungstate

11. Phosphotungstic acid is dissolved in distilled water.
12. The solution is neutralized to pH 7 using 5 M NaOH or KOH solution.
13. The solution can be kept for a long time with proper protection.

2.3.4 Preparation of Lanthanide Acetate

14. Dissolve lanthanide salts (samarium acetate [SmAc], gadolinium acetate [GdAc], thulium acetate [TmAc], or erbium acetate [ErAc]) at 1–2% (w/v) in ultrapure water.

2.3.5 Preparation of Ammonium Molybdate

15. Dissolve ammonium molybdate at 1–3% (w/v) in ultrapure water.
16. The pH needs to be adjusted to 7.0 using 5 M NaOH for dissolution.

Note: The staining solution needs to be filtered before every use to ensure its purity and no precipitations or interference from staining agents.

2.4 Phage–Nanoparticle Synthesis

1. Depending upon the nature of synthesized conjugates.
2. Metal salt (0.1 M $HAuCl_4 \cdot 3H_2O$, 1 M $AgNO_3$).

3. Reducing agent ($NaBH_4$).
4. Capping agent (sodium citrate, polymeric substance, peptides, detergents as CTAB).
5. Imidazole (imid).
6. Polyethylene glycol (PEG).
7. Succinimidyl S-acetylthiopropionate (SATP).
8. Stirrer with hot plate or water bath.

2.5 Purification of Nanoconjugates

1. Centrifuge with 25,000 g power.
2. Centrifuge tube of proper size depending upon the total volume of synthesis.
3. Washing buffer of particular pH (depends on the nature of the phage).
4. Oven with variable temperature settings.
5. Gooch-type filtering crucible.

3 Methods

3.1 Phage Purification

1. Remove host cells by centrifugation at 4500 g for 10 min. Transfer supernatant to a fresh tube. Repeat the centrifugation.
2. Transfer top clear supernatant to a new tube and add 2.5 M NaCl/20% PEG-8000 (w/v). For proper precipitation of the phages, 1:4 volume ratio of the PEG/NaCl mixture to supernatant solution provides the best results. Briefly mix. Precipitate phage overnight at 4 °C.
3. The NaCl/PEG solution acts as precipitating agent of the phage and electrostatic interaction helps phage precipitation. This method is best suited for M13 phage [5].
4. Pellet phage by centrifugation at 12000 g for 15 min. Decant supernatant. Resuspend pellet in 1 mL Tris-buffered saline (TBS) used to wash phages. The buffer solution needs to be selected depending upon the nature of the phage. TBS is used for the M13 phage. One should consider to carefully choose wash buffer depending upon the nature of the phage and optimize the concentration and other parameters [6].
5. Transfer supernatant to a fresh tube. Add 200 μL of 2.5 M NaCl/20% PEG-8000. Incubate on ice for 15–60 min. Spin at 12000–14000 rpm in a benchtop centrifuge for 10 min. Discard supernatant. Spin again briefly and remove remaining supernatant with pipette. Resuspend pellet in 200 μL TBS. For long-term storage at −20 °C, add 200 μL sterile glycerol.
6. To scale up the above protocol, use multiple culture flasks. Alternatively, after incubating 20 mL culture for 2 h, add the

entire culture to 1 L LB. Incubate the large culture for 4 h, then scale up the protocol, and repeat it to remove cells and purify phage.

3.2 Staining Method

A carbon-coated EM grid is placed on the tip of a reverse tweezer with carbon-coated side on top. Five microliters of purified phage sample is placed using a 10-μL pipette and wait for 1–3 min. The grid is blotted with filter paper and the excess solution is removed. The grid is washed with 5 μL of ultrapure water and blotted. Then the appropriate staining material (for most cases 2% aq UA) is placed on the grid for 30 s to 1 min. The solution is blotted with filter paper. The grid is washed with 5 μL ultrapure water for 30 s, two times. The grid is then air-dried for 15 mins and consequently imaged in TEM [7].

3.3 Phage–Nanoparticle Synthesis

The synthesis of nanoparticle–phage conjugates depends on many factors. The two different approaches mostly taken into consideration are as follows:

Previously synthesized nanoparticle coated with different kinds of ligands is attached to phage either using a chemical bond formation or via electrostatic interaction [8, 9].

Phage acts as capping agent and/or reducing agent and interacts with metal ions to form metal nanoparticles [10].

The amount of phage to be used and the ratio of phage to metal ions plays a major role in the structure of nanoconjugates. It should be properly optimized and checked for repeatability.

To explain two of the synthesis strategies for nanoparticle–phage conjugates, examples of metal–phage conjugates from published articles are summarized here:

1. Gold Nanoparticle–Phage Network (Au NP–Phage).

An Au nanoparticle solution (average size of 45 nm) is synthesized following the citrate-reduction procedure (mass ratio of 0.8 sodium citrate: 1 Au(III) chloride). Eight serial dilutions of fd-tet phage (10^7 pfu/mL) in nanopure water is mixed with an equal volume of Au nanoparticle solution (normalized to 4.2 a.u. using extinction measured at 530 nm) and allowed to stand for 12 h at room temperature. This facilitates Au–phage hydrogel formation and subsequent addition and mixing of 10^9 phage units, 1.0 M imidazole (metal binding molecule), with an equal volume of Au nanoparticles produces Au–phage–imid networks. Finally, the networks were purified by three consecutive centrifugation cycles (20,800 g, for 30 min) in glass sterile tubes [8, 11] (Fig. 1).

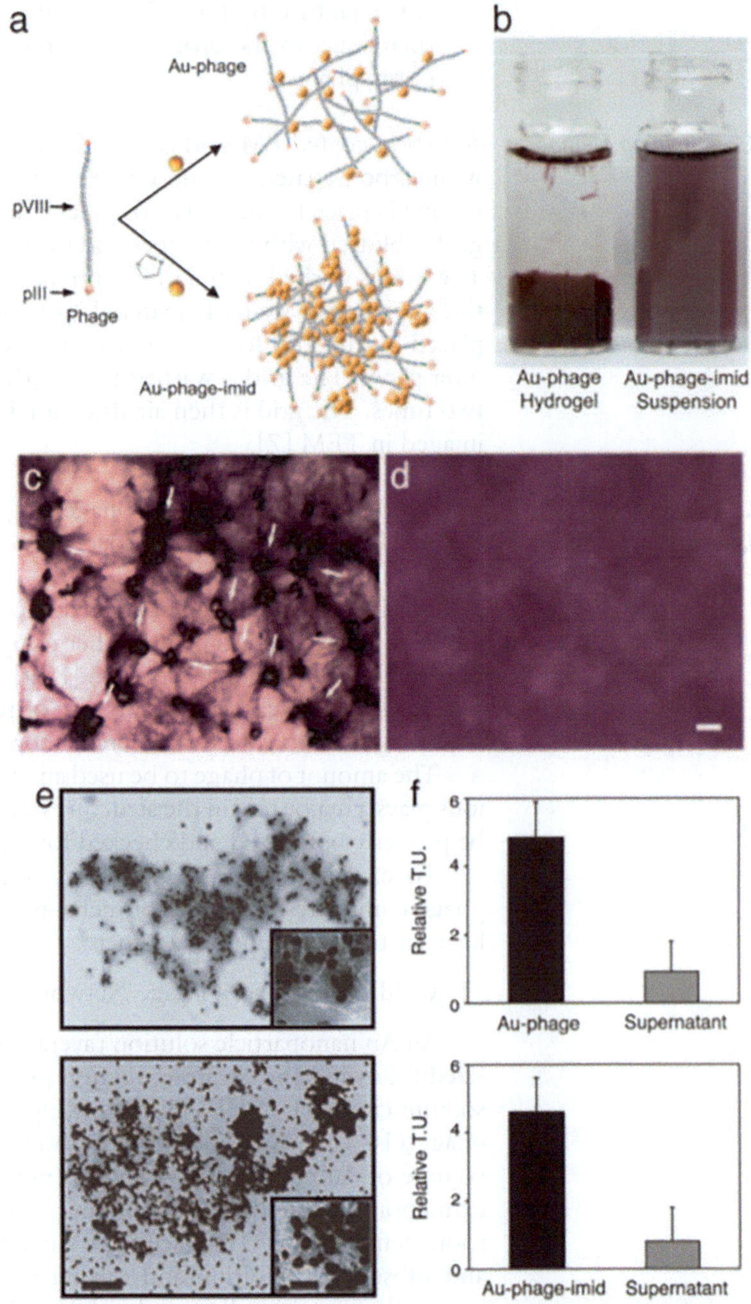

Fig. 1 (**a**) Concept and biological structural characterization of Au–phage and Au–phage–imid networks. Strategy for Au assembly onto phage nanoparticles. Imid and the yellow spheres (Au nanoparticles [not drawn to scale]). (**b**) Vials of nanoparticle solutions: Au–phage hydrogel (left) and suspension of purified Au–phage–imid (right; suspended from hydrogel precursor). (**c** and **d**) Hydrogel formed with RGD-4C-displaying phage. (Scale bar, 20 μm.) (**c**) C17.2 murine neural stem cells cultured within hydrogel for 24 h. Cell accumulation followed by cell-induced network displacement is shown (arrows point to cells within the network). (**d**) Control hydrogel (no cells). (**e**) TEM of purified networks: Au–phage

2. Gold nanorods–Phage Conjugates (Au NR–Phage).

Au NRs are synthesized following a seed-mediated protocol. The details can be found in published articles [12, 13]. Surface modification of gold nanostructures under aqueous conditions has been widely reported [14]. Au NR of uniform shape and size (with an average aspect ratio of 3.8) has been used to chemically attach modified phage. M13 phage was modified with SATP to introduce thiol groups to primary amines. Modified M13 phage is conjugated to AuNRs by formation of gold–sulfur bonds at room temperature at pH 3.0. Interaction between AuNRs and phages during bioconjugation was promoted by the positive charge from trace CTAB on the AuNRs and the negatively charged capsid protein of phage particles [7] (Fig. 2).

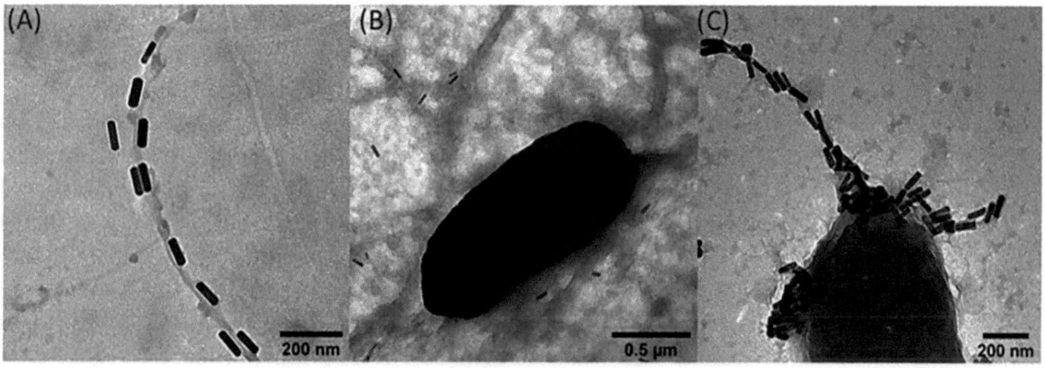

Fig. 2 Interaction between M13KE, AuNR, and *E. coli* cells. (**a**) TEM image of M13KE–AuNR, illustrating the conjugation of filamentous phage and AuNRs. When *E. coli* cells were mixed with M13KE and HOOC-PEG–AuNR (nonconjugated), no aggregation or localization of AuNRs to the cells was seen (**b**), but HS-PEG-COOH-modified M13KE–AuNR bioconjugates attached to *E. coli* cells did result in visible aggregation of AuNRs on the cell surface (**c**). Aggregation at one end of the bacterium (inC) presumably occurs near the position of the F pilus; stimulation of retraction by phage attachment may cause accumulation at the root of the pilus. [Note: CC BY-NC-ND licensed]
Reference [7]

Fig. 1 (continued) (upper) and Au–phage–imid (lower). (Scale bar, 500 nm; inset scale bar, 100 nm.) (**f**) Bacterial infection with purified Au–phage (upper) and Au–phage–imid (lower) networks; TUs are shown for purified and functional Au–phage and Au–phage–imid solution and for unbound phage present in the supernatant from centrifuged network solutions. [Note: Copyright (2006) National Academy of Sciences, U.S.A.]
Reference [8]

3. Silver–Phage Conjugates.

Silver nanoparticles were synthesized using M13 phage as both the reducing agent and capping agent.

Ag-M13

0.05 mL of M13 phage (10^{13} pfu/ mL) is added to 5 mL of Milli-Q water (pH maintained at 8–8.5). Then, 0.05 mL of 100 mM $AgNO_3$ solution is dropwise added with constant stirring. The total concentration of Ag ions in solution is 1 mM. Appearance of an initial turbid solution implies the start of the nucleation process. The solution is stirred at room temperature in the dark for 24 h. The change of color signifies formation of Ag NP. The solution was centrifuged at 7200 g for 15 min and washed with water to remove excess phage or unreacted Ag ions. The washing step is continued for two times, and finally, the precipitate was resuspended in Milli-Q water and stored at 4 °C until further use [15].

In a different synthesis strategy, PEG is used as a catalyzing agent to increase the rate of interaction between the phage and Ag ions. Moreover, 0.05 mL of the M13 phage (10^{13} pfu/ mL, separated after first PEG reprecipitation step) is added to 5 mL of Milli-Q water (pH -7). Then, 0.05 mL of 100 mM $AgNO_3$ solution is dropwise added with constant stirring. The solution is stirred at room temperature in the dark for 2 h. Further, 1.25 mL of 2.5 M NaCl/20% PEG-8000 is added to the solution and briefly mixed. The solution is stored at 4 °C overnight without any disturbance. The precipitate is then centrifuged at 14000 rpm for 10 mins and washed with water to remove excess phage, unreacted Ag ions, or PEG. The washing step is continued for two times, and finally, the precipitate is resuspended in Milli-Q water and stored at 4 °C until further use.

3.4 Purification of Nanoconjugates

It is very important to purify the nanoconjugates thoroughly to get artifact-free EM images. After optimizing the synthesis process, the conjugate is centrifuged at 14000 rpm for 10 mins and washed with water or the buffer used, to remove any excess phage or unreacted metal ions. The washing step is continued for at least two more times. Washing solution needs to be selected depending upon the nature of the phage. To verify complete removal of unattached phages, it is a good idea to check the absorbance spectra of the supernatant solution as an indication of residual protein or DNA [16].

3.5 Sample Preparation and Dilution

To obtain clear EM images, the conjugate needs to be properly cleaned. As metal nanoparticles are electron-dense and phages are electron translucent, the staining process should be more thorough.

The sample should first be imaged without any staining. It is expected that metal nanoparticles are easily identifiable due to their high contrast. In some cases, phages appear as a faint background. To use negative staining, one should pay attention with the timing of UA incubation. The researcher can prepare different grids incubated with UA for various times (30 s, 1 min, 2 min, 3 min) and compare the contrast difference between different images to finally optimize negative staining. The other parameter is concentration of UA. Generally, 1–2% UA in water serves the purpose well.

The relative concentration of phage and nanoparticle also plays a role in imaging. Researchers should start with three to five different concentrations of metal/phage ratio and, depending upon the obtained structure, finalize the synthesis protocol. If the stock solution is a very densely packed structure, then serial dilution should provide an overall perspective of the formed conjugate.

A carbon-coated EM grid is placed on the tip of a reverse tweezer with carbon-coated side on top. Place 5 μL of the nanoparticle–phage conjugate sample using a 10-μL pipette and wait for 10–15 min. The grid is blotted with filter paper and the excess solution is removed. The grid is then cleaned with 5 μL of ultrapure water and blotted. Then the appropriate staining material (for most cases 2% aq UA) is placed on the grid for 30 s to 1 min. The solution is bloated with filter paper. The grid is washed with 10 μL ultrapure water for 30 s, two times. The grid is then air-dried for 15 min to 2 h and consequently imaged in TEM. As a negative control, grids without any staining agents need to be used. Alternatively, drops of 1% ammonium molybdate in distilled water (pH 7.0, for 60 s) are used as staining material [8]. The EM image should provide a clear distinction between the nanoparticle and phage in terms of contrast and the ratio of nanoparticles to phage particles can be estimated. This ratio should be validated by other methods such as inductively coupled plasma mass spectrometry (ICP-MS) to measure the amount of nanoparticle and qPCR to measure the amount of phage in a sample.

3.6 Alternative Methods

The use of carbon-coated copper grids is standard for nanobiomaterials. Sometimes formavar-coated nickel grids are also used. For low pH samples, UA or UF interferes with samples. Lanthanide-based stains such as TmAc or ErAc may be more appropriate [17], although the overall pH of the preparation must be kept below the isoelectric point of the sample protein to help prevent positive staining. Especially for low pH-sensitive samples, anionic tungstate or molybdate stains may be more effective.

4 Notes

1. Phage purification is one of the most important steps for getting accurate EM images. The PEG/NaCl precipitation and reprecipitation step plays a vital role and the supernatant from the centrifuged solution should be carefully transferred for the next steps. If there is confusion of mixing, repeat the centrifugation step.
2. The pH of the washing buffer and the nature of the surface charge of phage particles or conjugates should be taken into consideration. Washing of the sample prior to staining may be necessary if the buffer in which the specimen is maintained has a high salt or phosphate component. In many cases, washing can be performed with ultrapure water but for more sensitive samples, which may degrade or undergo structural changes when exposed to water alone, washing may need to be performed with a volatile buffer of low ionic strength.
3. The concentration of phage needs to be in a certain range (generally 10^7–10^9 pfu/mL) to obtain good EM images. A higher concentration of phages tends to generate an aggregated ultrastructure which gives inaccurate information. Similarly, for nano-phage conjugates, the concentration should be properly optimized by doing a serial dilution study. Surface charge depending upon the nature of capping agent, nanoparticle, or chemical bonds can easily contribute to the agglomeration of a structure.
4. The grid needs to be cleaned before drop casting the sample. Plasma glow discharge is used for 10–30 s for etch cleaning of the organic residue from the grid surface.
5. Purity of the staining material also plays a vital role. Any staining material that will be used needs to be filtered freshly with a 0.2-μm filter. If there is any confusion, one should prepare a negative control grid using the staining material only.
6. It is good idea to air-dry the EM grid. Overnight drying results in best images; however, a few hours of drying or using an external air source increases the rate of the drying process.
7. During imaging, charging from the electron source, lower contrast, and not-dried sample can create a drifting problem. One should wait and interaction between the electron beam and sample stabilizes over time. If the problem continues after a certain time, the sample needs to be removed from the column and needs to be properly dried and stained. Sometimes using a higher concentration of staining materials helps to reduce the drifting issue.

8. The method by which a grid is prepared in terms of sample adsorption, blotting, and staining can also significantly affect what is observed. The most appropriate method is thus, again, highly sample dependent. So, imaging should be repeated within the same sample of a particular batch and for different batches as well.
9. Staining depth is influenced by multiple factors such as hydrophilicity of the grid surface, evenness of the carbon layer, the amount of stain applied to the grid, the length of time stain is in contact with the grid prior to blotting, the extent of blotting, and the time it takes for the grid to completely dry. A grid will never have an even distribution of stain across the entire area, and therefore, areas of the grid appropriate for imaging need to be selected carefully. Indeed, grids often vary in quality even when prepared on the same day under the same conditions.
10. Phage and nanoparticle–phage conjugates should be evaluated with EM in a time-dependent manner to check their stability. Firstly, the EM protocol needs to be standardized by optimizing the abovementioned parameters and then the EM should be used as a technique to ensure structural integrity of the samples used for further studies with progression of time.
11. Negative controls including individual component and staining materials under identical experimental conditions need to be properly evaluated.
12. For a detailed video description of EM sample preparation and negative staining process, refer to the video corresponding to reference [4].

References

1. Brenner S, Horne RW (1959) A negative staining method for high resolution electron microscopy of viruses. Biochim Biophys Acta 34:103–110
2. Ackermann HW (2009) Basic phage electron microscopy. Method Mol Biol (Clifton, NJ), 501:113–26
3. Lin DM, Koskella B, Lin HC (2017) Phage therapy: an alternative to antibiotics in the age of multi-drug resistance. World J Gastrointest Pharmacol Therap 83:162–73
4. Scarff CA, Fuller MJG, Thompson RF, Iadanza MG (2018) Variations on negative stain electron microscopy methods: tools for tackling challenging systems. JoVE 132:e57199
5. Peng H, Chen IA (2019) Rapid colorimetric detection of bacterial species through the capture of gold nanoparticles by chimeric phages. ACS Nano 132:1244–1252
6. Passaretti P, Khan I, Dafforn TR, Goldberg Oppenheimer P (2020) Improvements in the production of purified M13 bacteriophage bio-nanoparticle. Sci Rep 101:18538
7. Peng H, Borg RE, Dow LP, Pruitt BL, Chen IA (2020) Controlled phage therapy by photothermal ablation of specific bacterial species using gold nanorods targeted by chimeric phages. Proc Natl Acad Sci 1174:1951–1961
8. Souza GR, Christianson DR, Staquicini FI, Ozawa MG, Snyder EY, Sidman RL et al (2006) Networks of gold nanoparticles and bacteriophage as biological sensors and cell-targeting agents. Proc Natl Acad Sci 1035: 1215–1220
9. Peng H, Rossetto D, Mansy SS, Jordan MC, Roos KP, Chen IA (2022) Treatment of wound infections in a mouse model using Zn2+-releasing phage bound to gold Nanorods. ACS Nano 163:4756–4774

10. Scibilia S, Lentini G, Fazio E, Franco D, Neri F, Mezzasalma AM et al (2016) Self-assembly of silver nanoparticles and bacteriophage. Sens Bio Sens Res 7:146–152
11. Kannan P, Los M, Los JM, Niedziolka-Jonsson J (2014) T7 bacteriophage induced changes of gold nanoparticle morphology: biopolymer capped gold nanoparticles as versatile probes for sensitive plasmonic biosensors. Analyst 13914:3563–3571
12. Ye X, Jin L, Caglayan H, Chen J, Xing G, Zheng C et al (2012) Improved size-tunable synthesis of monodisperse gold Nanorods through the use of aromatic additives. ACS Nano 63:2804–2817
13. Pérez-Juste J, Pastoriza-Santos I, Liz-Marzán LM, Mulvaney P (2005) Gold nanorods: synthesis, characterization and applications. Coord Chem Rev 24917:1870–1901
14. Chen Y, Xianyu Y, Jiang X (2017) Surface modification of gold nanoparticles with small molecules for biochemical analysis. Acc Chem Res 502:310–319
15. Naik RR, Stringer SJ, Agarwal G, Jones SE, Stone MO (2002) Biomimetic synthesis and patterning of silver nanoparticles. Nat Mater 13:169–172
16. Elgiddawy N, Ren S, Yassar A, Louis-Joseph A, Sauriat-Dorizon H, El Rouby WMA et al (2020) Dispersible conjugated polymer nanoparticles as biointerface materials for label-free bacteria detection. ACS Appl Mater Interfaces 1236:39979–39990
17. Hosogi N, Nishioka H, Nakakoshi M (2015) Evaluation of lanthanide salts as alternative stains to uranyl acetate. Microscopy 646:429–435

Chapter 12

Visualization of Engineered M13 Phages Bound to Bacterial Targets by Transmission Electron Microscopy

Yanxi Yang and Irene A. Chen

Abstract

The filamentous phage M13 is one of the most well studied and characterized phages, particularly since it was introduced as a scaffold for phage display, a technique to express and evolve fusion proteins on the M13 phage's coat to study protein or peptide binding interactions. Since phages can be engineered or evolved to specifically bind to a variety of targets, engineered M13 phages have been explored for applications such as drug delivery, biosensing, and cancer therapy, among others. Specifically, with the rising challenge of antimicrobial resistance among bacteria, chimeric M13 phages have been explored both as detection and therapeutic agents due to the flexibility in tuning target specificity. Transmission electron microscopy (TEM) is a powerful tool enabling researchers to directly visualize and characterize binding of phages to bacterial surfaces. However, the filamentous phage structure poses a challenge for this technique, as the phages have similar morphology to bacterial structures such as pili. In order to differentiate between bacterial structures and the filamentous phages, here we describe a protocol to prepare TEM samples of engineered M13 phages bound to bacterial cells, in which the phage virions have been specifically labeled by decoration of the major capsid proteins with gold nanoparticles. This protocol enables clear visualization and unambiguous identification of attached filamentous phages within the context of bacterial cells expressing numerous pili.

Key words M13, Bacteriophage, Phage, Transmission electron microscopy, Filamentous phage, Virus

1 Introduction

Bacteriophages (or phages) are viruses that are specific to bacteria and harmless to humans [1]. Due to their biocompatibility and diagnostic and therapeutic potential, phages are increasingly studied and engineered for various applications such as drug delivery, biosensing, cancer therapy, and tissue regeneration [2, 3]. Phage capsids are generally composed of repetitions of a small number of capsid proteins [1, 4]. While they encompass a large diversity of morphologies, two major morphological classes are the tailed phages, which include an icosahedral head, and the filamentous, or rodlike, phages. One of the most well-characterized phages is

Huan Peng et al. (eds.), *Phage Engineering and Analysis: Methods and Protocols*,
Methods in Molecular Biology, vol. 2793, https://doi.org/10.1007/978-1-0716-3798-2_12,

M13, a filamentous phage that is about 1 μm in length and 6 nm in diameter. M13 has a single-stranded DNA genome packaged inside its capsid, which is composed of ~2700 copies of the major capsid protein pVIII (also called g8p), forming the bulk of the filamentous structure, and 3–5 copies of four minor capsid proteins, including the receptor-binding protein pIII (also called g3p) [5, 6]. M13 phage is a prominent tool in molecular biology as a system for phage display technology, in which protein variants can be rapidly selected, typically for binding activity. Major applications include display of a heterologous mutant library of antibody single-chain variable fragments (scFv) or display of a random library of peptides on the major (pVIII) or minor (pIII) capsid proteins [5]. The expression and selection of these proteins allows the discovery of M13-like phages engineered to bind to different targets. At the same time, since pVIII carries solvent-exposed carboxyl as well as primary amine groups, M13 phages have the potential for chemical surface modifications in order to achieve various functionalities [3].

An important application focus for phages is addressing antimicrobial resistance in bacteria, specifically for species that exhibit multidrug resistance and virulence. Prominent among these are *Enterococcus faecium, Staphylococcus aureus, Klebsiella pneumoniae, Acinetobacter baumannii, Pseudomonas aeruginosa,* and *Enterobacter* spp., which are collectively known as the ESKAPE pathogens and are a rising global challenge that poses a great risk to public health [7]. As bacteria develop resistance to existing antibiotics, the rate of development of new therapeutics for multidrug-resistant bacterial infections has been relatively low due to lack of financial incentives [8]. Phages and phage-derived products are promising alternatives to antibiotics, potentially providing high specificity and biocompatibility as well as substantial room for innovation [9].

Prior work from multiple laboratories, including ours, has demonstrated the potential for engineering M13 phages to bind different bacterial host species. In particular, exchanging the original M13 receptor-binding protein with homologs from other filamentous phages (with specificity for other bacterial species) creates chimeric M13 phages that are able to bind to alternative bacteria with high specificity [10, 11]. Engineered phages could be conjugated to gold nanoparticles, enabling new methods for bacterial detection as well as photothermal therapy of bacterial infections in both in vitro and in vivo models [11–14]. A foundational method in developing such technologies is the ability to characterize the ability of engineered phage candidates to bind to different bacterial species.

Transmission electron microscopy (TEM) is a powerful tool enabling researchers to directly visualize bacterial or phage samples at nanometer scale, using accelerated electron beams to interact with samples loaded on Formvar/carbon-coated mesh grids to obtain microscopy images [15]. TEM could be used to verify and characterize the attachment of a newly engineered phage to specific bacterial cells. However, the filamentous phage structure poses a challenge in interpreting these images, due to the difficulty in differentiating between bacterial structures, particularly pili, and filamentous phages, which both have long, filamentous morphologies with roughly similar dimensions (Fig. 1). Below, we describe a method to unambiguously identify the phages in a mixture of cells and phages using TEM after staining with uranyl acetate. The phages are differentiated from any bacterial cell structures through labeling with a primary antibody against the major coat protein (p. VIII) and a secondary antibody loaded with gold nanoparticles (Figs. 2 and 3).

2 Materials

If not otherwise specified, all reagents should be molecular biology grade. All solutions, media, and buffers should be prepared with ultrapure water, filtered through 0.22-μm pore size filters and equilibrated to room temperature prior to use.

1. Orbital shaker.
2. Benchtop centrifuge.
3. Spectrophotometer.
4. 50-mL volume Eppendorf tubes.
5. 1.5-mL volume Eppendorf tubes.
6. Reverse action tweezers.
7. Kimwipes.
8. Pipettes.
9. Pipette tips.
10. Petri dishes.
11. Formvar/Carbon 200 Mesh, Ni Grid (Electron Microscopy Sciences, SKU:FCF200-Ni-50).
12. TEM sample grid holder.
13. 1 × phosphate-buffered saline (PBS).
14. Washing solution: 1% w/v bovine serum albumin (BSA) in 1 × PBS buffer.
15. Blocking solution: 0.5% w/v gelatin solution in 1 × PBS buffer.
16. Primary antibody (mouse anti-M13 g8p IgG) working solution: 1:100 dilution from stock (Thermo Fisher Scientific, catalog number: MA1-06604) in 1 × PBS buffer.
17. Secondary antibody (donkey anti-mouse IgG preadsorbed to gold nanoparticles [6 nm dia.]) working solution: 1:20

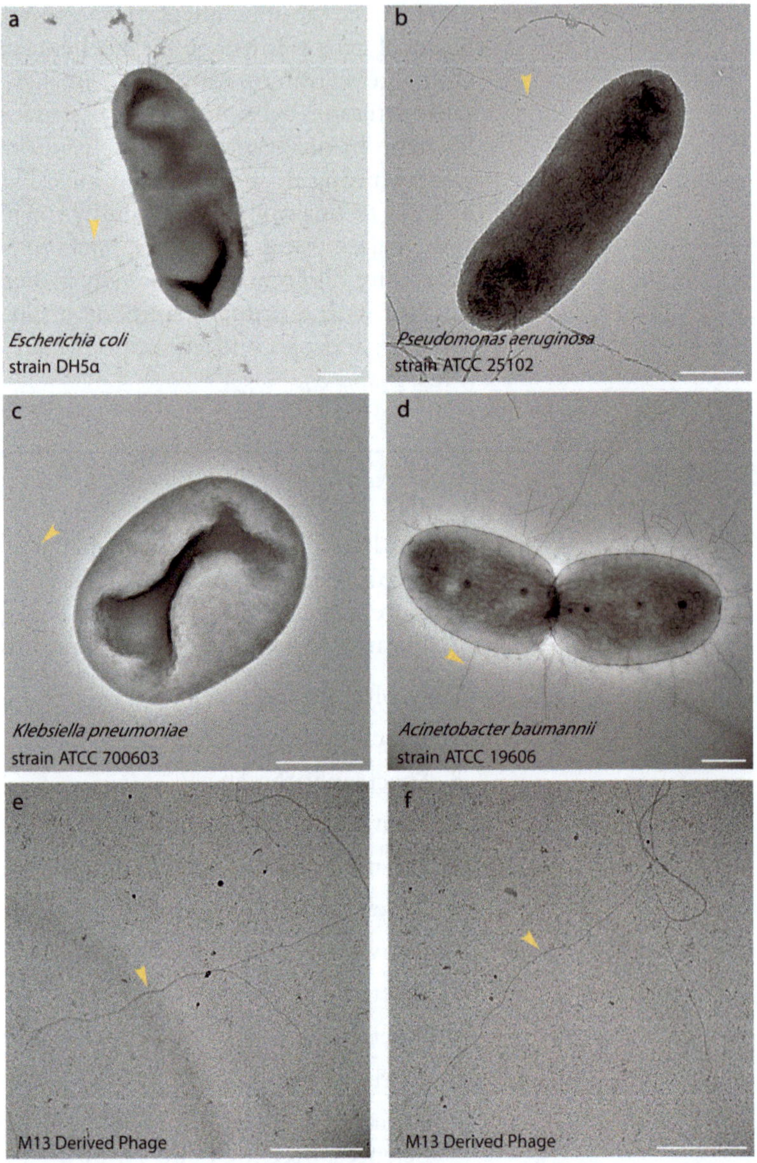

Fig. 1 TEM images of bacterial strains and M13-derived phages stained with 1% uranyl acetate. Pili can be observed in standard strains of (**a**) *Escherichia coli,* (**b**) *Pseudomonas aeruginosa,* (**c**) *Klebsiella pneumoniae,* and (**d**) *Acinetobacter baumannii.* Images of M13-derived phages (**e, f**) fixed on Formvar/carbon nickel grids demonstrate the morphological similarity between bacterial pili and M13-derived phages. All images have a scale bar of 500 nm. The yellow arrows indicate examples of pili on bacterial surfaces (**a–d**) or phages (**e–f**)

dilution from stock (Abcam, catalog number: ab105276) in 1 × PBS buffer.

18. 1% w/v uranyl acetate staining solution (*see* **Note 1**).
19. Purified chimeric M13 phage sample in × PBS buffer.

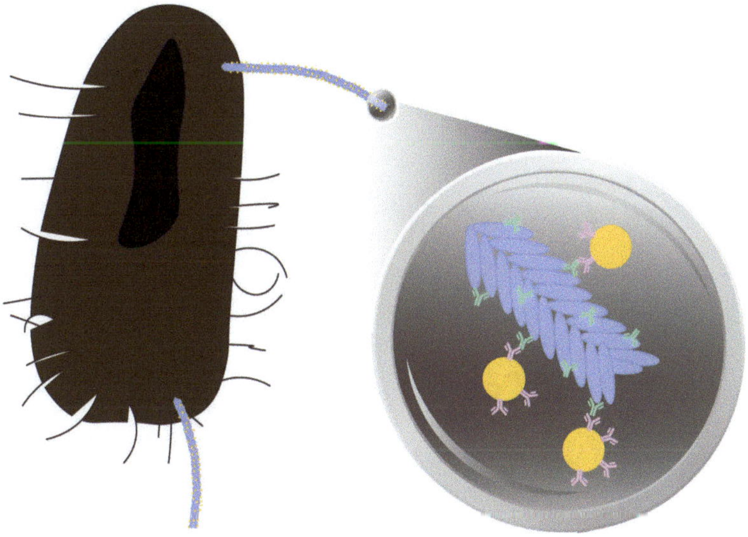

Fig. 2 Strategy for specific labelling of M13-derived phages with bacterial cells using antibody-bound gold nanoparticles. The phage coat is composed of repetitions of major capsid protein pVIII (purple) and is shown bound to a bacterial cell (dark oval) bearing pili (black curved lines). The phages are labeled with 6 nm gold nanoparticles through a 2-step procedure. The phage-cell sample is first incubated with mouse anti-pVIII primary antibodies (light green) and then incubated with 6 nm gold nanoparticles (yellow) preadsorbed with donkey anti-mouse secondary antibodies (pink)

20. Target bacterial cells on culture plates.
21. Luria–Bertani (LB) broth or other appropriate growth media.

3 Methods

All procedures should be done at room temperature unless specified otherwise. All bacteria should be spun down at 5000 rpm for 5 min using a benchtop centrifuge.

1. Grow an overnight culture from the appropriate bacterial species (*see* **Note 2**).
2. Subculture 100 μL of the overnight culture of bacterial cells into 5 mL fresh media and grow to log phase (*see* **Note 3**).
3. Spin down bacterial cells, remove supernatant of spent media, wash once with 5 mL of 1 × PBS buffer, and then resuspend bacterial cell pellet in 5 mL of 1 × PBS buffer (*see* **Note 4**).
4. Pipet 1 mL of cell suspension into a 1.5-mL Eppendorf tube, mix with appropriate amount of phage sample, and allow to bind for 30 min at room temperature (*see* **Note 5**).

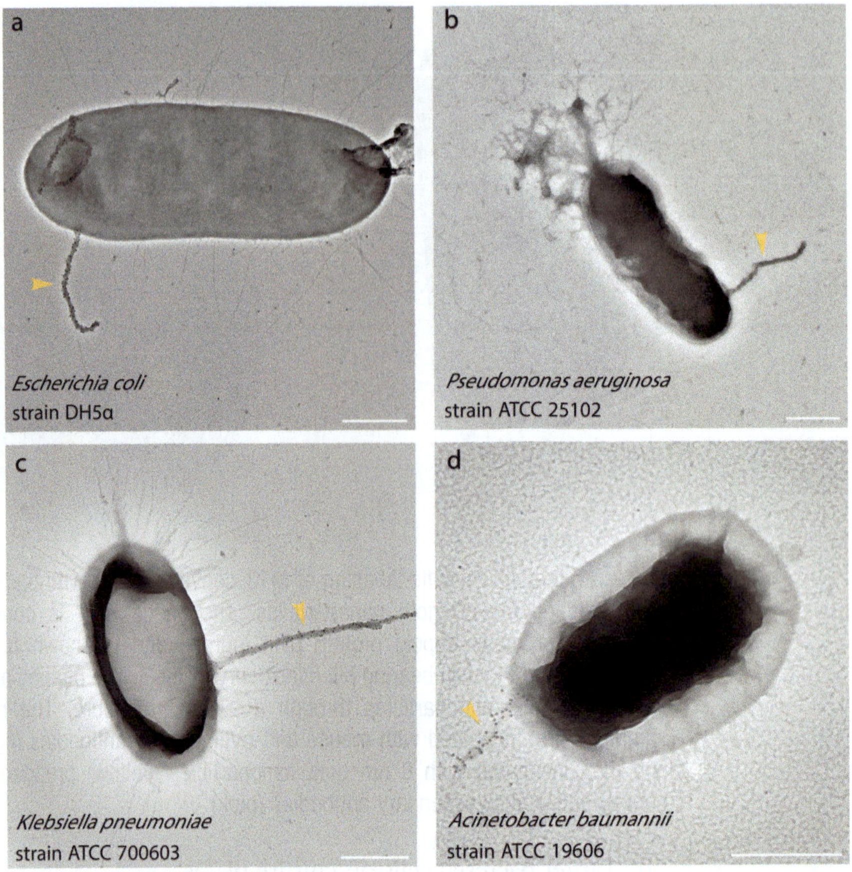

Fig. 3 TEM images of gold-labelled phages bound to cells stained by 1% uranyl acetate. All images show M13-derived phages expressing various receptor-binding proteins attached to the surface of bacterial cells. The cell–phage mixtures were fixed on formvar/carbon nickel grids. All images have a scale bar of 500 nm. The bacterial species are (**a**) *Escherichia coli,* (**b**) *Pseudomonas aeruginosa,* (**c**) *Klebsiella pneumoniae,* and (**d**) *Acinetobacter baumannii.* The yellow arrow in each image points to the phage labeled by the 2-step strategy illustrated in Fig. 2. Note that the width of the phage structure is increased compared to unlabeled phages, due to the bound antibodies and nanoparticles. Bacterial pili can be readily differentiated from the phages due to their lack of labeling and are prominently observed in (**a**) and (**c**)

5. Spin down bacteria (possibly with bound phage), remove supernatant containing unbound phage, and wash once with 1 mL of 1 × PBS buffer (*see* **Note 6**).
6. Resuspend cell pellet in 1 × PBS to obtain a final concentration with optical density at 600 nm (OD600) between 2 and 3.
7. Retrieve a TEM mesh grid with reverse action tweezers and position tweezers on the benchtop while holding, with the shiny side of the grid facing upward (*see* **Note 7**).
8. Load an 8 μL droplet of resuspended bacteria (with potentially bound phage) sample on the shiny side of the grid and incubate for 2 min.

9. Cut a piece of Parafilm with appropriate size and make 4 droplets of 500 μL washing solution on Parafilm for each sample. If not specified, all volumes of droplets for washing should be 500 μL in this protocol (*see* **Note 8**).
10. Using reverse action tweezers, move the sample grid to the surface of the first washing solution droplet with shiny side facing downward in contact with the washing solution. Wash sample by pipetting the droplet up and down for 50 times using a 200-μL pipette (*see* **Note 9**).
11. Move the sample grid to each of the other three washing solution droplets in order and repeat the washing step on each droplet.
12. In a Petri dish, make a droplet of 100 μL of blocking solution, transfer the sample grid to the surface of the droplet, and incubate for 1 h at room temperature (*see* **Note 10**).
13. Transfer the sample grid to a new droplet of washing solution and wash once.
14. In a Petri dish, make a droplet of 100 μL of primary antibody working solution, transfer the sample grid to the surface of the droplet with shiny side in contact with liquid surface, and incubate for 1 h at room temperature.
15. Repeat **steps 9–12** to wash and block.
16. In a Petri dish, make a droplet of 100 μL of secondary antibody (coated gold nanoparticles) working solution, transfer the sample grid to the surface of the droplet, and incubate for 1 h at room temperature (*see* **Note 10**).
17. Wash sample grid with ultrapure water droplets for five times.
18. Hold sample grid by its rim using reverse action tweezers with shiny side facing upward, add 8 μL of uranyl acetate staining solution to grid surface, and incubate for 1.5 min (*see* **Note 11**).
19. Dry sample by dabbing the grid gently on a Kimwipe tissue.
20. Store sample in a sample holder until ready to examine under transmission electron microscope (*see* **Note 12**).

4 Notes

1. Uranyl acetate is radioactive. Follow standard practice for radioactive material in the laboratory when handling. After resuspending powder in ultrapure water, fully dissolve the solid to make the 1% w/v solution, filter, and store solution at 4°C for short-term usage.

2. Sterility measures should be in place including using appropriate antibiotics to ensure quality of samples.
3. Assuming bacteria growing in log phase is ideal for phage attachment (e.g., for receptor expression), check culture optical density over time consistently to ensure bacteria growth is in log phase prior to proceeding to the next steps. This step may be modified if a different growth phase is preferred.
4. Washing of cells is performed by fully resuspending the cell pellet through careful pipetting up and down or vortexing at low speed. Then spin down again and remove supernatant. Be careful not to create too much shear stress during the wash step which may damage the cell surface.
5. Mix 10–100 times excess bacteriophages to bacteria (or other ratio as desired) to ensure sufficient binding.
6. Wash step done similarly as in **Note 4**. Avoid excessive stress to prevent cell damage.
7. Lay a piece of Kimwipe tissue on the benchtop to prevent the TEM mesh grid from dropping directly onto the benchtop. All handling of the grid with reverse action tweezers is done above Kimwipe tissues. In all steps involving transferring of the mesh grid, gently hold the rim of the grid without touching the inner mesh using reverse action tweezers, since the sharp tip of the tweezers may break the Formvar/carbon film. Avoid bending or other physical damage to the grid during transfers.
8. The size of Parafilm should be large enough to hold 4 of the 500 μL water droplets for each sample. In general, 4 grids of Parafilm in length should be enough for each sample. To make Parafilm stick to the surface of the benchtop without moving during the downstream wash steps, spray water onto the benchtop and push the water with one side of the Parafilm until it sticks to the benchtop due to water surface tension. Move the Parafilm around horizontally to ensure Parafilm is stuck to the surface uniformly, especially by 4 corners. To make the droplet, pipette 500 μL of the solution onto parafilm. Keep droplets well-separated to prevent them from combining together.
9. During the washing steps, it is important to make sure the grid floats on top of the liquid droplet without sinking to the bottom. If that happens, drying the sample with a piece of Kimwipe could potentially make the grid float again. Pipette up and down gently in the liquid droplet to avoid formation of air bubbles.
10. Keep the Petri dish covered with its lid during all incubation steps to reduce drying of the droplet. To prevent denaturing of the antibodies, only prepare primary antibody working

solution and secondary antibody (coated with gold nanoparticles) working solution prior to the step when they are required. It is not necessary to filter after diluting from stock solutions.

11. Do not stain samples with uranyl acetate solution longer than 4 min since this will overstain the cells.
12. The sample should be stable at room temperature for no less than 2 weeks.

Acknowledgements

We thank Dr. Chungyu Chang, University of California, Los Angeles, for her guidance on the development of this protocol. The authors acknowledge the use of instruments at the Electron Imaging Center for NanoMachines supported by NIH (1S10RR23057 to ZHZ) and CNSI at UCLA. Funding was provided by NIH (DP2GM123457 and R35GM148249) and the Camille Dreyfus Teacher-Scholar Program.

References

1. Ackermann HW (2007) 5500 phages examined in the electron microscope. Arch Virol 152(2):227–243
2. Chang C et al (2023) Engineered M13 phage as a novel therapeutic bionanomaterial for clinical applications: from tissue regeneration to cancer therapy. Mater Today Bio 20:100612
3. Ghosh D et al (2012) Refactored M13 bacteriophage as a platform for tumor cell imaging and drug delivery. ACS Synth Biol 1(12): 576–582
4. Hay ID, Lithgow T (2019) Filamentous phages: masters of a microbial sharing economy. EMBO Rep 20(6):e47427
5. Ledsgaard L et al (2018) Basics of antibody phage display technology. Toxins (Basel) 10(6):236
6. Conners R et al (2023) Cryo-electron microscopy of the f1 filamentous phage reveals insights into viral infection and assembly. Nat Commun 14(1):2724
7. Mulani MS et al (2019) Emerging strategies to combat ESKAPE pathogens in the era of antimicrobial resistance: a review. Front Microbiol 10:539
8. Klug DM et al (2021) There is no market for new antibiotics: this allows an open approach to research and development. Wellcome Open Res 6:146
9. Kim BO et al (2019) Phage-derived Antibacterials: harnessing the simplicity, plasticity, and diversity of phages. Viruses 11(3):268
10. Heilpern AJ, Waldor MK (2003) pIIICTX, a predicted CTXphi minor coat protein, can expand the host range of coliphage fd to include Vibrio cholerae. J Bacteriol 185(3): 1037–1044
11. Peng H, Chen IA (2019) Rapid colorimetric detection of bacterial species through the capture of gold nanoparticles by chimeric phages. ACS Nano 13(2):1244–1252
12. Peng H et al (2020) Chimeric phage nanoparticles for rapid characterization of bacterial pathogens: detection in complex biological samples and determination of antibiotic sensitivity. ACS Sens 5(5):1491–1499
13. Peng H et al (2020) Controlled phage therapy by photothermal ablation of specific bacterial species using gold nanorods targeted by chimeric phages. Proc Natl Acad Sci USA 117(4): 1951–1961
14. Peng H et al (2022) Treatment of wound infections in a mouse model using Zn(2+)-releasing phage bound to gold Nanorods. ACS Nano 16(3):4756–4774
15. Egerton RF (2005) Physical principles of electron microscopy: an introduction to TEM, SEM, and AEM. Springer, New York

Chapter 13

Purification of Single-Stranded RNA Bacteriophages and Host Receptors for Structural Determination Using Cryo-Electron Microscopy

Jirapat Thongchol and Junjie Zhang

Abstract

Single-stranded RNA bacteriophages (ssRNA phages) are small viruses with a compact genome (~3–4 kb) that infect gram-negative bacteria via retractile pili. These phages have been applied in various fields since their discovery approximately 60 years ago. To understand their biology, it is crucial to analyze the structure of mature virions. Cryo-electron microscopy (cryo-EM) has been employed to determine the structures of two ssRNA phages, MS2 and Qβ. This chapter presents a method for purifying these two phages and their receptor, the F-pilus, to allow examination using cryo-EM.

Key words ssRNA phages, MS2, Qβ, F-pilus, Cryo-EM

1 Introduction

In the 1960s, positive-sense, single-stranded RNA bacteriophages (ssRNA phages) were first discovered and isolated [1, 2]. These phages have a small genome of approximately 3–4 kb (Fig. 1), typically consisting of four genes, *mat*, *cp*, *rep*, and *lys*, which encode the maturation protein (Mat), the coat protein (Coat), the β-subunit of the replicase, and the single-gene lysis protein, respectively. The infection process of ssRNA phages begins with the attachment of Mat to its corresponding host-retractile pilus (Table 1). The mature virions of ssRNA phages typically contain 178 or 180 copies of Coat, a single copy of Mat, and a single strand of genomic RNA [3–6]. Since their discovery, ssRNA phages have been utilized in various applications, including RNA tracking [7, 8], delivery [9], protection [10], and peptide display [11].

Structural information on ssRNA phages has been reported for both culturable [2, 12–17] and unculturable ssRNA phages [18], in the form known as viruslike particles (VLPs). However, VLPs only provide the structure of the Coat and symmetry of these

Huan Peng et al. (eds.), *Phage Engineering and Analysis: Methods and Protocols*,
Methods in Molecular Biology, vol. 2793, https://doi.org/10.1007/978-1-0716-3798-2_13,

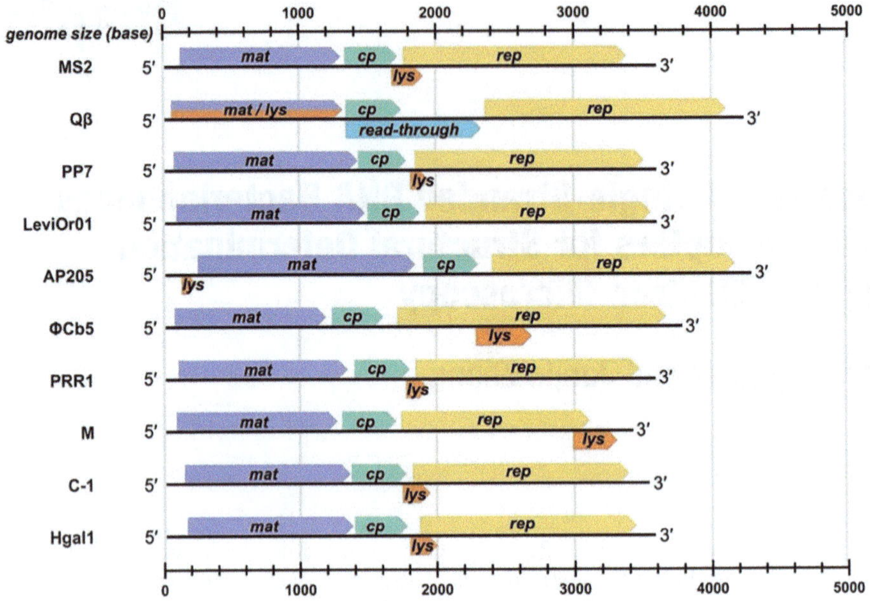

Fig. 1 The genome architecture of certain culturable ssRNA phages
The genome of ssRNA phages typically contain four open-reading frames. *mat* encodes the maturation protein. *cp* encodes the coat protein. *lys* encodes the single-gene lysis protein. *rep* encodes β-subunit of the replicase. *mat* and *lys* genes of Qβ are encoded from the same open-reading frame

Table 1
The host and receptor of certain culturable ssRNA phages

Phages	Hosts	Receptors
MS2	*E. coli*	F pili
Qβ	*E. coli*	F pili
PP7	*P. aeruginosa*	Type IV pili
LeviOr01	*P. aeruginosa*	Type IV pili
AP205	*Acinetobacter* spp.	Type IV pili
ϕCb5	*C. crescentus*	Type IV Tad pili
PRR1	*Pseudomonas, Salmonella, Vibrio, Escherichia*	IncP pili
M	*Escherichia, Salmonella, Klebsiella, Proteus, and Serratia*	IncM pili
C-1	*Escherichia, Salmonella, Proteus, and Serratia*	IncC pili
Hgal1	*Escherichia, Citrobacter, Klebsiella, Enterobacter*	IncH pili

particles, offering limited insights into the infection process or biological significance of ssRNA phages, as the visualization of their genome and Mat is not possible. Although the ssRNA phage genomic database has expanded significantly in recent times [19, 20], the structural information on mature virions is currently limited to Qβ and MS2, necessitating further exploration of other ssRNA phages.

To determine the structure of mature ssRNA phage virions, the most promising method is single-particle cryo-electron microscopy (cryo-EM) with asymmetric reconstruction [21]. This technique requires a relatively pure batch of phage particles. In this study, we present the procedure used for purifying ssRNA phages, specifically focusing on *E. coli* ssRNA phages MS2 and Qβ, which were utilized in cryo-EM studies [5, 22, 23]. Additionally, we provide the procedure for purifying the host receptor, the F-pilus, to investigate the structural complex formed between ssRNA phages and their receptors.

2 Materials

2.1 Culture Media

1. LB broth (BD Difco): 10 g tryptone, 5 g yeast extract, 10 g NaCl.
2. LB agar (1.5% agar): 10 g tryptone, 5 g yeast extract, 10 g NaCl, 15 g agar.
3. LB top agar (0.75% agar): 25 g LB broth, 7.5 g agar.
   - The formula per 1 L contains:
     - LB broth (BD Difco) 25 g
     - Agar (BD Bacto) 7.5 g

2.2 Bacterial Strains

1. *E. coli* (DH5α) harboring pMS2000 (*see* **Note 1**), Km^R (identifier: #JTB1H2).
2. ER2738, Tet^R (identifier: #JJZ0025).
3. *E. coli* (MC4100) harboring pOX38, Km^R (identifier: #JTB1D6).

2.3 Solution Buffers

1. MS2 buffer: 50 mM Tris–HCl pH 8, 150 mM NaCl.
2. MS2-1 M buffer: 50 mM Tris–HCl pH 8, 1 M NaCl.
3. Saline sodium citrate (SSC) buffer: 15 mM of sodium citrate pH 7.2, 150 mM of NaCl.
4. Buffer A: 50 mM Tris–HCl pH 8, 200 mM NaCl.
5. Buffer B: 50 mM Tris–Cl pH 8, 1 M NaCl.

2.4 Equipment and Reagents

2.4.1 ssRNA Phage MS2 Purification

1. Tetracycline hydrochloride, Tet.
2. Kanamycin sulfate, Km.
3. Chloramphenicol, Cm.
4. DNAse I.
5. RNAse A.

6. Tris(hydroxymethyl)aminomethane, Tris.
7. Sodium chloride, NaCl.
8. Ethylenedinitrilotetraacetic acid, EDTA.
9. Magnesium chloride, $MgCl_2$.
10. Ammonium sulfate (for ammonium sulfate precipitation).
11. Polyethylene glycol 8000, PEG8000 (for PEG sulfate precipitation).
12. Chloroform (for PEG sulfate precipitation).
13. Cesium chloride, CsCl.
14. 0.5–10 μL, 1–200 μL, and 100–1000 μL filtered tips (Corning).
15. Culture tube, 14 mL.
16. Conical tube, 50 mL.
17. Microcentrifuge tube, 1.5 mL.
18. Petri dish.
19. 0.22-μm syringe filter, PES membrane.
20. 0.45-μm filter units and bottle-top filters, PES membrane, sterile.
21. 20-kDa Slide-A-Lyzer Dialysis Cassettes, 30–70 mL.
22. 20-kDa Slide-A-Lyzer Dialysis Cassettes, 0.5–3 mL.
23. Quick-Seal round-top polypropylene tube (16 × 76 mm) (for CsCl isopycnic ultracentrifugation).
24. Open-top thinwall ultra-clear tube, 14 × 89 mm (for CsCl step-gradient ultracentrifugation).
25. Syringe, 5 mL.
26. Syringe, 50 mL.
27. Needle, 25G × 1½″.
28. Needle, 18G × 1″.
29. Pipetting needles with 90° blunt ends, metal hub, 18G × 4″.
30. Microwave.
31. 37 °C incubator.
32. 37 °C shaking incubator.
33. Spectrophotometer and cuvette.
34. Magnetic stir bar and stirrer.
35. Analytical balance (for CsCl ultracentrifugation).
36. Centrifuge with rotor A-4-62 rotor.
37. Centrifuge with rotor JLA-8.1000, JA-10, JA-17.

38. Ultracentrifuge with Ti70.1 rotor (for CsCl isopycnic ultracentrifugation) and SW41Ti (for CsCl step-gradient ultracentrifugation).
39. Polycarbonate centrifugal bottle with cap, 500 mL (for JA-10 rotor).
40. Polycarbonate centrifugal bottle with cap, 1 L (for JLA-8.1000 rotor).

2.4.2 F-Pilus Purification

1. Kanamycin sulfate, Km.
2. Polyethylene glycol 6000, PEG6000.
3. NaCl.
4. Tris(hydroxymethyl)aminomethane, Tris.
5. Sodium citrate, $C_6H_5O_7$. 3Na.
6. Culture tube, 14 mL.
7. Conical tube, 50 mL.
8. 0.5–10 μL, 1–200 μL, and 100–1,000 μL filtered tips (Corning).
9. Glass slide.
10. Needle, 23G × 1″.
11. Needle, 25G × 1½″.
12. Pipetting needles with 90° blunt ends, metal hub, 18G × 4″.
13. Syringe, 5 mL.
14. Syringe, 50 mL.
15. Petri dish.
16. Vortex mixer.
17. 10-kDa Slide-A-Lyzer Dialysis Cassettes, 3 mL.
18. Analytical balance (for CsCl ultracentrifugation).
19. Centrifuge with rotor JA-17.
20. Ultracentrifuge with rotor SW41Ti.

3 Methods

3.1 ssRNA Phage MS2 Purification

3.1.1 Making an ssRNA Phage Pickate

ssRNA phages are known to have high mutation rates and exist in quasi-species due to the error-prone nature of their genome replication [24, 25]. Therefore, it is advisable to isolate the phage from a single origin. To initiate the propagation and purification of ssRNA phage, we typically begin by selecting a single phage plaque. This involves choosing a plaque from the agar plate, similar to the process of selecting a single bacterial colony. The protocol described below focuses on the purification of MS2 but was also successfully used to purify Qβ [5, 22, 23, 26]. It is also highly recommended to use sterile filter pipette tips and autoclave, if possible, all the glass bottles, centrifuge bottles, flasks, and buffers to prevent contamination by other phages.

(Day 1, 5–10 min)

1. Inoculate *E. coli* (DH5α) harboring pMS2000 plasmid (JTB1H2) in 5 mL of LB broth media supplemented with 50 μg/mL kanamycin and 30 μg/mL chloramphenicol in 14-mL culture tube.
2. Incubate at 37 °C overnight (16 h).
3. Inoculate ER2738 in 5 mL LB broth media in 14-mL culture tube supplemented with 10 μg/mL tetracycline overnight. ER2738 is an F^+ cell, a host for MS2, and will be used the following day to propagate and determine phage titer.

(Day 2, 20–30 min)

4. Place the 1.5% LB agar plates supplemented with 50 μg/mL tetracycline in the 37 °C incubator for 10–20 min prior to usage to warm and dry the plates, especially if the plates have been stored at 4 °C for some time. Make sure to label each plate from 10^{-1} to 10^{-10}. These plates will be later used in **step 18**.
5. Melt the 0.75% sterile LB top agar with the microwave and let it cool down until temperature is below 42 °C. You can place the molten top agar in the 42 °C water bath with periodical swirling.
6. While waiting for plates and top agar in **steps 4** and **5**, centrifuge overnight JTB1H2 from day 1 at 4,000 rpm for 10 min, with A-4-62 rotor.
7. Filter the supernatant from **step 6** with 0.22-μm syringe filter with 5-mL sterile syringe. This supernatant should contain MS2 from an overnight expression (*see* **Note 2**).
8. Make serial dilution for lawn plates to make pickates (Fig. 2). Prepare ten 1.5-mL microcentrifuge tubes for ten serial dilutions.
9. Label each tube from 10^{-1} to 10^{-10} corresponding to each dilution.
10. Pipette 900 μL of MS2 buffer into each tube.
11. Carefully pipette 100 μL of filtered MS2 lysate from **step 7** into the first microcentrifuge tube labelled 10^{-1} without mixing it yet. Try not to submerge the entire pipette tip into the solution. Instead, allow the tip to touch only the surface of the solution and discard the tip.

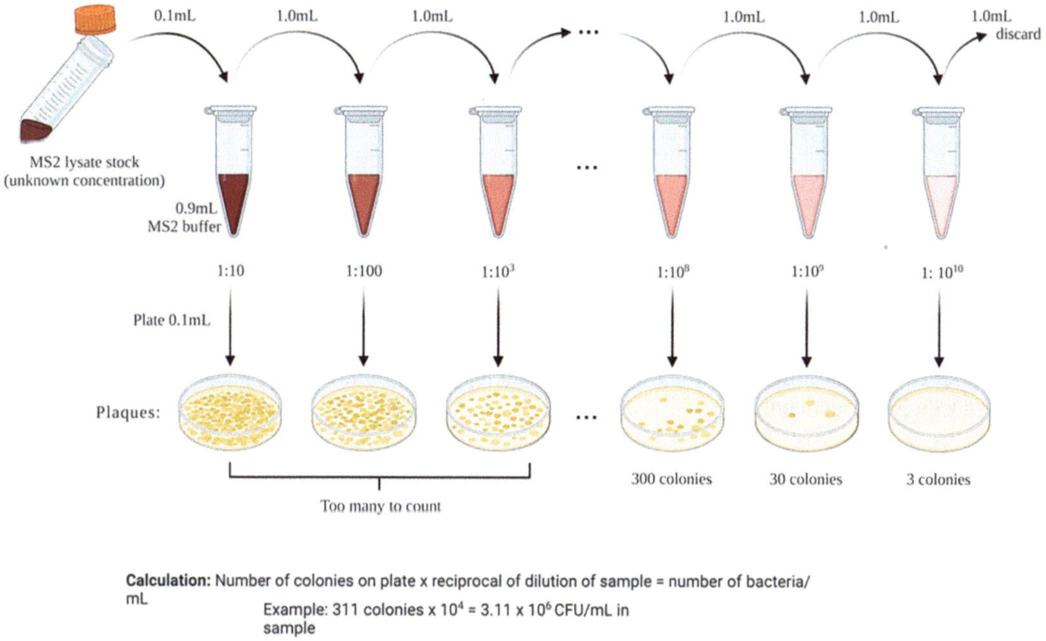

Fig. 2 The cartoon illustrates the procedure of making the lawn plaques described in Subheading 3.1.1, day 2, **step 5**. (Created with BioRender.com)

12. Mix the dilution by pipetting up–down slowly with a new pipette tip 15 times.
13. Using the same tip transfer 100 μL of well-mixed 10^{-1} dilution to the next tube, labelled 10^{-2}, by pipetting as described in **step 11**.
14. Continue repeating **steps 12–13** until you reach the 10^{-10} dilution.
15. When 10^{-10} dilution is reached, pipette 100 μL from the dilution and discard after mixing.
16. You should end up with a series of tenfold increment dilutions with each tube containing 900 μL of phage dilution.
17. In 14-mL culture tube, mix 100 μL of the 10^{-1} dilution and 100 μL of ER2738 inoculated from Subheading 3.1.1, day 1, **step 3**.
18. Pipette 5 mL of warm top agar into the culture tube from **step 17** and gently mix by swirling tube in circular motion (avoid creating bubbles) before pouring it onto the LB plate from **step 4**. Make sure to tilt the plate while pouring to ensure that the top agar covers the entire plate.
19. Let the plate stand for 10 s to allow the molten agar to solidify.
20. Repeat **steps 17–19** with all the dilutions. You should end up with 10 plates. Each plate corresponds to each dilution.

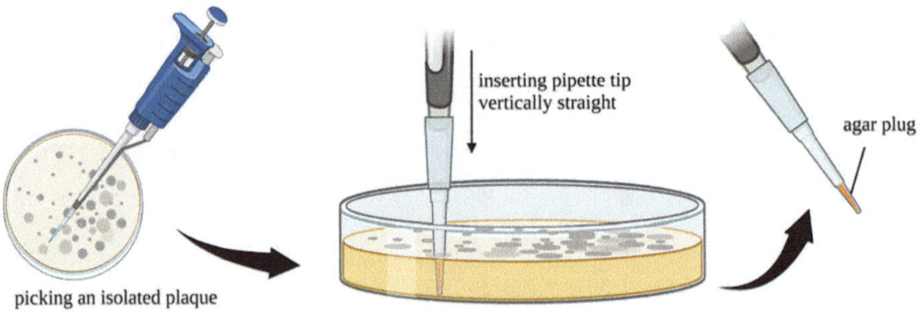

Fig. 3. The cartoon illustration showing how to pick an agar plug from a single plaque on the lawn plate. (Created with BioRender.com)

21. Incubate the plates in 37 °C incubator.
22. Before concluding the day, make sure to inoculate a fresh overnight culture of ER2738 from a single colony, as already described in Subheading 3.1.1, day 1, **step 3**, to be used later in day 3.

(Day 3, 5–10 min)

After day 2, you should have ten plates for each dilution, with the MS2 plaques decreasing in a tenfold increment from 10^{-1} to 10^{-10}. For example, if there are 300 plaques in the 10^{-8} dilution, you should expect approximately one-tenth of that number in the 10^{-9} dilution plate, resulting in around 30 plaques (Fig. 2). This serves as a good indicator of how accurately the serial dilutions were performed. Once we have these plates with lawn plaques, it is advisable to select the plate that exhibits well-isolated MS2 plaques, as this makes the picking process easier. Ideally, we will choose a plate that contains approximately 10–50 plaques.

23. In the 14-mL culture tube, add 5 mL LB supplement with 10 μg/mL tetracycline.
24. Use a P10 tip to pick the agar from an isolated single plaque of MS2 (Fig. 3).
25. Pipette up–down slowly in 5 mL LB from **step 23** to release the agar plug from **step 24** into the media.
26. Inoculate 50 μL of fresh overnight ER2738 host cell into the tube.
27. Incubate the tube for 5–6 h at 37 °C at 200 rpm.

28. Spin down the cell at 4,000 rpm for 10 min with an A-4-62 rotor.
29. Filter the supernatant through a 0.22-μm filter with a 5-mL sterile syringe.
30. Determine the phage titer using a spot titer assay. Routinely, the titer for 5 mL lysate is ~10^{10} PFU/mL.

3.1.2 Spot Titer Assay

Estimated hands-on time ~20–30 min.

1. Warm one 1.5% LB agar plate supplemented with 10 μg/mL tetracycline and melt the top agar, as described before in Subheading 3.1.1, day 1, **steps 1** and **2**.
2. Perform the tenfold serial dilutions as described in Subheading 3.1.1, day 2, **steps 8–16**.
3. In 14-mL culture tube, add 100 μL of fresh overnight ER2738 cell and 5 mL of warm top agar.
4. Gently swirl the tube and pour onto the warm plate from **step 1**.
5. Allow the plate to sit for 10 s to allow the top agar to solidify.
6. Spot 10 μL of the phage dilutions from **step 2** onto the plate. To increase accuracy, ensure that you change the tip for each dilution and maintain consistent mixing for each dilution. Using a larger volume for each spot will yield greater accuracy. However, it is important to ensure that the spots remain separate and do not merge with one another. The recommended volume for each spot is 10 μL. You can spot the dilutions as illustrated in Fig. 4.

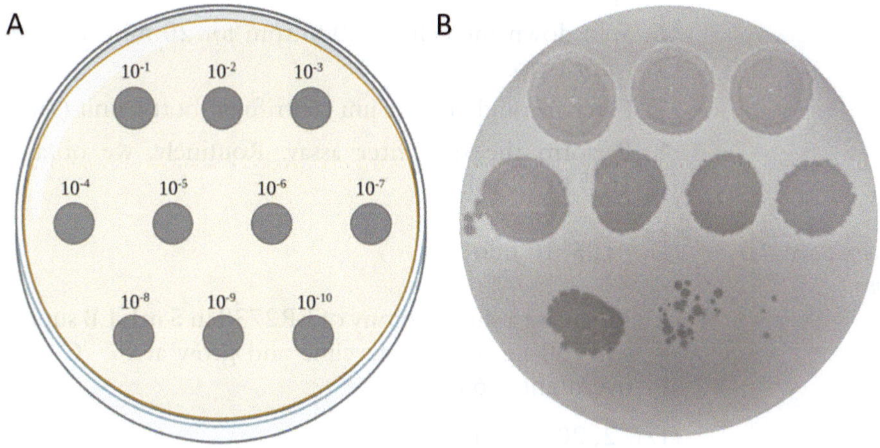

Fig. 4 Plate spot titer of bacteriophages
(**a**) A schematic representation of how to spot the phage dilution of the bacterial lawn for the spot titer test. (**b**) An image illustrating the result from the Qβ spot titer test. (Created with BioRender.com)

7. Allow the spots to air-dry inside the biosafety cabinet before incubation at 37 °C overnight (16–18 h).
8. Calculate the phage titer as follows:

$$\text{PFU/mL} = \frac{\text{number of plaques} \times \text{dilution factor}}{\text{volume of the spot}} \times \frac{1000\ \mu\text{L}}{1\ \text{mL}}$$

For example, Fig. 4b shows an example of the spot titer results of Qβ using 10 μL volume for each dilution. By following the calculation above, the titer of Qβ lysate is ~3×10^{12} PFU/mL (see calculation below).

$$\text{PFU/mL} = \frac{3 \times 10^{10}}{10\ \mu\text{L}} \times \frac{1000\ \mu\text{L}}{1\ \text{mL}} = 3 \times 10^{12}$$

3.1.3 Propagation of 200 mL Starting MS2 Lysate

(Day 1, 5–10 min)

1. Inoculate a single colony of ER2738 in 5 mL LB supplemented with 10 μg/mL tetracycline and grow at 37 °C at 200 rpm overnight (16 h).

(Day 2, 15–30 min)

2. Inoculate ER2738 bacterial host cell into 200 mL LB broth supplemented with 50 μg/mL tetracycline and incubate at 37 °C at 200 rpm until OD_{600} is around 0.5–0.6.
3. Add the entire 5 mL phage lysate from Subheading 3.1.1, day 3, into the culture.
4. Shake the culture at 37 °C, 200 rpm for 6–8 h.
5. Split the culture into 50-mL conical tubes.
6. Spin down the cell at 4,000 rpm for 20 min with an A-4-62 rotor twice.
7. Filter through a 0.45-μm centrifugal-bottle unit (*see* **Note 3**).
8. Perform the spot titer assay. Routinely, we obtain ~10^{10}–10^{11} PFU/mL.

3.1.4 Large-Scale (3 L) Propagation

(Day 1, 5–10 min)

1. Inoculate a single colony of ER2738 in 5 mL LB supplemented with 50 μg/mL tetracycline and grow at 37 °C at 200 rpm overnight (16 h).

(Day 2, 20–30 min)

2. Grow 200 mL starting culture of ER2738 by inoculating 4 mL of ER2738 overnight into 200 mL LB supplemented with 10

μg/mL tetracycline at 37 °C 200 rpm until OD_{600} reaches 0.5–0.6.

3. Subculture 30 mL of the starting ER3728 into 500 mL LB supplemented with 10 μg/mL tetracycline in a 2-L flask for total of 6 flasks (total volume 3 L). This helps ensure a good aeration of the culture during the growth. Grow at 37 °C at 200 rpm until OD_{600} is around 0.5–0.6.
4. Add 30 mL phage from 200 mL lysate propagated on Subheading 3.1.3 into each flask.
5. Shake at 37 °C at 200 rpm for 6–8 h.
6. Collect the supernatant by centrifugation at 4,000 rpm for 20 min with JLA-8.1 rotor twice.
7. Carefully collect and store the supernatant in a 4-L autoclaved beaker with a sterile magnetic stir bar.
8. Perform phage titer and store the lysate at 4 °C overnight before performing the next step. Routinely, the titer is ~10^{10}–10^{11} PFU/mL.

3.2 Concentrating the Phage Lysate

This section describes the common procedures used to concentrate a large volume of phage lysate for further purification processes. Generally, for MS2 and Qβ, the phage particles can be concentrated by ammonium sulfate precipitation and resuspended in a small volume of buffer (**method 1**). Alternatively, for some ssRNA phages, a high concentration of salt might affect the particle stability; therefore, PEG precipitation can also be used (**method 2**). We prefer to use method 1 rather than method 2 for MS2 and Qβ due to the simplicity and lower hands-on time requirement.

3.2.1 Method 1: Ammonium Sulfate Precipitation

Estimated hands-on time ~30 min.

Estimated wait time ~45 h.

1. To the 3 L lysate, add ammonium sulfate 280 g per liter of lysate by adding ~20 g at a time and dissolving with a magnetic stirrer before adding another portion (*see* **Note 4**).
2. Incubate at 4 °C overnight or at least 4 h.
3. Transfer the 500-mL polycarbonate centrifugal bottle.
4. Spin down the lysate at 7,000 rpm (~9,000× *g*) for 45 min with a Backman JA-10 rotor. Ensure to balance the bottle properly. The white pellet will be formed on the side and at the bottom of bottle.
5. Resuspend the pellet using ~10 mL MS2 buffer per 1 L lysate. This will bring the final volume to ~30–40 mL. The pellet might be packed. You can add the buffer and gently shake with the buffer at 4 °C for some time to loosen the pellet.

6. Transfer resuspended sample to 20-kDa Slide-A-Lyzer Dialysis Cassettes, 30–70-mL size capacity.
7. Dialyze against 4 L MS2 buffer overnight twice at 4 °C with vigorous stirring of the buffer with magnetic stirrer to increase the diffusion rate during the dialysis process.
8. Collect the dialyzed sample from the cassette and put into the 50-mL conical tube.
9. Add $MgCl_2$ to the final concentration of 10 mM (0.08 g per 40 mL sample) and dissolve completely by manual mixing.
10. Add DNase I and RNase A to a final concentration of 1 unit/mL and incubate for 1 h at 37 °C.
11. Clarify the lysate by centrifuging at 15,000× *g* (JA-17 rotor) for 30 min at 4 °C.
12. Collect the supernatant and titer the lysate. Routinely, the titer should be ~10^{12} PFU/mL.

3.2.2 Method 2: PEG Precipitation

Estimated hands-on time ~1 h.

Estimated wait time ~35 h.

1. To the 3 L lysate, add NaCl to a final concentration of 0.5 M and dissolve completely.
2. Add PEG8000 to a final concentration of 10%g/v (100 g per 1 L of lysate) and dissolve completely.
3. Incubate at 4 °C overnight or at least 4 h.
4. Spin down the lysate at 9,000 rpm (14,334× *g*) for 40 min with a Backman JA-10 rotor. Ensure to balance the bottles properly and the max volume is ~450 mL. The white pellet will be formed on the side and at the bottom of the bottle.
5. Resuspend the pellet using ~10 mL MS2 buffer per 1 L lysate. This will bring the final volume to ~30–40 mL.
6. Dialyze the lysate against 4 L MS2 buffer using 20-kDa Slide-A-Lyzer Dialysis Cassettes overnight.
7. Treat the dialyzed sample with 10 mM $MgCl_2$ (0.06 g per 30 mL sample), DNAse I and RNAse A for 1 h at room temperature.
8. Perform chloroform extraction by splitting ~30 mL lysate into two conical tubes with equal volumes of lysate (~15 mL) and add an equal volume of chloroform (*see* **Note 5**).
9. Rotate with rotator for 10 min until chloroform is mixed with lysate uniformly. The solution should become milky.
10. Centrifuge at 4,000 rpm for 10 min to promote phase separation (Fig. 5).
11. Carefully aspirate the aqueous phase into new conical tubes.

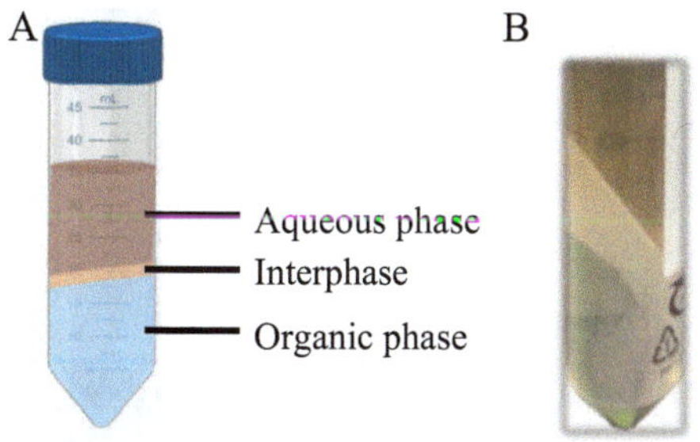

Fig. 5 The result from the centrifugation of chloroform-mixed lysate
(**a**) A cartoon representation of the phase separation resulting from centrifugation during chloroform extraction. (**b**) An image illustrating the actual phase separation resulting from the early step of chloroform extraction of ssRNA phage after centrifugation. (Created with BioRender.com)

12. Repeat **step 8–11** until the white interphase is very faint or none. Typically, this will take at least ~4–5 rounds.
13. Perform the spot titer test and store samples at 4 °C.

3.3 Purification of ssRNA Via CsCl Centrifugation

Usually after ammonium sulfate precipitation, the titer of MS2 is routinely ~10^{12} PFU/mL. Further purify the phage from possible contaminations such as host ribosomes and membrane liposomes. The CsCl ultracentrifugation is performed. We determined the CsCl ultracentrifugation method based on the titer of the phage we obtained. If the titer is higher than 10^{11} PFU/mL, the CsCl isopycnic ultracentrifugation is performed. If the titer is higher than 10^{7} PFU/mL, but lower than 10^{11} PFU/mL, the CsCl step gradient ultracentrifugation is performed. However, if the titer of the phage is lower than 10^{7} PFU/mL, we do not recommend pursuing CsCl ultracentrifugation since the phage bandage will be very faint or not formed at all. You should restart 3 L propagation to obtain more phage particles.

3.3.1 CsCl Isopycnic Centrifugation

Estimated hands-on time ~1–2 h.

Estimated wait time ~35 h.

1. Add CsCl crystal into the phage solution until the density is reaches 1.4 g/mL (19 g into 30 mL). Add a small portion of CsCl gradually, ensuring it dissolves completely before adding more.
2. Transfer the sample into 13.5-mL Quick-Seal round-top polypropylene tubes (16 × 76 mm) using an 18G × 1″ needle/syringe. With 30 mL, this will be enough to transfer into two Quick-Seal tubes. Make sure to balance the tubes properly and seal the tube.
3. Spin at 45,000 rpm for 24 h at 4 °C with a fixed-angle Ti70.1 rotor (*see* **Note 6**).

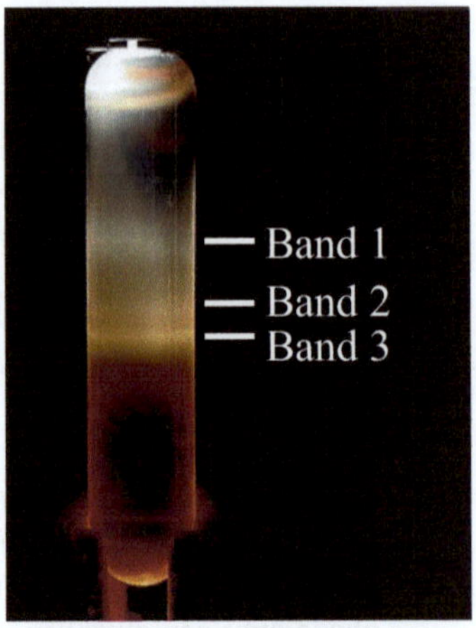

Fig. 6 The example result from isopycnic CsCl ultracentrifugation shows bands formed
The bandages formed from isopycnic CsCl ultracentrifugation. Band 3 corresponds to the phage band that produced the highest titer from the spot titer test. To collect all the bands, you should start by collecting band 1, band 2, and band 3 in order

4. To collect the sample from the Quick-Seal tube, puncture the tube with the 18G × 1″ needle to release the vacuum inside the Quick-Seal tube.
5. Collect the band with a 25G × 1½″ needle/syringe by puncturing through the side of the tube slightly below (1–2 mm) the band you want to collect. Collect all the bands starting from the topmost band to the bottom (*see* **Note** 7). The phage band should appear around the middle of the tube (Fig. 6).
6. Typically, the volume for each band should not exceed 3 mL. Transfer each band into separate 20-kDa Slide-A-Lyzer Dialysis Cassettes, 0.5–3 mL.
7. Dialyze against 2 L of MS2-1M buffer for at least 4 h.
8. Change buffer to 4 L MS2 buffer and dialyzed overnight. Repeat this step twice.
9. Check titer and analyze by SDS-PAGE for all the bands.
10. Routinely, titer should be ~10^{12}–10^{13} PFU/mL with the volume ~3–5 mL.

3.3.2 CsCl Step-Gradient Centrifugation

Estimated hands-on time ~1–2 h.

Estimated wait time ~35 h.

It is recommended to perform the CsCl step gradient when the phage titer is higher than 10^7 PFU/mL but lower than 10^{10} PFU/mL. The volume of the lysate from **step 3.1.4** should be around 30–40 mL. You need to concentrate the lysate down to around 2–4 mL, so that the phage bandage can be easily observed after

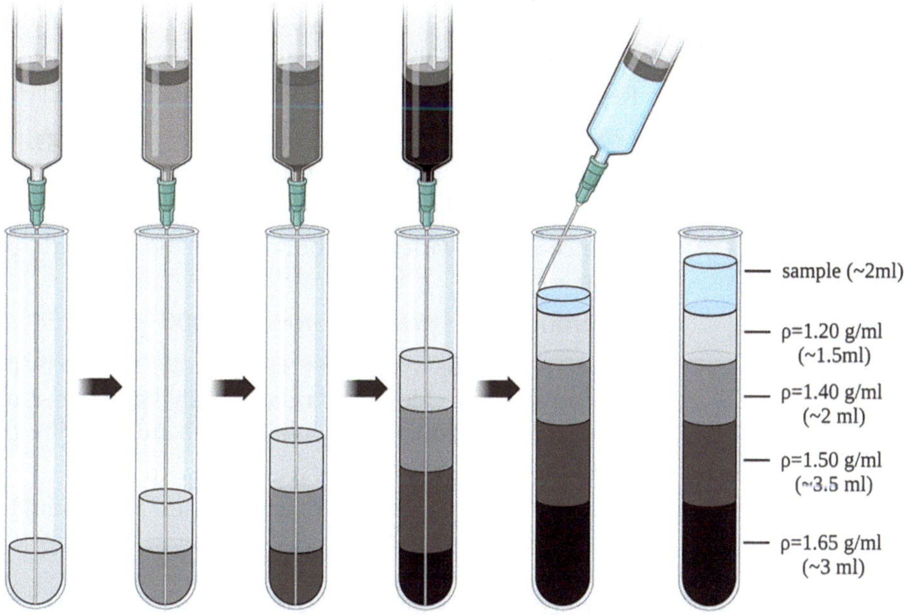

Fig. 7 The cartoon illustrating the process of layering the CsCl density steps for CsCl step-gradient ultracentrifugation. (Created with BioRender.com)

centrifugation and that all the phage particles can be collected. The concentration can be done using 30- or 100-kDa MWCO Amicon™ Centrifugal Filter Units.

1. Start by making four different CsCl solutions at different densities (ρ):

   ρ = 1.2 g/mL in MS2 buffer

   ρ = 1.4 g/mL in MS2 buffer

   ρ = 1.5 g/mL in MS2 buffer

   ρ = 1.65 g/mL in MS2 buffer

   It is a good idea to check again if the density is right by simply weighing the solution with scale since the density of the MS2 buffer is roughly 1 g/mL. After adding CsCl, the weight of the solution should increase, e.g., 1 mL of ρ = 1.2 g/mL CsCl solution should weigh 1.2 g.

2. Set up the CsCl density steps by (*see* Fig. 7) preparing the 13.2-mL open-top thin wall ultra-clear tube, 14 × 89 mm.
3. Layer the CsCl steps starting from the lowest density (ρ = 1.2 g/mL) using 90° blunt-end needles. Try to avoid bubbles by pushing the tip of the needle against the bottom of the tube and injecting the solution slowly.
4. Set up the next density (ρ = 1.4 g/mL) by using the blunt-point needle and slowly submerging the needle until it reaches

the bottom of the tube. Inject the solution slowly against the bottom of the tube. The higher-density CsCl solution pushes the lower density up.

5. Repeat **steps 3–4** until you have four layers of CsCl densities.
6. Load 1.5–2 mL of sample on top by slowly injecting the sample against the wall of the tube. You can use a 25G × 1½″ needle or pipette with 1000-μL filter tips if you prefer.
7. Balance it well with rotor bucket and the cap with analytical balance.
8. Centrifuge using SW41Ti rotor at 40,000 rpm for 24 h at 4 °C.
9. Collect the bands with 25G × 1½″ needle/syringe by puncturing through the side of the tube slightly below (1–2 mm) the band you want to collect. Collect the bands starting from the topmost to the bottom.
10. Perform procedures as described in Subheading 3.3.1, **steps 6–9**.

3.4 F-Pilus Purification

We described below a method previously used to purify F-pilus for cryo-EM study [23] which was previously described elsewhere [27–29]. This method has proved to be useful to us for the purification of other types of pili by changing the buffer.

(Day 1, 5–10 min)

1. Inoculate a single colony of JTB1D6 (*see* **Note 8**) from a freshly streaked plate into 5 mL LB supplemented with 50 ug/mL kanamycin and incubate overnight (16 h) at 37 °C with agitation at 200 rpm.

(Day 2, 1 h)

2. Inoculate 500 μL of overnight JTB1D6 culture into 50 mL LB supplemented with 50 ug/mL kanamycin.
3. Incubate at 37 °C, 200 rpm, until OD_{600} is ~0.6.
4. Spread 150 μL onto 100 plates of 1.5% LB agar supplemented with 50 ug/mL kanamycin.
5. Incubate at 37 °C overnight.

(Day 3, 2–3 h)

6. After overnight incubation, scrape the cells from the surface of agar plates using a glass slide. It is recommended to perform this step on ice or in the cold room, if possible, to slow down the retraction of the F-pili. To help with the scraping, you may add 1 mL of SSC buffer onto the plate beforehand (*see* **Note 9**).

7. For every 30 plates (*see* **Note 10**), the SSC buffers should not be exceeded 30 mL.
8. Vortex vigorously for 1 min followed by incubation on ice for another 1 min.
9. Assemble a 23G × 1″ needle on a 50-mL syringe.
10. Add the cell to the syringe and pass the cell suspension through the needle for three times. This will help ensure the additional shearing of the F-pili from the cell surface.
11. Combine all the pass-through cells into sterile flask with sterile stir bar and add SSC buffer to adjust volume to 1 L.
12. Stir the pass-through at 4 °C for 2–4 h.
13. Centrifuge twice at 10,800 × *g* for 20 min and collect the supernatant.
14. Add NaCl to a final concentration of 500 mM and stir to dissolve.
15. Add PEG6000 to a final concentration of 5%g/v and stir to dissolve.
16. Incubate at 4 °C for 4 h.
17. Collect the pellet by centrifugation at 15,000 *g* and 4 °C for 40 min twice and then resuspend the pellet in small volume (1–2 mL) of buffer A (*see* **Notes 11** and **12**).

(Day 4, 1–2 h)

18. Make three different CsCl solutions at different densities (ρ):
    (i) ρ = 1.1 g/mL in buffer A
    (ii) ρ = 1.2 g/mL in buffer A
    (iii) ρ = 1.3 g/mL in buffer A
19. Set up the CsCl steps and add the sample (*see* **step 3.3.2**, Fig. 7). Layer 1.1, 1.2, and 1.3 g/mL CsCl solution for 4, 3, and 3 mL volume, respectively.
20. Balance the tubes properly with analytical balance.
21. Centrifuge at 192,000× *g* for 19 h at 4 °C with the swing-bucket SW40.1Ti rotor.

(Day 5, 1–2 h)

22. Collect the bands with a 25G × 1½″ needle/syringe by puncturing through the side of the tube slightly below (1–2 mm) the band you want to collect. Collect all the bands from the topmost band to the bottom.
23. Dialyze with 10K dialyzer cassette against 1 L of buffer B at 4 °C with stirring for 5 h.

24. Change the buffer to 4 L buffer A and dialyzed further overnight twice.
25. Store the sample at 4 °C.
26. Analyze by SDS-PAGE or negative-stain TEM.

4 Notes

1. JTB1H2 is an *E. coli* strain that also harbors an additional arabinose-inducible plasmid (pBAD33) containing a bacteriophage lambda (λ) CI repressor and open-reading frame expressing a single-chain MS2 coat dimer. λ CI repressor will bind to the pL promotor on pMS2000 suppressing the transcription of MS2 cDNA at 30 °C. The expression of λ CI is induced by 0.2% arabinose. The single-chain MS2 coat dimer acts as a transcriptional repressor.
2. Alternatively, you could pause at this step and store the filtered lysate at 4 °C and continue with the following steps later.
3. The filtration in this step can be skipped, but it is recommended to do so since the culture might have some resistant cells emerging and might interfere/outgrow during 3 L expansion. If you choose to skip the filtration, we recommend going to **step 3.1.5** directly after the collection of supernatant.
4. This helps to ensure that the osmotic pressure increases gradually.
5. Chloroform is toxic if swallowed or inhaled. Short-term exposure to high levels of chloroform can affect the central nervous system. When working with chloroform, use appropriate personal protective equipment and perform the operation in the fume hood. Please refer to safety data sheet and laboratory procedures.
6. Spinning for 24 h might not bring the CsCl gradient to the equilibrium yet. Alternatively, you can spin for 48–72 h to ensure it reaches equilibrium. However, this will take longer time and we find out that 24-h spinning is sufficient for cryo-EM data collection without major contaminations of host proteins. When you purify other ssRNA phages and do not see bands forming after 24 h, you can allow the centrifugation to go further for another 24–48 h.
7. This will ensure that the bands below will not be disturbed before the collection.
8. This is *E. coli* (MC4100) strain harboring pOX38. The pOX38 is the conjugal F-derived plasmid allowing cells to express the F-pilus.

9. It is recommended to make sure not to add too much of SSC buffer. We do not want to dilute the F-pilus concentration. It will also make the next step easier if you have a small volume. For every 30 plates, the SSC buffers should not be exceeded ~30 mL.
10. To make **steps 7–10** easier, it is recommended to do ~30 plates at a time.
11. You would want to use a small volume of buffer to ensure that we can load all the sample into one tube. This ensures that the sample gets concentrated.
12. You can directly proceed with the procedures in day 4 for CsCl ultracentrifugation if you prefer or conclude the day at this point and store the sample at 4 °C.

Acknowledgements

We would like to express our gratitude to Prof. Ry Young at the Texas A&M University, TX, for graciously granting us the pMS2000. We extend our sincere thanks to Prof. Peter J. Christie at McGovern Medical School, Houston, TX, for generously providing us with pOX38 as a kind gift. Additionally, we would like to acknowledge Zachary Lill, a lab member, for his invaluable assistance in proofreading. We are grateful for the support received from TAMU T3 and X-grants, the Welch Foundation grant A-1863, NSF grant MCB-1902392, and NIH grants R21AI156846 and R01GM141659, as well as funding from the Center for Phage Technology, jointly sponsored by Texas AgriLife and Texas A&M University.

References

1. Loeb T, Zinder ND (1961) A bacteriophage containing RNA. Proc Natl Acad Sci USA 47(3):282–289. https://doi.org/10.1073/pnas.47.3.282
2. Shishovs M et al (2016) Structure of AP205 coat protein reveals circular permutation in ssRNA bacteriophages. J Mol Biol 428(21): 4267–4279. https://doi.org/10.1016/j.jmb.2016.08.025
3. Dai X et al (2017) In situ structures of the genome and genome-delivery apparatus in a single-stranded RNA virus. Nature 541(7635):112–116. https://doi.org/10.1038/nature20589
4. Cui Z, Gorzelnik KV, Chang J, Langlais C, Jakana J, Young R, Zhang J (2017) Structures of Qβ virions, virus-like particles, and the Q-β-MurA complex reveal internal coat proteins and the mechanism of host lysis. Proc Natl Acad Sci USA 114:11697
5. Gorzelnik KV, Cui Z, Reed CA, Jakana J, Young R, Zhang J (2016) Asymmetric cryo-EM structure of the canonical Allolevivirus Qβ reveals a single maturation protein and the genomic ssRNA in situ. Proc Natl Acad Sci USA 113(41):11519–11524. https://doi.org/10.1073/pnas.1609482113
6. Koning RI et al (2016) Asymmetric cryo-EM reconstruction of phage MS2 reveals genome structure in situ. Nat Commun 7:12524. https://doi.org/10.1038/ncomms12524
7. Pichon X, Robert MC, Bertrand E, Singer RH, Tutucci E (2020) New generations of MS2 variants and MCP fusions to detect single mRNAs in living eukaryotic cells. Methods

Mol Biol 2166:121–144. https://doi.org/10.1007/978-1-0716-0712-1_7

8. Tutucci E, Vera M, Biswas J, Garcia J, Parker R, Singer RH (2018) An improved MS2 system for accurate reporting of the mRNA life cycle. Nat Methods 15(1):81–89. https://doi.org/10.1038/nmeth.4502
9. Fu Y, Li J (2016) A novel delivery platform based on Bacteriophage MS2 virus-like particles. Virus Res 211:9–16. https://doi.org/10.1016/j.virusres.2015.08.022
10. Li J, Sun Y, Jia T, Zhang R, Zhang K, Wang L (2014) Messenger RNA vaccine based on recombinant MS2 virus-like particles against prostate cancer. Int J Cancer 134(7):1683–1694. https://doi.org/10.1002/ijc.28482
11. Zhao L, Kopylov M, Potter CS, Carragher B, Finn MG (2019) Engineering the PP7 virus capsid as a peptide display platform. ACS Nano 13(4):4443–4454. https://doi.org/10.1021/acsnano.8b09683
12. Valegård K, Liljas L, Fridborg K, Unge T (1990) The three-dimensional structure of the bacterial virus MS2. Nature 345(6270):36–41. https://doi.org/10.1038/345036a0
13. Golmohammadi R, Fridborg K, Bundule M, Valegård K, Liljas L (1996) The crystal structure of bacteriophage Q beta at 3.5 A resolution. *Structure* 4(5):543–554. https://doi.org/10.1016/s0969-2126(96)00060-3
14. Tars K, Bundule M, Fridborg K, Liljas L (1997) The crystal structure of bacteriophage GA and a comparison of bacteriophages belonging to the major groups of Escherichia coli leviviruses. J Mol Biol 271(5):759–773. https://doi.org/10.1006/jmbi.1997.1214
15. Tars K, Fridborg K, Bundule M, Liljas L (2000) The three-dimensional structure of bacteriophage PP7 from Pseudomonas aeruginosa at 3.7-A resolution. Virology 272(2):331–337. https://doi.org/10.1006/viro.2000.0373
16. Persson M, Tars K, Liljas L (2008) The capsid of the small RNA phage PRR1 is stabilized by metal ions. J Mol Biol 383(4):914–922. https://doi.org/10.1016/j.jmb.2008.08.060
17. Plevka P et al (2009) The structure of bacteriophage phiCb5 reveals a role of the RNA genome and metal ions in particle stability and assembly. J Mol Biol 391(3):635–647. https://doi.org/10.1016/j.jmb.2009.06.047
18. Rūmnieks J et al (2020) Three-dimensional structure of 22 uncultured ssRNA bacteriophages: flexibility of the coat protein fold and variations in particle shapes. Sci Adv 6(36). https://doi.org/10.1126/sciadv.abc0023
19. Neri U et al (2022) Expansion of the global RNA virome reveals diverse clades of bacteriophages. Cell 185(21):4023–4037.e18. https://doi.org/10.1016/j.cell.2022.08.023
20. Callanan J, Stockdale SR, Shkoporov A, Draper LA, Ross RP, Hill C (2020) Expansion of known ssRNA phage genomes: from tens to over a thousand. Sci Adv 6(6):eaay5981. https://doi.org/10.1126/sciadv.aay5981
21. Goetschius DJ, Lee H, Hafenstein S (2019) CryoEM reconstruction approaches to resolve asymmetric features. Adv Virus Res 105:73–91. https://doi.org/10.1016/bs.aivir.2019.07.007
22. Cui Z et al (2017) Structures of Qβ virions, virus-like particles, and the Qβ-MurA complex reveal internal coat proteins and the mechanism of host lysis. Proc Natl Acad Sci USA 114(44):11697–11702. https://doi.org/10.1073/pnas.1707102114
23. Meng R et al (2019) Structural basis for the adsorption of a single-stranded RNA bacteriophage. Nat Commun 10(1):3130. https://doi.org/10.1038/s41467-019-11126-8
24. Kopsidas G et al (2007) RNA mutagenesis yields highly diverse mRNA libraries for in vitro protein evolution. BMC Biotechnol 7:18. https://doi.org/10.1186/1472-6750-7-18
25. Novella IS, Presloid JB, Taylor RT (2014) RNA replication errors and the evolution of virus pathogenicity and virulence. Curr Opin Virol 9:143–147. https://doi.org/10.1016/j.coviro.2014.09.017
26. Strauss Jr JH, Sinsheimer RL (1963) Purification and properties of bacteriophage MS2 and of its ribonucleic acid. J Mol Biol 7:43–54 [Online]. Available: https://www.ncbi.nlm.nih.gov/pubmed/13978804
27. Costa TRD et al (2016) Structure of the bacterial sex F pilus reveals an assembly of a stoichiometric protein-phospholipid complex. Cell 166(6):1436–1444.e10. https://doi.org/10.1016/j.cell.2016.08.025
28. Gonzalez Rivera AK, Forest KT (2017) Shearing and enrichment of extracellular type IV pili. Methods Mol Biol 1615:311–320. https://doi.org/10.1007/978-1-4939-7033-9_25
29. Helmuth R, Achtman M (1978) Cell-cell interactions in conjugating Escherichia coli: purification of F pili with biological activity. Proc Natl Acad Sci USA 75(3):1237–1241. https://doi.org/10.1073/pnas.75.3.1237

Part V

Probing Phage Biology

Chapter 14

Obtaining Detailed Phage Transcriptomes Using ONT-Cappable-Seq

Leena Putzeys, Danish Intizar, Rob Lavigne, and Maarten Boon

Abstract

Detailed transcription maps of bacteriophages are not usually explored, limiting our understanding of molecular phage biology and restricting their exploitation and engineering. The ONT-cappable-seq method described here brings phage transcriptomics to the accessible nanopore sequencing platform and provides an affordable and more detailed overview of transcriptional features compared to traditional RNA-seq experiments. With ONT-cappable-seq, primary transcripts are specifically capped, enriched, and prepared for long-read sequencing on the nanopore sequencing platform. This enables end-to-end sequencing of unprocessed transcripts covering both phage and host genome, thus providing insight on their operons. The subsequent analysis pipeline makes it possible to rapidly identify the most important transcriptional features such as transcription start and stop sites. The obtained data can thus provide a comprehensive overview of the transcription by your phage of interest.

Key words Bacteriophage, Transcriptomics, Nanopore sequencing, Promoter identification, Terminators, Operons

1 Introduction

Comprehensive insights in phage transcription are imperative to understanding their core biology and can enable us to extract useful tools and parts for synthetic biology, as well as provide major benefits in their successful engineering as this is done with little insight on how changes could impact transcription of other genes (e.g., by disruption of promoters, operons, or other transcriptional features) [1, 2]. The limited number of existing transcriptional studies generally makes use of short-read RNA sequencing [3–6]. This generates information on transcript presence and abundance, but misses out on key transcriptional features such as where transcripts initiate and terminate. Some short-read-based methods partially alleviate this by specialized library prep methods to look at one specific feature such as transcription start sites (TSS) [7]. By contrast, long-read RNA sequencing approaches have the

Huan Peng et al. (eds.), *Phage Engineering and Analysis: Methods and Protocols*,
Methods in Molecular Biology, vol. 2793, https://doi.org/10.1007/978-1-0716-3798-2_14,

advantage that they can capture the entire RNA transcript in a single read, thus providing significantly more information on transcriptional events. One major hurdle is the extensive RNA processing that occurs in the bacterial cell, which makes it difficult to distinguish true TSS from processing sites. In SMRT-cappable-seq this hurdle is overcome by combining an enrichment of primary, non-processed transcripts with PacBio long-read sequencing [8]. Reads are enriched by enzymatic capping of the triphosphate group typically present at the 5′ end of unprocessed transcripts. The combination with a long-read approach thus results in a complete end-to-end sequencing of the transcriptome, enabling identification of TSS, transcription termination sites (TTS), operon structures, and other intricate details of transcription in one go. However, due to the size selection step that is inherent to the PacBio library prep, it is less suited for phage transcription, which is notorious for its high density in small ORFs [9].

The ONT-cappable-seq method described here is an adaptation of SMRT-cappable-seq, to fit the more accessible long-read sequencing platform of Oxford Nanopore Technologies (ONT). For this sequencing platform, no stringent size selection is used in the library prep, preventing that shorter phage reads would be filtered out. This makes it ideal for obtaining a comprehensive overview on the transcriptional landscape of phages and their host, though it does not provide the quantitative comparative gene expression profiles of RNAseq [10]. In ONT-cappable-seq, primary transcripts are enriched by capping their 5′-triphosphates and selecting for capped RNA (Fig. 1). The enriched transcripts are reverse transcribed, amplified, and prepared for loading on a nanopore flow cell. This chapter also offers an automated analysis pipeline to pinpoint the main transcription hallmarks such as promoters and terminators from the sequencing data.

2 Materials

2.1 RNA Preparation

1. HiScribe T7 High Yield RNA synthesis kit.
2. NanoDrop 1000 or similar spectrophotometer.
3. Heat block or thermomixer.
4. 1 U/μL DNase I, RNase-free.
5. 10× reaction buffer with $MgCl_2$ (supplied with DNase I enzyme).
6. 40 U/μL RNaseOUT.
7. PCI mix: Mix phenol, chloroform, and isopropanol, at a ratio of 25:24:1 (e.g., 25 mL phenol, 24 mL chloroform, and 1 mL of Isopropanol). Mix by shaking and let it settle (*see* **Note 1**).

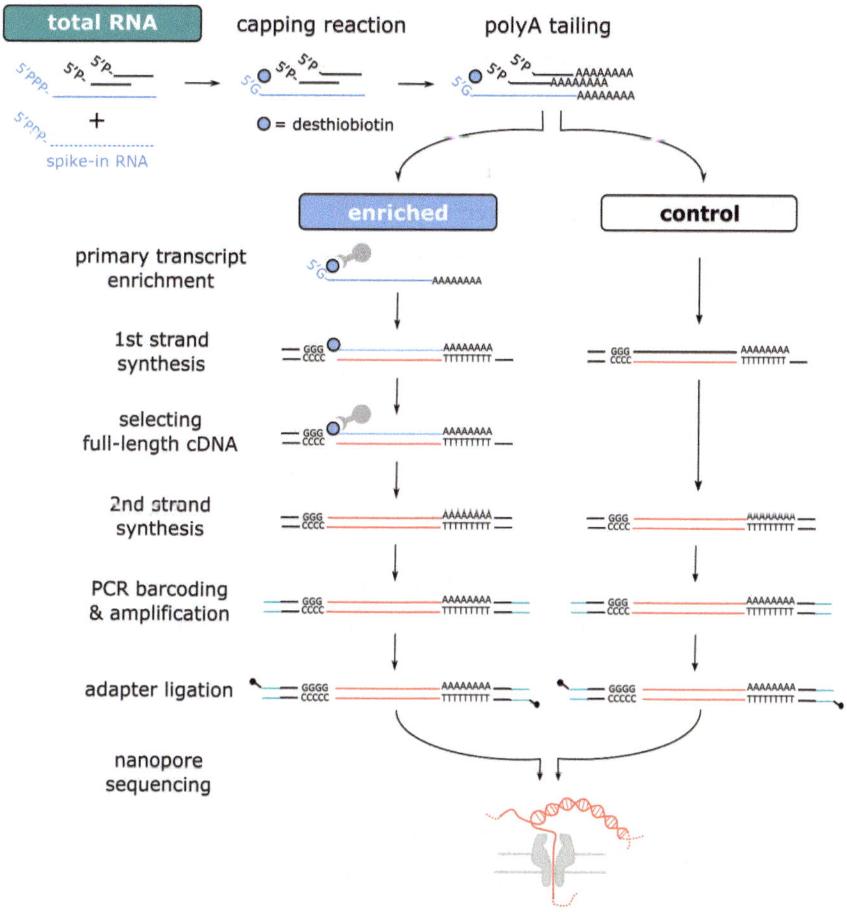

Fig. 1 Schematic overview of the ONT-cappable-seq experimental workflow. Starting from a total prokaryotic RNA sample supplemented with an RNA control spike-in, primary transcripts (blue) are enzymatically labelled with a desthiobiotin tag on their 5′-triphosphate group. After polyA tailing, the sample is split in an enriched and a control sample. In the enriched sample, the primary transcripts are specifically captured during a streptavidin enrichment procedure. Afterwards, both samples are reverse transcribed and the enriched sample undergoes another round of enrichment to select full-length cDNA. Second-strand synthesis is carried out, followed by PCR barcoding and amplification. Finally, the samples are pooled together, adapters are ligated, and the library is ready to load on a nanopore sequencing platform

8. Ethanol–3 M NaOAc mix: Combine 30 parts ethanol with 1 part 3 M NaOAc solution, pH 5.2. Make this mix fresh right before use.
9. Cooled centrifuge.
10. Nuclease-free water (NFW).
11. RNA Clean & Concentrator-5 columns.
12. Absolute ethanol.
13. Qubit fluorometer.

14. Qubit RNA High Sensitivity (HS) kit.
15. 1.5-mL DNA LoBind Tubes.
16. 0.5-mL DNA LoBind Tubes.

2.2 ONT-Cappable-Seq Library Prep

1. SQK-PCB109 PCR-cDNA sequencing/barcoding kit.
2. Nuclease-free water.
3. RNA Clean & Concentrator-5 columns.
4. Cooled centrifuge.
5. NanoDrop 1000 or similar spectrophotometer.
6. 1.5-mL DNA LoBind Tubes.
7. 0.5-mL DNA LoBind Tubes.
8. Heat block or thermomixer.
9. Vaccinia Capping System (containing vaccinia capping enzyme and buffer).
10. 5 mM 3′-desthiobiotin-GTP.
11. 0.1 U/μL yeast inorganic pyrophosphatase (YIPP).
12. 5 U/μL *E. coli* poly(A) polymerase.
13. Poly(A) polymerase reaction buffer (supplied with *E. coli* poly (A) polymerase).
14. 10 mM ATP (supplied with *E. coli* poly(A) polymerase).
15. 4 mg/mL hydrophilic streptavidin magnetic beads.
16. Tube rack with magnet.
17. Washing buffer: 10 mM Tris–HCl (pH 7.5), 1 mM EDTA, 250 mM NaCl (*see* **Note 2**).
18. 2× binding buffer: 10 mM Tris–HCl (pH 7.5), 1 mM EDTA, 2 M NaCl.
19. Biotin buffer: 10 mM Tris–HCl (pH 7.5), 1 mM EDTA, 500 mM NaCl, 1 mM biotin; store at 4 °C.
20. Qubit fluorometer
21. Qubit RNA High Sensitivity (HS) kit.
22. Agilent Bioanalyzer.
23. Agilent RNA 6000 Pico Kit
24. 10 mM dNTP mix.
25. Maxima H Minus Reverse Transcriptase.
26. 5× RT buffer (supplied with Maxima H Minus Reverse Transcriptase).
27. 40 U/μL RNaseOUT.
28. 50 U/μL RNase I_f.

29. AMPure XP reagent.
30. 80% ethanol in NFW solution (make fresh before use).
31. 70% ethanol in NFW (make fresh before use).
32. Absolute ethanol.
33. Thermocycler.
34. Low TE buffer: 1 mM Tris–HCl (pH 7.5), 0.1 mM EDTA.
35. Overhead rotator.
36. 2.5 U/μL LongAmp Hot Start *Taq* DNA Polymerase.
37. LongAmp Hot Start *Taq* Reaction buffer (supplied with the DNA polymerase).
38. 20 U/μL exonuclease I.
39. Qubit dsDNA HS Assay kit.

2.3 Sequencing

1. Flow Cell Priming kit (EXP-FLP002).
2. SQK-PCB109 PCR-cDNA sequencing/barcoding kit.
3. R9.4.1 flow cell (MinION or PromethION; *see* **Note 3**).
4. Vortex mixer.
5. Microfuge.
6. MinION or PromethION sequencing platform (*see* **Note 3**).
7. MinKNOW.

2.4 ONT-Cappable-Seq Data Analysis

1. UNIX environment. For this chapter a Linux machine running Ubuntu 20.04.6LTS with an Intel Core i7-9700X 3.00 GHz 8 core CPU and 64 GB RAM (though 32 GB would suffice) was used.
2. Conda or equivalent package environment manager.
3. Integrative Genomics Viewer (IGV) [11].
4. Snakemake (v7.26.0) [12].
5. Pychopper [13].
6. Cutadapt [14].
7. Minimap2 [15].
8. Samtools [16].
9. Samclip [17].
10. Bedtools [18].
11. Termseq-peaks [19, 20].
12. R dplyr package [21].
13. Complete reference genome sequences of phage and optionally bacterial host.

3 Methods

3.1 RNA Preparation

3.1.1 Preparation of RNA Spike-In

The RNA spike-in RNA control is used as an internal validation of the library prep. It enables verification of the enrichment steps, RNA quality, or biases that may arise during library preparation and can be used to normalize the resulting data. Here, a single 1.8-kb RNA transcript was used. However, the inclusion of additional transcripts of varying lengths enables the detection of additional biases or allows troubleshooting (*see* **Notes 2** and **4**).

1. Synthesize the 1.8-kb spike-in control with the HiScribe T7 High Yield RNA Synthesis kit using 250 ng of the included FLuc control template and standard reaction conditions as indicated by the manufacturer.
2. Measure the RNA concentration by NanoDrop 1000 (*see* **Note 5**).
3. Treat up to 40 μg of resulting RNA with DNase I. To do this, first denature the RNA by incubation at 65 °C for 5 min and place it directly on ice. After letting it cool for 5 min, transfer up to 40 μg of in vitro transcribed RNA 1.5-mL tube and add 5 μl Dnase I buffer with $MgCl_2$, 0.5 μL RNaseOUT, and 4 μL DNase I (4 U). Incubate the reaction for 30 min at 37 °C.
4. Perform an ethanol precipitation: Add 50 μL nuclease-free water and 100 μL phenol/chloroform/isopropanol mix (25: 24:1 ratio) to the DNase I-treated sample and centrifuge for 12 min at 10,000 *g* and 15 °C (*see* **Note 6**).
5. Transfer the top layer to a new tube and add 2.5 volumes of 30: 1 (EtOH/3M NaOAc) mix and invert the tubes until they form a homogeneous solution (*see* **Note** 7). Precipitate the sample overnight at -20 °C.
6. Centrifuge tubes for 1 h at 10,000 *g* and 4 °C and remove the supernatant carefully (*see* **Note 8**). Air-dry for approximately five minutes.
7. Resuspend the pellet in 50 μL nuclease-free water (NFW). To improve resuspension, incubate the tube at 65 °C for 5 min with regular or continuous shaking/mixing.
8. To remove residual nucleotides, clean up the resulting RNA product using RNA Clean & Concentrator-5 columns according to the manufacturer's instructions. To ensure sufficient cleanup, perform the wash step with RNA wash buffer (see kit) four times. Perform all centrifugation steps at 10,000 *g*.
9. Measure the concentration using Qubit to get a more accurate measurement. Dilute the sample further to 1 ng/μL and make small aliquots for use during the ONT-cappable-seq library prep. Store at −80 °C until use.

3.1.2 Preparation of RNA Samples

While high-quality RNA samples can be prepared in many different ways (e.g., TriZOL-based extraction [22]), a number of criteria are essential for this application.

Sampling

1. To sample phage (and host) RNA, keep in mind the research question. In case you are interested in strict temporal resolution of transcription features, infection at high multiplicity of infection (MOI; ranging from 5 to 100 depending on your phage) is required to ensure near-synchronous infection of the entire culture. You can then sample at the timepoints of interest. However, this choice increases the number of samples for library prep and sequencing and, subsequently, the overall cost of the experiment.

In many cases a synchronized infection is less important, for example, when the temporal aspects can be distinguished in data analysis (e.g., identification of phage promoter motifs that can be associated to different infection stages). While a high MOI can still help to ascertain the culture is mostly infected and therefore contains phage RNA at the same stage of infection, it is less essential since coverage across the entire infection cycle is desired. After infection, sample at different timepoints to cover the complete infection cycle until right before lysis. These samples can then be pooled in equal amounts before or after extraction and treated as a single sample for library preparation (*see* **Notes 3** and **9**).

Extraction

1. Phages tend to contain a lot of smaller transcription units and resulting small RNAs in their transcriptome. It is therefore best to choose an RNA extraction protocol or kit that retains these small RNAs. While chemical extraction methods such as hot phenol or TriZOL are laborious, they have the advantage that they capture the full spectrum of RNA species in the sample. By contrast, kits are faster, but many lose the smaller RNAs. As such, the chemical extraction methods are recommended, but when a kit is opted for, verify its specifications in terms of which RNA species are retained.
2. Since ONT-cappable-seq is a long-read-oriented protocol, it is important to select a method that does not introduce excessive shearing of RNA. At the very least, protocols should be adapted to prevent shearing, e.g., by avoiding vortexing and intensive pipetting steps, by mixing by inverting or flicking the tubes). Check your RNA quality with Bioanalyzer and aim for RIN values >9 (*see* **Notes 2** and **10**).

Additional Preparation

1. Ensure your RNA samples are free of DNA by treating with DNase (repeat this step if necessary). DNA presence can be checked with PCR on a small ~100–300 bp genomic region.
2. rRNA depletion is not required for starting ONT-cappable-seq library prep. The enrichment procedure will remove a large portion of rRNAs.
3. Store RNA at −80 °C until ready for library prep.

3.2 ONT-Cappable-Seq Library Prep

The ONT-cappable-seq library prep is a modification of the SMRT-cappable-seq pipeline, switching it from PacBio to the Oxford Nanopore Technologies (ONT) sequencing platforms. It is a hybrid protocol between SMRT-cappable-seq for primary transcript enrichment and the ONT PCR-cDNA kit (SQK-PCB109) for cDNA synthesis and sequencing ([8, 23]; *see* **Note 11**). Make sure you have made all the necessary preparations in advance and allow sufficient time for the library prep to prevent having to freeze-thaw the samples too often (*see* **Note 2**).

Capping and A-Tailing

In the capping step, 5′-triphosphates (unprocessed transcripts) are specifically capped with a desthiobiotin-tagged GTP to enable enrichment at a later stage. A-Tailing at the 3′ end of RNA ensures compatibility with the PCR-cDNA Nanopore kit, as this is required for first-strand synthesis.

1. Thaw your samples and one aliquot of the spike-in on ice.
2. Transfer 10–15 μg of each sample in separate tubes and bring each to a total of 50 μL with NFW.
3. Clean up your RNA samples with RNA Clean & Concentrator-5 columns according to the manufacturer's protocol, but elute in 31 μL NFW (*see* **Note 12**).
4. Use 1 μL to prepare a 1:10 dilution in NFW and assess the concentration using NanoDrop.
5. Transfer 5 μg of each cleaned sample to separate tubes and bring each to a total volume of 29 μL with NFW.
6. Add 1 μL of the prepared 1 ng/μL spike-in RNA to each sample.
7. Denature each sample by incubating at 65 °C for 5 min and directly place samples on ice.
8. Set up the capping reaction by adding 5 μL 10× VCE buffer, 5 μL of 5 mM 3′DTB-GTP, 5 μL (50 U) Vaccinia Capping Enzyme, and 5 μL (0.5 U) YIPP to each sample. Mix by flicking the tube and incubate at 42 °C for 40 min.

9. Clean up the sample using RNA Clean & Concentrator-5 columns, according to the manufacturer's guidelines. However, perform the wash step with RNA wash buffer four times (*see* **Note 13**). Elute each sample in 36 μL NFW.
10. Set up the polyA-tailing reaction by adding the following to each 36 μL DTB-capped RNA sample (add polymerase last): 5 μL of 10× *E. coli* Poly(A) Polymerase reaction buffer, 5 μL of 10 mM ATP, and 4 μL *E. coli* Poly(A) Polymerase. Mix the components by flicking and incubate the reaction at 37 °C for 15 min.
11. Directly proceed to the cleanup step using RNA Clean & Concentrator-5 columns according to manufacturer's guidelines and elute in 33 μL NFW.
12. Transfer 3 μL of each sample to a separate tube and keep it aside on ice (control RNA). Use the remaining 30 μL to proceed with primary transcript enrichment.

3.2.1 Primary Transcript Enrichment

1. For each sample (not control samples), prepare 30 μL hydrophilic streptavidin magnetic beads. Bring the beads into a 1.5-mL tube and place them on a magnet. Remove the supernatant and remove the tubes from the magnet. Add 200 μL washing buffer and place back on the magnet. Repeat this washing step three times. After removal of the last washing buffer supernatant, remove the tubes from the magnet and resuspend each pellet in 30 μL 2× binding buffer (*see* **Note 14**).
2. Add 30 μL of the prewashed streptavidin beads to 30 μL of the capped-tailed RNA and add 47 μL of NFW to the control RNA. Incubate both control and capped-tailed samples at room temperature (RT) for 45 min on an overhead rotator (*see* **Note 15**).
3. Pellet the capped-tail samples on a magnetic rack. Meanwhile, keep the control samples at RT to mimic similar temperature conditions.
4. Remove and keep the supernatant (= flow-through) on ice. Wash the beads three times with 200 μL washing buffer (*see* also **Note 16**).
5. Elute the RNA by resuspending in 50 μL biotin buffer and incubate at 37 °C for 30 min on an overhead rotator. Also incubate the control samples at this temperature.
6. Take out all samples from 37 °C and pellet the magnetic beads on a magnetic rack. Transfer the supernatant to a new tube (= enriched RNA).
7. Clean up both enriched and control RNA samples using RNA Clean & Concentrator-5 columns according to manufacturer's instructions, but use four RNA wash steps (*see* **Note 13**).

8. Remove the tubes from the magnet and elute the RNA/DNA hybrid by adding 10 μL NFW.
9. Pellet the beads on a magnet and transfer the supernatant to a fresh tube. Take care not to transfer any of the beads.

3.2.2 Quality Control Analysis

1. Take 1 μL of control RNA to make a 1:5 dilution in NFW and measure the concentration on a Qubit. If needed, dilute the 1:5 dilution further in NFW to ~1–2 ng/μL (*see* **Note 17**).

   Optional: Also measure concentrations of the flow-through samples from ***step 4*** *in Subheading 3.2.1 and bring them to ~1–2 ng/μL.*
2. Prepare an Agilent RNA 6000 Pico chip according to the manufacturer's instructions and load the samples. Enriched samples can be loaded directly on the chip. For control samples (and flow-throughs; *see* **Note 18**) take the ~1 ng/μL dilutions. Proceed to first-strand synthesis when all checks are ok (*see* **Note 19**).

3.2.3 First-Strand Synthesis

1. For each of the control and enriched samples, prepare the following strand synthesis mix in a PCR tube: 1–2 ng of RNA (*see* **Note 20**), 1 μL of 2 μM VN primer (supplied with PCB109 kit), and 1 μL of 10 mM dNTP mix. Bring the mix to a total volume of 11 μL with NFW.
2. Denature the mix for 5 min at 65 °C, spin down briefly, and place on ice promptly.
3. Meanwhile, mix together the following in a separate tube: 4 μL of 5× RT buffer, 1 μL RNaseOUT, 1 μL NFW, and 2 μL strand-switching primer (SSP at 10 μM supplied with PCB109 kit). Mix by flicking the tube and spin down briefly.
4. Combine this mix with the denatured RNA mix and preincubate at 42°C for 2 min.
5. Add 1 μL of Maxima H Minus Reverse Transcriptase and mix by flicking the tube (spin down if needed).
6. Incubate at 42 °C for 90 min, followed by heat inactivation at 85 °C for 5 min. Hold at 4 °C until you proceed (*see* also **Note 21**).

3.2.4 Selecting Full-Length cDNA

1. Add 1 μL RNase I_f (50 U) to the enriched samples (not the controls) and incubate at 37 °C for 30 min. In parallel incubate the controls also at 37 °C.
2. Clean up both control and enriched samples with AMPure XP beads (*see* **Note 22**). For this, add 1.8 volumes of AMPure beads to the eluted RNA volume and add 1.5 volume of absolute ethanol to the resulting volume of the AMPure/RNA mix (e.g., if you have 20 μL of sample, add 36 μL

AMPure beads and 84 μL ethanol). Incubate the bead/RNA mix on the bench for 5 min, and then expose to a magnet and wash beads two times with 250 μL 80% ethanol while confined to the magnet (*see* **Note 23**). Dry briefly (<5 min) until the pellet gives a matte appearance. Remove the tube from the magnet and elute the enriched RNA in 30 μL low TE buffer. Elute the control RNA in 20 μL of low TE buffer.

3. For each enriched sample, prepare 30 μL hydrophilic streptavidin magnetic beads by washing them three times with 200 μL washing buffer. Next, resuspend in 30 μL 2× binding buffer (*see* also **step 1** in Subheading 3.2.1 and **Note 14**).
4. Add 30 μL of cDNA/RNA duplex (only for enriched samples) to 30 μL of the prewashed streptavidin beads and incubate at RT for 45 min on an overhead rotator. In parallel, incubate the control samples at RT for 45 min.
5. Place the enriched samples on a magnetic rack, allow the beads to settle, and remove the supernatant. Take the tubes from the rack and add 200 μL washing buffer. Incubate for 1 min and place the tubes back in the rack. Once the beads are settled, remove the supernatant and repeat this wash step twice more.
6. After removal of the final washing buffer, resuspend the washed beads in 20 μL low TE buffer (*see* **Note 24**). Place both enriched and control samples on ice.

3.2.5 Second-Strand Synthesis

1. Set up two of the same PCR reactions per sample as follows: In a PCR tube combine 10 μL of 5× LongAmp *Taq* reaction buffer, 1.5 μL of 10 mM dNTP mix, 1.5 μL barcode primer (*see* **Note 25**), 5 μL template cDNA, and 2 μL LongAmp Hot Start *Taq* DNA Polymerase (5 U) and top up the mix to 50 μL total volume with NFW (*see* **Note 26**).
2. Place all the reaction tubes in a thermocycler and use the PCR settings outlined in Table 1 (*see* **Note 27**).

Table 1
PCR settings for second-strand synthesis

Step	Temperature	Time	Cycles
Initial denaturation	95 °C	30 s	–
Denaturation	95 °C	15 s	16
Primer annealing	62 °C	15 s	
Primer extension	65 °C	15 min	
Final extension	65 °C	6 min	–
Pause	4 °C	∞	–

3. Add 1 μL exonuclease I (20 U) to each PCR tube and mix by flicking the tube. Incubate at 37 °C for 15 min, followed by 80 °C inactivation for 15 min.
4. Combine replica PCRs in one tube per barcode and clean up the samples using AMPure XP beads. Add 80 μl of AMPure beads (0.8× sample volume) and mix by flicking (*see* also **Note 22**). Incubate on an overhead rotator for 5 min at RT. Briefly spin down the sample and pellet on a magnet. Once the solution is clear, remove the supernatant (*see* **Note 28**) and wash twice with freshly prepared 70% ethanol without disturbing the pellet while the tube stays confined to the magnet (*see* also **Note 23**). After washing and removal of the ethanol, briefly spin down the tube and place it back on the magnet. Pipette off residual 70% ethanol and shortly air-dry the pellet (*see* **Note 29**).
5. Remove the tube from the rack and resuspend in 12 μL elution buffer (EB in PCB109 kit).
6. Incubate on an overhead rotator mixer for 10 min at RT and pellet the beads on a magnet until the eluate is clear and colorless.
7. Remove and retain 12 μL of the eluate in a 1.5-mL LoBind tube (*see* **Note 30**). Keep the samples on ice.
8. Measure the concentrations of each sample using a Qubit dsDNA HS Assay kit (*see* **Note 31**).
9. Pool together equal amounts of barcoded library samples to a total of ~100 fmol for the combined library in a final volume of 23 μL (*see* **Note 32**). Keep this library at 4 °C (or −20 °C for long term) until ready for sequencing.

3.3 Nanopore Sequencing

3.3.1 Flow Cell Priming and Loading

In this part, the flow cell is prepared and set up for sequencing. It follows the standard PCR-cDNA protocol by Oxford Nanopore Technologies and the reader is advised to check the manufacturer's website for the latest updates and fine details for correct setup. What follows is an excerpt from the latest SQK-PCB109 protocol update for PromethION (version PCB_9092_v109_revK_10Oct2019) [23].

1. Add 1 μL of rapid adapter to the cDNA library mix and incubate for 5 min at RT (*see* **Note 33**).
2. Spin down briefly and store the library on ice until ready for loading on your flow cell.
3. Thaw the sequencing buffer (SQB), loading beads (LB), flush tether (FLT), and one tube of flush buffer (FB) at RT.
4. Mix SQB, FB, and FLT tubes separately by vortexing, and spin down at RT. Mix LB by gentle pipetting.

5. Prepare the flow cell priming mix by adding 30 μL of the thawed and mixed FLT directly to the tube of the thawed and mixed flush buffer (FB). Mix by vortexing.
6. Take out the flow cells from the 4 °C storage, let it get to RT for ~20 min, and insert it into the PromethION sequencer (*see* **Note 34**). Make sure the R9.4.1 PromethION flow cell is properly docked and recognized by the PromethION sequencer.
7. Turn the cover over the inlet port (port 1) to the open position.
8. Remove air from the inlet port by placing a P1000 pipette, set at 200 μL, and turn the dial until it shows 220–230 μL or until the liquid enters the tip.
9. Using a P1000 pipette, flush 500 μL of the priming mix into the inlet port of the flow cell, avoiding the introduction of air bubbles (*see* **Note 35**). Let the priming mix incubate on the flow cell for five minutes. Repeat the flush and incubation steps. Meanwhile, your library can be prepared.
10. Thoroughly mix the contents of the loading bead (LB) tube by pipetting. Make sure all the beads are resuspended and you get a homogenous, milky solution.
11. Prepare the library by combining 75 μL SQB, 51 μL LB, and 24 μL of library (*see* **Note 36**).
12. Load your 150 μL of prepared library on the flow cell through the inlet port, close the valve to seal it, and close the PromethION lid when ready (*see* **Note 37**).

3.3.2 Sequencing Settings

1. Data acquisition and basecalling can be performed in real time using the operating software MinKNOW according to flow cell and library preparation kit version. Start the run with the appropriate parameters and make sure to disable barcode trimming. The MinKNOW-run parameters outlined in Table 2 are an example of a typical ONT-cappable-seq experiment on a PromethION 24 device:
2. Monitor the progression of the sequencing run after ~24 h to assess pore availability, translocation speed, and sample distribution. If necessary, the run can be paused to refuel and/or reload the flow cell to increase and balance the sequencing yields of your samples.
3. Allow your sequencing experiment to continue until all pores are exhausted and/or a sufficient number of reads are obtained. In general, a minimum 2–5 million raw reads of sufficient quality ($Q > 9$) is advisable for each sample (enriched and control).

Table 2
Recommended run parameters for ONT-cappable-seq sequencing on PromethION

Parameters	Setting
Flow cell type	R9.4.1
Kit	SQK-PCB109
Run duration	72h
Minimum read length	20 bp (unfiltered)
Basecalling	On
Basecalling model	High accuracy
Barcoding	On
Trim barcodes	Off
Mid-read barcode filtering	Off
Barcode both ends	Off
Override minimum barcoding score	Off
Output format	.fast5 (raw read signal) .fastq (basecalled reads)
Filtering	On (Qscore: 9)

3.4 ONT-Cappable-Seq Data Analysis

3.4.1 Installation and Requirements

1. The automated ONT-cappable-seq pipeline (version 2) is available from GitHub [24]. Type in the following command to download it (*see* **Note 38**):

```
$ git clone https://github.com/LoGT-KULeuven/ONT-
cappable-seq2
```

2. The automated ONT-cappable-seq data analysis tool (v2) is implemented in the workflow management system Snakemake [12]. Using Conda, create a separate environment (here called ONT-cappable-seq) where we will install the tools and run the pipeline.

```
$ conda create -n ONT-cappable-seq python=3.9
$ conda activate ONT-cappable-seq
```

3. The pipeline uses eight external tools (pychopper, cutadapt, minimap2, samtools, samclip, bedtools, termseq-peaks, r-dplyr [for references, *see* Subheading 2]). All tools, except for termseq-peaks and dplyr, are automatically installed in their

own environment upon running the Snakemake pipeline. In the ONT-cappable-seq environment, navigate to the ONT--cappable-seq2 project folder and manually install the termseq-peaks package using the following commands:

```
$ cd ONT-cappable-seq2/
$ git clone https://github.com/NICHD-BSPC/term-
seq-peaks
$ cd termseq-peaks
$ conda install -c bioconda --file requirements.
txt
$ python setup.py install
```

Check if the installation worked by typing:

```
$ termseq_peaks
```

4. Afterwards, install the Snakemake tool and the rdplyr package in the same environment:

```
$ conda install -c bioconda -c conda-forge snake-
make
$ conda install -c conda-forge r-dplyr
```

3.4.2 Project Organization and Data Preparation

1. After downloading, the ONT-cappable-seq project folder is organized as follows:

```
ONT-cappable-seq2/
├── workflow/
| ├── envs/
| ├── rules/
| ── snakefile
├── termseq-peaks/
├── config/
| ── config.yaml
├── input/
| ├── fastq_data/
| ── genome_data/
── peak_clustering.r
```

2. Retrieve high-quality FASTA files of the reference genomes of your organisms of interest and store them in the input/genome_data directory. For clarity, we recommend to rename the FASTA files to the respective IDs of the phage (or host)

(=sampleID) (*see* **Note 39**). The sequence of the FLuc control plasmid used for in vitro transcription of the RNA control spike-in can be downloaded from the manufacturer's website [25].

```
ONT-cappable-seq2/
...
├── input/
| ── genome_data/
| ├── phageID.fasta
| ├── hostID.fasta
| ── in-vitro-RNA.fasta
```

3. Place the raw sequencing files in the input/fastq_data directory and rename them to ensure they contain information on sampleID and treatment (enriched or control sample). For each individual sample, it is important to have a single decompressed .fastq file that contains sufficient reads with an appropriate mean read length (>200 bp) (*see* **Note 40**).

```
ONT-cappable-seq2/
...
├── input/
| ├─ fastq_data/
| ├── sampleID_enriched.fastq
| ── sampleID_control.fastq
| ── genome_data/
| ├── phageID.fasta
| ├── hostID.fasta
| ── in-vitro-RNA.fasta
```

3.4.3 Pipeline

1. Overview of the Pipeline

The automated ONT-cappable-seq pipeline (v2) can be divided in three general steps: read processing, read mapping, and transcript boundary detection (Fig. 2). During read processing, full-length cDNA reads are identified, trimmed, and oriented by Pychoppper, based on the SSP and VNP primers used during cDNA synthesis. Afterwards, cutadapt removes the polyA tail and remnant adapter sequences from the reads. The processed reads are subsequently used for mapping to the reference genome of interest using minimap2, after which samclip discards reads with more than 10 clipped bases at either side. Samtools removes low-quality mapped read from the alignment file and converts it to a sorted and indexed BAM file.

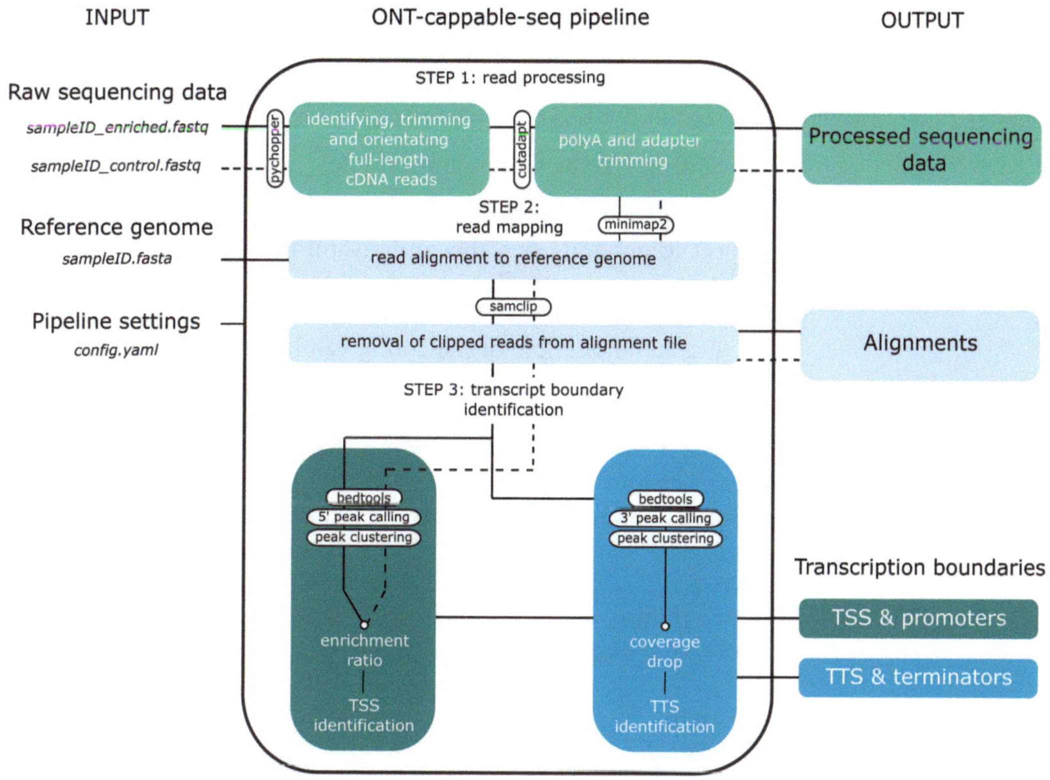

Fig. 2 Schematic overview of the automated ONT-cappable-seq data analysis pipeline. The ONT-cappable-seq tool requires two raw sequencing data files (enriched and control), a reference sequence and a configuration file that specifies the settings of the pipeline. As a first step, the raw read files will be oriented and trimmed to obtain high-quality full-length cDNA reads without adapter remnants (output directory = processed sequencing data). In the second step, the processed reads are mapped to the reference sequence. Upon removal of low-quality mapped reads, the alignment files are sorted and indexed (output directory = alignments). Finally, the alignment files are used to annotate TSS and TTS by identifying genomic positions with local maxima of 5′ and 3′ ends using a peak calling algorithm, followed by the calculation of an enrichment ratio (TSS) or coverage drop (TTS) at the respective position (output directory = transcript_boundaries)

Next, transcript boundary identification starts by generating strand-specific BED files that contain the number of 5′ and 3′ read ends at each position of the genome, which is used as an input for the termseq-peaks program to identify local maxima of 5′ and 3′ termini. Peak positions within a specified distance are clustered, retaining only the peak position with the highest number of reads and assigning all other reads to this cluster representative. To discriminate between TSS and processed 5′ ends, relative read counts of 5′ peak positions in the enriched sample and the control sample are used to calculate an enrichment ratio (see formula below). In case this enrichment ratio is above a specified cutoff value, and the 5′ peak position is annotated as a TSS.

Finally, 3′ peak positions are annotated as TTS in case the average read coverage drop across this position exceeds the specified threshold. Threshold values and other parameters can be customized in the ONT-cappable-seq tool configuration file.

1. Setting Up the Configuration File

Navigate to the config directory in the ONT-cappable-seq project folder and open the config.yaml file. This file contains the paths to your input files and the different parameters used for annotation of the transcriptional boundaries.

A general example of the config.yaml file is displayed below:

```
Sample name: sampleID
fasta file: input/genome_data/sampleID.fasta
 enriched fastq: input/fastq_data/sampleID_en-
riched.fastq
 control fastq: input/fastq_data/sampleID_control.
fastq
ID: group
termseq alpha: 0.001

cluster width:
 TSS: 15
 TTS: 30

minimum coverage:
 enriched:
 TSS: 25
 TTS: 25
 control:
 TSS: 2
 TTS: 2

peak alignment error: 2
TSS Threshold: 1.5
TTS threshold: 0.25

TSS sequence extraction:
 upstream: 40
 downstream: 0

TTS sequence extraction:
 upstream: 30
 downstream: 30
```

where

- **Sample name**: contains the name of your organism of interest (sampleID, e.g., LUZ19).
- **Fasta file**: contains the path to your reference genome of interest (input/genome_data/sampleID.fasta, e.g., input/genome_data/LUZ19.fasta).
- **Enriched fastq**: contains the path to the raw sequencing file of your enriched sample (input/fastq_data/sampleID_enriched.fastq, e.g., input/fastq_data/LUZ19_enriched.fastq).
- **Control fastq**: contains the path to the raw sequencing file of your control sample (input/fastq_data/sampleID_control.fastq, e.g., input/fastq_data/LUZ19_control.fastq).
- **ID**: additional identifier to classify your samples (cannot be empty), e.g., timepoint, specific condition, replicate, etc.
- **Termseq alpha:** peak calling threshold value used by the termseq-peak algorithm (-t flag).
- **Cluster width:** peak positions within the specified distance are clustered. The position with the highest number of reads is taken as the representative of the cluster.
- **Minimum coverage:** absolute number of reads required at this position to be considered as candidate TSS/TTS position. This can be specified for the enriched and the control sample separately.
- **Peak alignment error:** positional difference (n) allowed between peak positions identified in the enriched and the control dataset used to calculate the enrichment ratio at a specific genomic position i, based on the read count per million mapped reads (RPM) at that peak position, as calculated by:

$$\text{Enrichment ratio }(i) = \frac{\text{RPM}_{\text{enriched sample}}(i)}{\text{RPM}_{\text{control sample}}(i \pm n)}$$

- **TSS threshold:** enrichment ratio value that needs to be surpassed to annotate 5′ peak position as a TSS.
- **TTS threshold:** minimum read reduction to annotate 3′ peak position as TTS, determined by calculating the coverage drop across the putative TTS, averaged over a 20-bp region up- and downstream the TTS.
- **TSS/TTS sequence extraction**: selection of promoter and terminator region for which the user wants to extract the DNA sequence, defined by the number of up- and downstream nucleotides relative to the TSS and TTS.

It is important that the sampleIDs and names of the FASTA and FASTQ files in the config file are modified according to the input files provided. The default settings for TSS and TTS identification can be adjusted to optimize for your specific organism of interest (*see* **Note 41**).

2. Running the Pipeline

In the terminal, change directory to the ONT-cappable-seq project folder and execute the Snakemake pipeline, which will run according to the settings specified in the configuration file (*see* **Note 42**).

```
$ cd ONT-cappable-seq2
$ snakemake --cores N --use-conda
```

where N is the number of cores available on your computer to run the pipeline. In-house, generally 8 cores are used. All command line options can be printed by calling Snakemake -h.

3.4.4 Result Interpretation

After running the pipeline, you can find the results in the newly created Results directory, which contains the subdirectories processed_fastq, alignments, and transcript_boundaries. The directory tree is displayed below and indicates which files are generated in each folder.

```
ONT-cappable-seq2/
...
├── results/
| ├── processed_fastq/
| ├── pychopper/pychopper_sampleID_ID/
| ├── sampleID_enriched_ID_full_length_output.fq
| ├── sampleID_control_ID_full_length_output.fq
| ├── report_enriched.pdf
| ├── report_control.pdf
| ├── statistics_enriched.tsv
| ── statistics_control.tsv
| ├── cutadapt/
| ├── sampleID_enriched_ID_cutadapt.fq
| ── sampleID_control_ID_cutadapt.fq
| ├── alignments/
| ├── BAM_files_sampleID/
| ├── sampleID_enriched_ID.sorted.bam
| ├── sampleID_control_ID.sorted.bam
| ├── sampleID_enriched_ID.sorted.bam.bai
| ── sampleID_control_ID.sorted.bam.bai
| ├── transcript_boundaries/
| ├── TSS_sampleID/TSS_sampleID_ID/
| ├── enr_ratios_sampleID_ID.plus.csv
| ├── enr_ratios_sampleID_ID.minus.csv
```

```
| ├── sampleID_enriched_ID_peaks.bed
| ── TSS_seq_sampleID_enriched_ID.fa.out
| ├── TTS_sampleID/TTS_sampleID_ID/
| ├── eff_ratios_sampleID_ID.plus.drop.coverage
| ├── eff_ratios_sampleID_ID.minus.drop.coverage
| ├── TTS_sampleID_ID.bed
| ── TTS_seq_sample_ID.fa.out
```

1. The raw sequencing files of the enriched and control sample are processed in two stages, in which they are first oriented with pychopper and trimmed using cutadapt. The processed sequencing FASTQ (.fq) files and tool-specific metadata can be retrieved in their respective processed_fastq folder (*see* **Note 43**).
2. The sorted alignment files (.BAM) together with their index files (.BAI) of the enriched and the control samples can be found in the alignments/BAM_files_sampleID folder. These files can be uploaded in Integrative Genomics Viewer (IGV) to visually inspect the full-length transcriptome of the input genome/sequence you provided. Ideally, the transcriptional landscape shows reads with a broad-length distribution (~0.1–7 kb) and sufficient sequencing coverage across the entire reference sequence. For additional quality checks of the mapped reads, *see* **Notes 44** and **45**.
3. For each sampleID and additional ID provided, TSS identification results are deposited in their respective TSS_sampleID_ID subfolder within the transcript_boundaries folder. This folder contains a .BED file and FASTA (.fa) file with the promoter regions and sequences associated with the identified TSS, respectively, as specified in the config file. In addition, one can find two strand-specific .CSV files that specify the absolute and relative read counts of each annotated TSS position on the specified strand, as well as the enrichment ratio. More specifically, each column in the 15-column file returns the info displayed in Table 3.
4. Similarly, TTS identification results are deposited in their respective TTS_sampleID_ID subfolder within the transcript_boundaries folder. This folder contains a .BED file and FASTA (.fa) file that respectively contain the terminator regions and sequences associated with the TTS positions defined by ONT-cappable-seq, as specified in the config file. In addition, two strand-specific .CSV files can be found, which indicate the read count reduction across each annotated TSS position on the specified strand. Details for each column in the 9-column file can be found in Table 4.

Table 3
Info on the strand-specific TSS .CSV output files. An explanation for each column in the file is given

Column ID	Value
1	Reference genome (same as fasta header ID)
2	Start position of peak cluster identified in the enriched sample
3	End position of peak cluster identified in the enriched sample
4	TSS position identified in the enriched sample
5	Strand of TSS position in the enriched sample
6	Absolute read count at TSS in the enriched sample
7	Relative read count at TSS in the enriched sample (RPM)
8	Reference genome
9	Start position of peak cluster identified in the control sample
10	End position of peak cluster identified in the control sample
11	TSS position identified in the control sample
12	Strand of TSS position in the control sample
13	Absolute read count at TSS in the control sample
14	Relative read count at TSS in the control sample (RPM)
15	Enrichment ratio at TSS (RPM(enriched)/RPM(control))

5. To evaluate TSS and TTS calling performance, it is recommended to upload the .BED files of the promoter and terminator positions in IGV alongside their corresponding alignment files (and, if available, annotation files [.GFF or .GTF]). Manual inspection of the identified positions can also serve as a visual guide to tailor the ONT-cappable-seq pipeline parameters to your experimental conditions.
6. Optional: Depending on the user's interests, the full-length transcriptional landscape, together with the identified TSS and TTS, can be used in follow-up analyses. For example, the information can be used to improve genome annotations, gain insights in promoter and terminator architectures, delineate 5′- and 3′ untranslated regions (UTR), and elucidate of operon structures. Examples of these can be found in some recent publications that use ONT-cappable-seq [1, 10, 26].

4 Notes

1. Use a low pH (<5.5) buffered phenol, and make sure to take the bottom (nonaqueous) layer.

Table 4
Info on the strand-specific TTS .CSV output files. An explanation for each column in the file is given

Column ID	Value
1	Reference genome (same as fasta header ID)
2	TTS position
3	TTS-20 position
4	Average read coverage in 20-bp upstream window
5	TTS position
6	TTS+20 position
7	Average read coverage in 20-bp downstream window
8	Readthrough (downstream/upstream coverage)
9	Coverage drop (1- readthrough)

2. As a general note for RNA and library preparations: To ensure that long transcripts do not get fragmented, avoid shearing forces as much as possible. Minimize freeze-thaw steps, do not vortex for mixing and keep pipetting to a minimum. When possible, flick the tubes for mixing or invert them in case of larger volumes. Also work with RNase-free materials and solutions and work on ice as much as possible to prevent enzymatic degradation. Use LoBind tubes (or similar) for all steps to prevent RNA/DNA loss due to binding of nucleic acids to the plastic.
3. In our experience, up to six samples for PromethION and two for MinION will provide sufficient read depth in the sequencing run. Exceeding six samples is not recommended as you will exceed the number of barcodes (12) of the SQK-PCB109 kit, due to the sample splitting between control and enriched samples in a later step.
4. When using other spike-in RNAs, it is important they contain a 5′-triphosphate end so enrichment can also be tracked. Using a transcription kit like the one used here will ensure 5′-triphosphate ends.
5. Yields can be quite high (>100 μg), so diluting the sample in nuclease-free water might be required to bring it within the measurable range of your device.
6. The PCI mix consists of two layers; ensure you take the bottom layer.
7. Take care not to take up any of the bottom layer or interphase layer (if present/visible). Using Phase Lock Gel (PLG) tubes

for the centrifugation or cutting off part of the tip to create a wider opening can make this easier.

8. Check on which side the pellet should be and do not disturb it when removing the supernatant. Ensure that all residual ethanol is gone but do not dry for too long, as this will make dissolving the pellet more difficult.
9. Sampling after lysis occurs can result in more fragmented RNA species as RNA is released from the cell into an environment where it is easily degraded.
10. RIN values are calculated for a large part based on 16S and 23S rRNA peaks. For some bacteria these peaks are different, resulting in an incorrect RIN value. In that case, check whether the peaks are sharp and have a flat baseline. This indicates limited degradation products.
11. ONT sequencing is continuously evolving and improving its sequencing technologies. At the time of writing there is a shift taking place from the R9.4.1 flow cells (and accompanying library prep kits such as the SQK-PCB109 used here) to more performant R10.4.1 flow cells and a new set of library prep kits, with the former kits gradually becoming unavailable. The reader is advised to check for changes in the PCR-cDNA library preparation protocol and flow cell priming/loading, and adapt the ONT-cappable-seq protocol accordingly.
12. This step is important to ensure removal of nucleotides that tend to co-extract and co-precipitate together with RNA. Insufficient removal of contaminating nucleotides will prevent tagging primary RNAs with DTB-GTP and will consequently result in poor enrichment of primary transcripts. The use of RNA Clean & Concentrator-5 kits is strongly advised for cleanup as it is one of the few kits that retains smaller RNAs (>17 nt) and has high recovery rates.
13. To save time, perform the first three washes with centrifuge for only 30 s and only spin down the final wash centrifuge for two minutes to make sure that all ethanol is removed. With each wash step, add the RNA wash buffer via the sides of the column that may have come into contact with the capping reaction to reduce carryover of DTB-GTP. Sufficient washing is important for DTB-GTP removal, which will otherwise inhibit primary transcript enrichment.
14. For the streptavidin precipitate fast, mix the beads well by vortexing and directly proceed on taking the needed aliquots. During washing, do not allow the beads to dry to the point of cracking.
15. Rotating the sample will prevent precipitation of the magnetic beads. Alternatively, mix the samples regularly by flicking the tubes.

16. Remove the samples from the magnetic rack for each wash step. Carefully add the washing buffer (no need to completely resuspend the beads), incubate for ~1 min, and place the sample back on the magnet. Remove the washing buffer and repeat. Flow-through RNA can be used in quality checks later on (optional).
17. It is better not to measure enriched samples with Qubit to save the sample. Rather, check estimated concentrations in the Bioanalyzer results. Enriched samples will be very low in concentration (typically 0.25–2 ng/μL in a 10 μL volume) since the majority of the crude RNA is processed (>95%) and washed away during enrichment.
18. Flow-through samples still contain salts from the elution buffer which will affect peak intensity and position. It is best that these are diluted in NFW or even cleaned up separately with, e.g., RNA Clean & Concentrator-5 columns.
19. For quality checks there are a couple of things to check. First, the RNA should still be intact. This will mainly be visible in the control samples as they contain large amounts of rRNA. Oftentimes the rRNA peaks are shifted and broadened due to the library preparation causing RIN values from the Bioanalyzer to be absent or wrong. In our experience, samples are of good quality if any rRNA peaks are present in control samples. If these are missing, it may be useful to check the flow-through samples as well, as these should also contain rRNA. Second, check for successful enrichment. This can be quite tricky to estimate from Bioanalyzer results as it also depends on the concentration that was achieved for the enriched samples. In general, upon good enrichment rRNA peaks are either completely absent in enriched samples or present together with/located within a "bulge" of other RNA lengths, indicating a larger relative amount of other transcripts compared to rRNA transcripts. Third, check the estimated concentrations of the samples. High concentrations (as a guideline >15 ng/μL for *P. aeruginosa*) in the enriched sample could indicate poor enrichment, especially when also rRNA peaks are strongly present. However, this can be organism dependent.
20. For enriched RNA one can go up to 5 ng if the RNA concentration allows, since yields of cDNA synthesis on enriched RNA tend to be a lot lower.
21. This is a safe break point to leave the sample overnight at 4 °C and continue the next day. Since DNA is a lot more stable than RNA, it is recommended to continue the library prep until this point, once you have done the primary transcript enrichment. The small amount of RNA in the enriched samples makes that even low levels of RNase activity can affect sample quality.

22. Magnetic beads tend to precipitate, so ensure proper mixing before taking them from the vial.
23. During washing, beads just need to have been well in contact with ethanol, no mixing is required. Take off the supernatant and repeat the wash.
24. This will not elute the RNA from the beads, so it is crucial not to remove them from the solution. Poor PCR yields may be observed when eluting and cleaning the sample to remove biotin, likely due to RNA losses that are inherent to these steps.
25. Use a different BP (the kit offers up to 12 barcodes) for each sample, but use the same for between the two PCR replicates. Make note of which sample is associated with each barcode. Two PCR replicates are generally sufficient to ensure sufficient yield.
26. Ensure that for the enriched samples beads are included in the PCR. Flick the tubes to bring the beads completely in suspension before taking the RNA sample. Keep any remainder of the sample at 4 °C. Freezing the samples is not recommended since this is detrimental for the magnetic beads and will complicate their downstream removal.
27. Use 11–18 cycles, depending on the application (14 is recommended by Oxford Nanopore Technologies). From experience, 16 cycles ensure sufficient yields, while less cycles are better to decrease the bias towards smaller transcripts. The extension time of 15 min ensures that even the longest reads will be captured (polymerase operates at 50 s/kb).
28. For enriched samples, both the streptavidin beads and AMPure beads will migrate to the magnet. Streptavidin beads tend to migrate a lot faster to the magnet, so be sure to wait until also the AMPure beads have pelleted and solution is clear.
29. Dry until the pellet appears matte. Try to avoid that the pellet reaches the stage of cracking.
30. Take care not to pick up any of the beads. In some cases, a small amount of beads can co-migrate with the meniscus of the elution buffer. With the removal of the streptavidin beads, any remaining capped RNAs from the sample are also cleared from the sample.
31. From the enriched samples, use 1 μL directly. Control samples generally have a higher concentration and it is recommended to make a ½ dilution first, measure 1 μL of this and adjust the dilution factor in case the sample concentration is out of range (too high). Absorbance-based concentration measurements such as with NanoDrop are not recommended as they will result in inaccurate concentration estimates for low-concentration samples.

32. This volume is for PromethION library preparation, for MinION preparations consult the manual provided on the Oxford Nanopore Technologies website. The nanogram quantity required to reach 100 fmol of cDNA depends on the average fragment length. One may generally aim for ~54.68 ng, corresponding to 0.9-kb average fragment lengths. Alternatively, verify the average fragment length on a Bioanalyzer. If the yield is insufficient to reach this quantity, redoing the second-strand synthesis PCR (Subheading 3.2.5) with an increased number of cycles is advisable.
33. After starting this step, continue the protocol immediately to library loading and the start of the sequencing run. The RAP adapter contains an enzyme that is sensitive to freezing or 4 °C storage in the elution buffer conditions.
34. Check the gold pins for condensation and wipe off with a Kimwipe if present. Also make sure the heat pad (black pad) is present on the bottom side of the flow cell.
35. Insert the tip into the inlet port and turn the wheel to insert the volume in the flow cell in a much more controlled way. Make sure not to place the entire volume in (leave a few μL in your tip), to avoid introducing air bubbles.
36. Do not wait too long to load this library on the flow cell, the enzyme of your RAP adapter starts using sequencing fuel as soon as it is combined with SQB.
37. Wait for 10 min after loading the flow cell before starting the sequencing run. This increases the sequencing output.
38. A nonautomated version of the workflow is also available on https://github.com/LoGT-KULeuven/ONT-cappable-seq [27]. There, a step-by-step workflow is presented where each operation is explained in more detail. However, this workflow requires manual curation of the output files.
39. In case your phage has terminal repeats, manually remove one of the repeats from the FASTA file to avoid ambiguous mapping.
40. The sequencing yields, quality, and read lengths of the raw nanopore sequencing data can be evaluated using NanoPlot (individual sample) or NanoComp (compare samples) [28], or a similar tool.
41. We recommend to try out different settings to optimize TSS and TTS calling for your organism of interest. In general, adjustment of peak clustering widths can strongly influence the number of 5′ and 3′ peaks. Afterwards, for TSS identification, increasing the minimum coverage (enriched and control) and enrichment ratio values (>1.5) allows for a more stringent TSS annotation. Similarly, increasing the coverage drop

threshold (>0.25) will only return the most certain transcription terminators.

42. The tool can only be carried out for one reference sequence at the time, exemplified here for a phage. If there is interest in the transcriptional features of another sequence included in your sample (phage genome, host genome, plasmid, etc.), the config file can readily be adapted. Instead of the name of the phage, provide the appropriate sampleID and adjust the fasta file directory to point to the reference sequence of interest. Likewise, the procedure can be carried out on the in vitro RNA control spike-in to evaluate successful enrichment and read lengths.
43. It can be useful to evaluate and compare the processed FASTQ files in terms of yields, mean read length, and quality. These reads will ultimately be used for mapping. Recommend tools are provided by NanoPack (NanoPlot, NanoComp) [28].
44. Use samtools flagstat to assess the number of reads that were mapped to your reference sequence. In addition, the NanoPack tools also accept input data in BAM format and offers insights in the percent identity.
45. If annotation data of your host bacterium is available, a read summary tool such as FeatureCounts [29] can be used to evaluate and compare the fraction of reads that are assigned to coding sequences and processed RNA species (rRNA and tRNA) in the enriched and the control sample. In case primary transcript enrichment was successful, a significant reduction in the number of rRNA and tRNA species should be observed.

Acknowledgements

This work was supported by the European Research Council (ERC) under the European Union's Horizon 2020 research and innovation programme [819800].

References

1. Lammens E-M, Putzeys L, Boon M, Lavigne R (2023) Sourcing phage-encoded terminators using ONT-cappable-seq for SynBio applications in pseudomonas. ACS Synth Biol 12: 1415–1423
2. Mahler M, Costa AR, van Beljouw SPB, Fineran PC, Brouns SJJ (2023) Approaches for bacteriophage genome engineering. Trends Biotechnol 41:669–685
3. Blasdel BG, Chevallereau A, Monot M, Lavigne R, Debarbieux L (2017) Comparative transcriptomics analyses reveal the conservation of an ancestral infectious strategy in two bacteriophage genera. ISME J 11: 1988–1996
4. Adéla F, Ivana M, Reuter BGB, Doškař J, Götz F, Pantůček R (2022) Global transcriptomic analysis of bacteriophage-host interactions between a kayvirus therapeutic phage and Staphylococcus aureus. Microbiol Spectr 10:e00123–e00122
5. Brandão A, Pires DP, Coppens L, Voet M, Lavigne R, Azeredo J (2021) Differential transcription profiling of the phage LUZ19

infection process in different growth media. RNA Biol 18:1778–1790
6. Lood C, Danis-Wlodarczyk K, Blasdel BG, Jang HB, Vandenheuvel D, Briers Y et al (2020) Integrative omics analysis of Pseudomonas aeruginosa virus PA5oct highlights the molecular complexity of jumbo phages. Environ Microbiol 22:2165–2181
7. Wicke L, Ponath F, Coppens L, Gerovac M, Lavigne R, Vogel J (2021) Introducing differential RNA-seq mapping to track the early infection phase for Pseudomonas phage πKZ. RNA Biol 18:1099–1110
8. Yan B, Boitano M, Clark TA, Ettwiller L (2018) SMRT-cappable-seq reveals complex operon variants in bacteria. Nat Commun 9: 3676
9. Fremin BJ, Bhatt AS, Kyrpides NC, Sengupta A, Sczyrba A, Maria da Silva A et al (2022) Thousands of small, novel genes predicted in global phage genomes. Cell Rep 39: 110984
10. Putzeys L, Boon M, Lammens E-M, Kuznedelov K, Severinov K, Lavigne R (2022) Development of ONT-cappable-seq to unravel the transcriptional landscape of Pseudomonas phages. Comput Struct Biotechnol J 20:2624–2638
11. Thorvaldsdóttir H, Robinson JT, Mesirov JP (2013) Integrative Genomics Viewer (IGV): high-performance genomics data visualization and exploration. Brief Bioinform 14:178–192
12. Mölder F, Jablonski KP, Letcher B, Hall MB, Tomkins-Tinch CH, Sochat V et al (2021) Sustainable data analysis with Snakemake. F1000Res 10
13. Pychopper: cDNA read preprocessing. https://github.com/epi2me-labs/pychopper. Accessed 12 June 2023
14. Martin M (2011) Cutadapt removes adapter sequences from high-throughput sequencing reads. EMBnet.journal 17:10–12
15. Li H (2018) Minimap2: Pairwise alignment for nucleotide sequences. Bioinformatics 34: 3094–3100
16. Li H, Handsaker B, Wysoker A, Fennell T, Ruan J, Homer N et al (2009) The sequence alignment/map format and SAMtools. Bioinformatics 25:2078–2079
17. Samclip: filter SAM file for soft and hard clipped alignments. https://github.com/tseemann/samclip. Accessed 12 June 2023
18. Quinlan AR, Hall IM (2010) BEDTools: a flexible suite of utilities for comparing genomic features. Bioinformatics 26:841–842
19. Dar D, Shamir M, Mellin JR, Koutero M, Stern-ginossar N, Cossart P et al (2016) Term-seq reveals abundant ribo-regulation of antibiotics resistance in bacteria. Science 352
20. Termseq-peaks: peak-calling for term-seq data in bacteria. https://github.com/nichd-bspc/termseq-peaks. Accessed 12 June 2023
21. dplyr: a grammar of data manipulation. https://github.com/tidyverse/dplyr. Accessed 12 June 2023
22. Blasdel B, Ceyssens P-J, Lavigne R (2018) Preparing cDNA libraries from lytic phage-infected cells for whole transcriptome analysis by RNA-seq. Methods Mol Biol 1681:185–194
23. Oxford Nanopore Technologies (2019) PCR-cDNA-barcoding_SQK-PCB109 protocol. Accessed 12 June 2023
24. ONT-cappable-seq2. https://github.com/LoGT-KULeuven/ONT-cappable-seq2. Accessed 12 June 2023
25. NEB DNA Sequences and Maps Tool. https://international.neb.com/tools-and-resources/interactive-tools/dna-sequences-and-maps-tool. Accessed 12 June 2023
26. Putzeys L, Poppeliers J, Boon M, Lood C, Vallino M, Lavigne R (2023) Transcriptomics-driven characterization of LUZ100, a T7-like Pseudomonas phage with temperate features. mSystems 8:e0118922
27. ONT-cappable-seq. https://github.com/LoGT-KULeuven/ONT-cappable-seq. Accessed 12 June 2023
28. De CW, D'Hert S, Schultz DT, Cruts M, Van Broeckhoven C (2018) NanoPack: visualizing and processing long-read sequencing data. Bioinformatics 34:2666–2669
29. Liao Y, Smyth GK, Shi W (2014) FeatureCounts: an efficient general purpose program for assigning sequence reads to genomic features. Bioinformatics 30:923–930

Chapter 15

Multidisciplinary Methods for Screening Toxic Proteins from Phages and Their Potential Molecular Targets

Xing Wan and Mikael Skurnik

Abstract

This chapter presents a comprehensive methodology for the identification, characterization, and functional analyses of potentially toxic hypothetical proteins of unknown function (toxHPUFs) in phages. The methods begin with in vivo toxicity verification of toxHPUFs in bacterial hosts, utilizing conventional drop tests and following growth curves. Computational methods for structural and functional predictions of toxHPUFs are outlined, incorporating the use of tools such as Phyre2, HHpred, and AlphaFold2. To ascertain potential targets, a comparative genomic approach is described using bioinformatics toolkits for sequence alignment and functional annotation. Moreover, steps are provided to predict protein–protein interactions and visualizing these using PyMOL. The culmination of these methods equips researchers with an effective pipeline to identify and analyze toxHPUFs and their potential targets, laying the groundwork for future experimental confirmations.

Key words Bacteriophages, Hypothetical proteins of unknown function (HPUFs), Antimicrobial resistance, Next-generation sequencing, Functional prediction

1 Introduction

Antimicrobial resistance is one of the most critical challenges confronting contemporary medicine. Antibacterial development, often carried out by small- and medium-sized enterprises, typically employs traditional R&D strategies based on chemical modifications of existing molecules [1]. The development and introduction of new antibiotic classes have substantially decelerated over the years, with only two new classes introduced to the market since 1962 [2]. This slowdown has caused a significant concern, as the capacity for analogue development from existing antibiotic classes is reaching saturation. The urgency for novel antibiotic classes is escalating, particularly to combat pathogens classified under the ESKAPE acronym—*Enterococcus faecium*, *Staphylococcus aureus*, *Klebsiella pneumoniae*, *Acinetobacter baumannii*, *Pseudomonas aeruginosa*, and *Enterobacter* species [3].

Huan Peng et al. (eds.), *Phage Engineering and Analysis: Methods and Protocols*,
Methods in Molecular Biology, vol. 2793, https://doi.org/10.1007/978-1-0716-3798-2_15,

An emerging avenue to address this problem is through the study of bacteriophages, particularly lytic phages. These bacterial viruses have deeply intertwined evolutionary history with bacteria, which have equipped them to develop highly specific mechanisms to hijack bacterial cellular metabolism for their own propagation [4]. Their unique ability to reprogram bacterial machinery has opened up new possibilities for drug discovery. Phage endolysins, holins, polysaccharide depolymerases, and virion-associated peptidoglycan hydrolases, which phages naturally use to infect or lyse bacteria, have already been extensively studied for their potential use as antibiotics [5]. Given the fact that a large proportion of phage gene products have entirely unknown structures and functions, screening these hypothetical proteins of unknown function (HPUFs) from bacteriophages for toxic activity against bacteria can potentially unlock new strategies to counter bacterial infections [6, 7].

In our earlier studies, we utilized a conventional plating assay to screen 94 HPUFs of *Yersinia* phage φR1-RT, which led to the discovery of four HPUFs that exhibited toxicity against *E. coli* [8]. Following this, a systematic examination of the HPUFs within the genome of phage fHe-Kpn01, a bacteriophage known to infect extended-spectrum beta-lactamase (ESBL)-producing *K. pneumoniae*, revealed three toxic antibacterial HPUFs (toxHPUFs) demonstrating cross-species activity towards *E. coli* [9]. Later, we developed a high-throughput screening method exploring next-generation sequencing (NGS) for detecting phage-encoded toxic proteins using fHy-Eco03 phage as an example [10]. This approach significantly shortens the screening process compared with the plating-based method. After latest refinements, the current streamlined NGS-based screening approach not only provides results comparable to conventional plating-based toxicity screening methods, but also demonstrates superior accuracy, efficiency, and reliability [11].

This book chapter describes an innovative phage genomics-based screening approach for toxHPUFs aiming to exploit these unique phage–bacteria interactions to identify new bacterial targets for drug discovery. Here, we present detailed workflows from a plating-based assay suitable for screening a small number of HPUFs to a high-throughput NGS-based approach optimized for analyzing a phage genome for toxHPUFs. Subsequent comparative genomics and protein–protein docking analysis will then offer insights into the molecular targets of these identified toxHPUFs.

By embracing the potency of phages and integrating modern genomic tools, we aim to introduce a robust pipeline to accelerate the discovery of novel antibiotic targets, shedding light on potentially transformative ways to confront the escalating crisis of antimicrobial resistance.

2 Materials

All equipment and media used in bacterial work should be sterile. Follow good laboratory practices to ensure septic working procedures. Obtain general plastic and glass labware from your own supply. Use ultrapure water for phage and DNA works.

2.1 Phage Purification

1. Amicon® Ultra-4 centrifugal filter units with molecular weight cutoff (MWCO) 100 kDa.
2. SM buffer:100 mM NaCl, 50 mM Tris-HCl pH 7.5, 10 mM Mg_2SO_4, 0.01% (w/v) gelatin.
3. Äkta Purifier.
4. Monolithic column: CIMmultus® QA 1 mL monolithic column (6 μm).
5. Injection and washing buffer: 350 mM NaCl.
6. Eluted buffer: 450 mM NaCl.
7. Discontinuous glycerol.
8. Ultracentrifuge.
9. SW55Ti swing-out rotor.
10. Sucrose.
11. TM buffer: 50 mM Tris pH 7.5, 10 mM Mg_2SO_4.

2.2 Phage Genomic DNA Isolation and Sequencing

1. Invisorb Spin Virus DNA Mini Kit (Stratec Biomedical, Birkenfeld, Germany).
2. Nextera sample prep kit (Illumina, San Diego, CA, USA).
3. Illumina MiSeq sequencer (Illumina) (optional).
4. Access to GenBank database (*see* **Note 1**).

2.3 Phage Particle Proteomes by LC-MS/MS (Optional)

1. Ultracentrifuge.
2. Tris(2-carboxyethyl) phosphine (TCEP).
3. Iodoacetamide.
4. Trypsin.
5. nLC1000 high-pressure liquid chromatography (HPLC) coupled with tandem Orbitrap Elite Electron-Transfer Dissociation (ETD) mass spectrometer.
6. C18 reversed-phase chromatography column.
7. Software: Xcalibur version 2.7.1, Proteome Discoverer 1.4, SEQUEST, usually incorporated to the HPLC/MS devices.

2.4 Preparation of toxHPUFs for Toxicity Screening

1. PCR device: T100™ or iCycler Thermal Cycler.
2. dNTP mix (10 mM each).
3. 5× Phusion HF buffer.
4. Phusion DNA Polymerase.
5. Cloning plasmid vectors under basal expression promoter or strains carrying those vectors (*see* **Note 2**).
6. T4 DNA ligase.
7. NucleoSpin Gel and PCR Clean-up XS kit (Machery-Nagel, North Rhine-Westphalia, Germany).
8. Electrocompetent *E. coli* cells (*see* **Note 3**). It is also possible to prepare electrocompetent cells in house, essentially following the protocol as described in [12].
9. SOC medium: 2% (w/v) tryptone, 0.5% (w/v) yeast extract, 10 mM NaCl, 2.5 mM KCl, 10 mM $MgCl_2$, 10 mM MgSO4, 20 mM glucose.
10. 2 mm electroporation cuvettes.
11. Gene Pulser II coupled with a Pulse Controller Plus.

2.5 Plating-Based Screening

1. Lysogeny broth (LB): 10 g/L Bacto™ tryptone, 5 g/L Bacto™ Yeast Extract, 10 g/L NaCl.
2. LA: LB supplemented with 1.5% Bacto agar.
3. Colony counter.

2.6 NGS-Based Screening

1. NucleoSpin Gel and PCR Clean-up XS kit (Machery-Nagel).
2. SOC broth (*see* Subheading 2.4).
3. NucleoBond™ Xtra Midi kit and NucleoBond™ Finalizers (Machery-Nagel).
4. Elution buffer: Tris–HCl pH 8.5.
5. Illumina HiSeq sequencer (Illumina) (optional).

2.7 Bioinformatic Analysis for NGS-Based Toxicity Screening

1. Puhti: high-performance computing system (*see* **Note 4**).
2. WinSCP 5.19.2 (Windows Secure Copy).
3. Microsoft Office.
4. Notepad.

2.8 Confirmation of Protoin Toxioity

1. Suitable inducible expression vector (*see* **Note 5**).
2. Suitable restriction enzymes based on the vector in use.
3. Access to a Sanger sequencing core facility.
4. Appropriate antibiotic stock solutions based on the vector in use.
5. Glucose.

6. Arabinose.
7. M9 minimal media: 3 g/L KH_2PO_4, 0.5 g/L NaCl, 6.78 g/L Na_2HPO_4, 1 g/L NH_4Cl, casamino acid 0.2% (v/v), $MgSO_4$ 2 mM, 0.1 mM $CaCl_2$, 1 mg/L thiamine.
8. Tryptic soy broth (TSB).
9. Tryptic soy agar (TSA).
10. Bioscreen Honeycomb 2.
11. Bioscreen C MBR.
12. Sterile phosphate-buffered saline (PBS): 8 g of NaCl, 0.2 g of KCl, 1.44 g of Na_2HPO_4, 0.24 g of KH_2PO_4 per liter of distilled water. Adjust the pH to 7.4 with HCl.
13. Anhydrotetracycline (ATc) stock solution.

2.9 Structural and Functional Analyses of toxHPUFs

For bioinformatic analyses in Subheadings 2.9, 2.10, and 2.11, ensure the accesses to the following software.

1. AlphaFold2 is an online tool for protein structure prediction (*see* **Note 6**).
2. Phyre2 is a protein fold recognition server for protein structure prediction (*see* **Note 7**).
3. HHpred provides protein function prediction online (*see* **Note 8**).
4. PyMOL is a molecular visualization system for analyzing predicted protein structures (*see* **Note 9**).

2.10 Comparative Genomics and Functional Annotation

1. Puhti (*see* **Note 4**).
2. eggNOG-mapper web server: a tool for functional annotation of proteins at genome scale (*see* **Note 10**).

2.11 Protein–Protein Interaction Analysis

1. AlphaFold2 (*see* **Note 6**).
2. HDock, a webserver for protein–protein and protein–DNA/RNA docking (*see* **Note 11**).
3. National Centre for Biotechnology Information (NCBI) Genome database (*see* **Note 12**).
4. PyMOL (*see* **Note 9**).

Buffer and media recipes provided here are for commonly used recipes and suitable for the conditions mentioned through the text. Depending on the bacterial strains, plasmid, and phages used, appropriate modifications and optimizations may need to be done. Always verify and ensure compatibility with your working materials.

3 Methods

3.1 Phage Purification

This chapter provides a stepwise approach for the purification of bacteriophages in general. The method described is flexible and can be adapted as required for a specific experimental setup.

1. Begin with the lysate of the phage to be purified. A high concentration of plaque-forming units (PFU/mL) is ideal for achieving optimal results.
2. Proceed with ultrafiltration of the lysate using a centrifugal filter unit suitable for your experimental setup (e.g., Amicon Ultra-4, MWCO 100 kDa). Continue this process until the initial volume is reduced to one-quarter.
3. Following the ultrafiltration, add three volumes of an appropriate buffer (SM buffer, or a similar buffer according to your specific phage) and repeat the ultrafiltration process.
4. Subsequently, the ultrafiltrated phage sample can be purified using ion-exchange chromatography (IEX) or via discontinuous glycerol density gradient ultracentrifugation.
   (a) For IEX, use an appropriate purifier (such as Äkta Purifier) and a suitable monolithic column. Inject and wash the sample into the column using 350 mM NaCl concentration and elute with 450 mM NaCl.
   (b) For discontinuous glycerol density gradient ultracentrifugation, conduct the ultracentrifugation at 40,000 rpm at 4 °C for an extended period (e.g., 4 h).
5. After the purification, phage-containing fractions should be pooled. Use a centrifugal filter unit to concentrate the product and to change the buffer to a suitable storage buffer, such as SM or TM buffer.
6. It is also recommended to add a cryoprotectant to the storage buffer (e.g., 8% sucrose).
7. Purified phage particles, as well as intermediate purification products, should be stored at 4 °C.

3.2 Phage Genomic DNA Isolation and Sequencing

Phage genomic DNA can be isolated from high-titer phage preparations either manually as described earlier [13] or using a suitable DNA isolation kit.

1. Extract phage DNA from high-titer phage lysate utilizing a kit such as the Invisorb Spin Virus DNA Mini Kit or an equivalent product, following the manufacturer's instructions.
2. Upon successful DNA isolation, proceed with the construction of the DNA library for next-generation sequencing. A kit like the Nextera sample prep kit can be used for this purpose.

3. Perform paired-end sequencing using a suitable sequencing platform, such as the Illumina MiSeq sequencer with a short read length (commonly 150 PE150 or 300 nucleotides PE300).
4. Use an integrated pipeline for de novo assembly of microbial genomes to obtain the genome sequence. The A5 (Andrew And Aaron's Awesome Assembly)-miseq pipeline is one example [14].
5. Finally, submit the resulting sequence to GenBank (*see* **Note 1**) for accessibility and future reference. Ensure that an accession number is received for your submitted sequence.

3.3 Phage Particle Proteomes by LC-MS/MS

Genome annotation of a newly sequenced phage can already identify many HPUFs. However, depending on the searching algorism and the database, some open reading frames (ORF) annotated as HPUFs could in reality be associated to the phage particles and have no effect on bacterial reprograming. Therefore, identification of phage particle associated proteins is needed.

1. Start with high-titer phage samples for the analysis. Concentrate the phages by centrifugation at 4 °C and at 16,000 × *g* for an appropriate duration (e.g., 2 h) depending on the phage of handling.
2. Prior to protein digestion, reduce the proteins in the samples with tris(2-carboxyethyl) phosphine (TCEP), followed by alkylation with iodoacetamide.
3. Digest the proteins into peptides using a suitable protease. Trypsin is a very common enzyme for this purpose.
4. Purify the resultant tryptic peptide digests using C18 reversed-phase chromatography column.
5. Perform the mass spectrometry (MS) analysis on an Orbitrap Elite Electron-Transfer Dissociation (ETD) mass spectrometer using Xcalibur version 2.7.1 (*see* **Note 13**), coupled to a nano-flow HPLC system (Thermo Scientific nLC1000).
6. Extract the peaks and identify the proteins using suitable software (e.g., Proteome Discoverer 1.4). Calibrated peak files should be searched against the phage and host bacteria protein sequences by a reliable search engine (e.g., SEQUEST).
7. Set error tolerances for precursor and fragment ions accordingly (e.g., ±15 ppm and ± 0.6 Da).
8. Apply a stringent cutoff for peptide identification to minimize false discoveries (e.g., 0.5% false discovery rate).
9. Set exclusion criteria for the proteins identified from LC-MS/MS analysis (for instance, a protein possessing more than 2 unique peptides and/or providing more than 5% coverage).

Those proteins meeting the criteria are deemed potential phage particle-associated proteins and excluded for further screening (*see* **Note 14**).

Ensure all LC-MS/MS analyses are performed in a suitably equipped facility, either in-house or at an outsourced proteomics unit.

3.4 Preparation of toxHPUFs for Toxicity Screening

1. Design primers for amplifying the true HPUF genes as per established guidelines [15].
2. Perform PCRs in either 0.2-mL thin-walled tubes or in a 96-well microtiter plate format. For the former, use 50 μL volumes containing 10 ng of DNA template; for the latter, use 30 μL volumes with 5 ng of DNA template.
3. For each reaction, add 0.5 μM of primers, 0.2 mM of dNTP mix, along with the appropriate volume of 5× Phusion Buffer and 0.02 U/μL Phusion DNA Polymerase.
4. Run the PCRs in a suitable thermal cycler, such as in a T100™ or iCycler Thermal Cycler, following the standard manufacturer protocol for the polymerases.
5. Restrict the amplified HPUF DNA fragments with appropriate restriction enzymes.
6. Select a suitable vector (*see* **Note 2**). Restrict the vector using suitable restriction sites same as individual HPUF fragment. Ensure a 3:1 molar ratio of the insert over the linearized vector for ligation with T4 DNA ligase (0.1 U/mL).
7. Purify the ligation mixtures using a suitable DNA fragment purification kit, such as NucleoSpin Gel and PCR Clean-Up XS kit.
8. Adjust the concentration of purified ligation mixtures to 10 ng/mL.
9. Transfer the ligation mixture to *E. coli* DH10B (*see* **Note 3**) using electroporation.
10. Use electrocompetent *E. coli* cells with a high transformation efficiency, preferably of 10^9 CFU per μg of intact vector (*see* **Note 15**). Preparation of the electroporation mixtures and thawing of electrocompetent cells should be performed on ice.
11. Use pre-chilled 2-mm electroporation cuvettes for the electroporation mixture. Follow the standard electroporation protocol as essentially described previously [16]. An example of parameters using Gene Pulser II coupled with a Pulse Controller Plus is shown as the following: 200-Ω resistance, 25-mF capacitance, and 2.5-kV voltage. This setting normally results in a time constant between 4.5 and 5.0 ms.

12. Immediately after the pulse, transfer the cells to 950 μL SOC medium and incubate at 37 °C for 1 h with shaking vigorously.
13. Spread every 50 μL of the recovered cells onto LB plates supplemented with suitable antibiotic selection (e.g., 100 μg/mL ampicillin). Incubate the obtained 20 plates at 37 °C overnight. Ensure single colonies are formed on the plates, and adjust the cell volume for plating if necessary.

3.5 Plating-Based Screening

For small-scale screening, it would be good to introduce known toxic and nontoxic controls in the preparation. For instance, RegB, a restriction endoribonuclease from phage T4 [17], has been shown to function as the toxic control [8, 9]. A nontoxic control can be the structural protein of your phage, like the phage capsid vertex protein Gp178 from phage fR1-RT [18].

1. Follow the same procedure as described previously in Subheading 3.4 for the control genes.
2. Count the colony-forming unit (CFU) from the obtained plates (*see* Subheading 3.4, **step 12**) using a colony counter.
3. Normalize the CFU counts, if necessary (*see* **Note 16**).

Biological replicates of each HPUF gene for electroporation can be used to achieve higher reliability of the experiments.

3.6 NGS-Based Screening

For large-scale screening, introduction of control genes is not necessary.

1. Initiate the NGS screening process by following Subheading 3.4 until **step 6**.
2. Following digestion and gene–vector ligation, pool ligation mixtures of HPUF gene and vector (*see* **Note 17**).
3. Concentrate each pool with a kit, such as NucleoSpin Gel and PCR Clean-Up XS kit. Use a small volume of sterile ultrapure water for each pool during the elution process to achieve high final DNA concentrate.
4. Adjust the DNA concentration to 200 ng/mL per pool.
5. Transfer 1 μL of each ligation pool into *E. coli* cells via electroporation, as described in Subheading 3.4, **step 9** onwards.
6. Collect all the transformation colonies of each pool from the plates by first adding 1 mL SOC. Resuspend the cells using, e.g., a cell scraper or a cell spreader.
7. Inoculate the harvested cell suspension into 100 mL SOC broth containing appropriate antibiotic for plasmid selection. Ensure the selection strength is sufficient (*see* **Note 18**).
8. Grow the cells in a 500-mL Erlenmeyer flask at 37 °C with 220 rpm for 3 h.

9. Isolate the plasmid mixture of each pool from the 3-h cell cultures with a plasmid purification kit, for example, NucleoBond™ Xtra Midi kit and NucleoBond™ Finalizers according to manufacturer's instruction. Elute the plasmid pools in 200 μL Tris–HCl pH 8.5 elution buffer.
10. Sequence the DNA samples from both the ligation pool and the plasmid pool using a 150-bp paired-end protocol on Illumina HiSeq, either in-house or at a sequencing facility (*see* **Note 19**).

3.7 Bioinformatic Analysis for NGS-Based Toxicity Screening

All the bioinformatics analyses are recommended to perform on a local or remote supercomputer for a high-performance computing. The following procedure (**steps 3–6**) is carried out on, for example, the Puhti supercomputer at the Finnish Centre for Scientific Computing (CSC), which provides suitable biotool packages for the sequence analysis. Data transfer between local personal computer and the remote supercomputer can be done using WinSCP 5.19.2. For bioinformatic procedures, remember to replace the files marked in <> with your own suitable files.

1. Design the 4 joint sequences for each HPUF–vector ligation. Ensure each ligation–joint sequence contains approximately 25–30 nucleotides covering the restriction site flanked by partial sequences from the vector and partial from the gene (Fig. 1).
2. Save the joint sequences each on its own line under the name in a text file <your_list.txt>. Upload the file to your working directory on Puhti containing the NGS raw data files.
3. Activate relative biotool packages $ module load biokit and $ module load velvet.
4. Uncompress the compressed fastq.gz NGS sequence read files $ gunzip <file_name.gz>
5. Combine the paired-end fastaq files into a single file. $ shuffleSequences_fastq.pl <read_file_1.fq> <read_file_2.fq> <file_-name.fastq>
6. Run the alignments as a batch job using the following script named <file_name.sh> with sbatch command $ sbatch <file_-name.sh>. Remember to change the required file names and paths to your own (*see* **Note 20**).

```
#!/bin/bash -l
#SBATCH -o std1.out
#SBATCH -e std1.err
#SBATCH -p small
#SBATCH --account=<your puhti account>
#SBATCH --ntasks=1
```

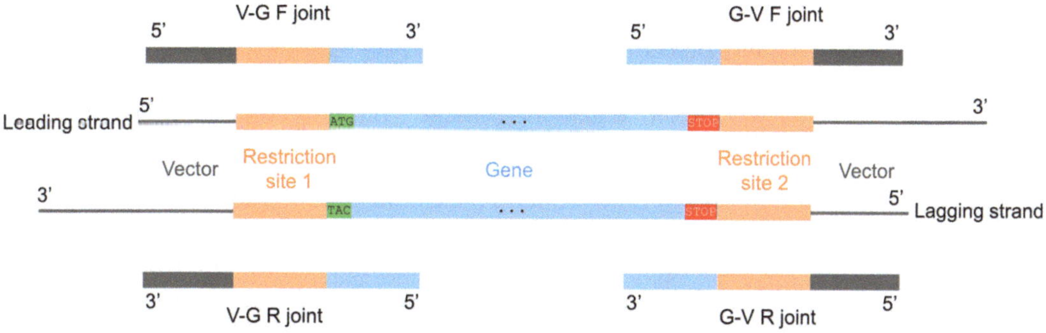

Fig. 1 Illustration of the four ligation-joint sequences used in the determination of sequence read coverage for each of the screened HPUFs (*V* vector, *G* gene fragment, *F* forward, *R* reverse). (Adapted from Kasurinen et al. [10])

```
#SBATCH --cpus-per-task=1
#SBATCH --nodes=1
#SBATCH -t 48:00:00
#SBATCH --mem=128000
module load biokit
##change directory to the one where you have the data
cd /<file_path>
##use the file name for your sequence segment text file
a=1
for pat in $(cat <your_list.txt>)
do
##use the fastq filename of your combined paired end reads
fuzznuc -pattern "$pat" <file_name.fastq> -rformat excel -fil-
ter | awk '{ if ( $1 != "SeqName") print $1}' | sort | uniq >
name_$a (( a = a + 1))
done
```

7. Calculate the total number of the four ligation–joint sequences for each HPUF-encoding gene, representing their total read coverage (N joint reads; *see* Formula 1).
8. Determine the relative number of joint-sequence reads for all genes in the pools by dividing the total read coverage of a single gene by the total number of joint-sequence reads for all genes in the pool and express this as a percentage (relative gene percentage, Formula 1).
9. Calculate the ratio between the relative joint-sequence reads of individual genes from plasmid pools and those from the corresponding ligation mixture as an indication of the presence of a toxHPUF gene (Formula 2).

$$\text{Relative gene percentage} = \frac{N \text{ joint reads of single gene}}{N \text{ joint reads of all genes in pool}} \times 100\% \quad (1)$$

$$\text{Ratio} = \frac{\text{Relative gene percentage from pooled plasmid DNA}}{\text{Relative gene percentage from ligation mixtures}} \quad (2)$$

A low ratio is due to the elimination of transformants carrying a toxic gene [10]. A ratio near or above 1 indicates a nontoxic gene, reflecting the successful replication of the recombinant plasmid. A HPUF could be regarded as bactericidal if this ratio was less than 1. For a very large screening pool, an even lower ratio can be applied to determine the toxicity, such as 0.5.

3.8 Confirmation of Protein Toxicity

Since the *lac* promoter in pUC19-based plasmid is notoriously known as leaky and the initial toxicity screening depends on the leaky expression, to confirm the toxicity of the HPUF subset, an all-or-none inducible promoter should be used.

1. Choose potentially toxic HPUFs obtained from the initial screening assay and clone these into an appropriate expression vector, such as pBAD30 or pBAD33 where the gene expression is under the control of the P_{BAD} promoter, or pRAB11 plasmid under *tet* promoter. Remember to select suitable restriction sites to avoid any internal presence of same sites in the genes.
2. Once cloning is done, purify the plasmids and confirm the presence of correct inserts by PCR and by Sanger sequencing at a trusted institute such as Finnish Institute for Molecular Medicine (FIMM). This step provides additional confirmation of the successful cloning and enables the following expression of the potentially toxic HPUFs.
3. Electroporate the plasmids into electrocompetent *E. coli*, such as strain DH5α (*see* Subheading 3.4, **steps 9–12**).
4. Plate the recovered cells on LB agar plates with suitable antibiotic selection using 10 μL, 100 μL, and the remaining of cells collected through centrifugation. Incubate the plates at 37 °C overnight to obtain single colonies.
5. For each gene, inoculate three single colonies into 1 mL LB medium supplemented with an appropriate antibiotic, such as ampicillin (100 μg/mL) or chloramphenicol (20 μg/mL). Ensure to add 2% (w/v) glucose to suppress any possible basal expression, when using pBAD plasmids.
6. Incubate the cultures overnight at 37 °C with shaking (200 rpm). Following incubation, collect the bacteria by centrifuging and replace the medium with the same volume of minimal medium M9 or TSB.
7. Prepare two portions of M9 medium with the appropriate antibiotic and add either 0.2% (w/v) glucose for repression or

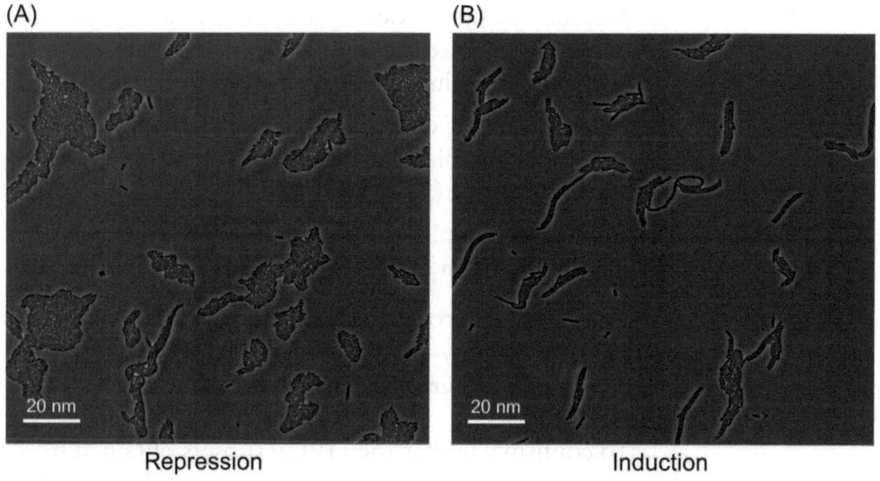

Fig. 2 Time-lapse microscopy of *Yersinia enterocolitica* YeO3-R1 carrying *g232* from fR1-RT. Gene expression under PBAD promoter in pHERD20T-based plasmid. (**a**) Gene under repression condition with 2% glucose. (**b**) Gene under induction condition with 0.2% arabinose. Images obtained 4 h after repression/induction

0.2% (w/v) arabinose for induction. When the pRAB11 vector is used, prepare TSB media and add 0.4 μM ATc for induction. Inoculate the media with 1% (v/v) inoculum of washed bacterial cells.

8. Transfer the bacterial dilutions in triplicate (300 μL/well) to Bioscreen Honeycomb 2 plates or similar. Monitor the OD at 600 nm every hour for a set period (between 16 and 20 h is recommended) using a suitable reader, such as the Bioscreen C MBR. Ensure the plates are shaken continuously with high amplitude and normal speed and that shaking is paused briefly (5–10 s) before each measurement.

9. For each measurement, consider the average values across the triplicate readings. From this, compute the overall mean values and standard deviations for the three biological replicates.

   As controls, consider using *E. coli* strains carrying plasmids containing known toxic or nontoxic genes, or strains carrying the empty vector.

   Combinational use of several plasmids with different stringent promoters is also recommended.

   As an alternative or supplementary confirmation of HPUF toxicity, an agar plate drop test can be employed. Follow these steps:

10. Follow the abovementioned **steps 1–3** to obtain single colonies of each HPUF-expressing clone.

11. Pick three colonies from each construct and suspend in 500 μL sterile phosphate-buffered saline (PBS) pH 7.4.

12. Measure the optical density at 600-nm wavelength (OD_{600}) and dilute the culture to an OD_{600} of 0.2. Subsequently, serially dilute the inoculum to create a series from 10^{-1} to 10^{-8}.
13. Pipette 5 μL of each dilution onto both inducing plates and non-inducing plates with suitable antibiotic and repression and/or inducer (*see* **steps 4 and 6**).
14. Allow the samples to absorb for 30 min at 22 °C before incubating them at 37 °C overnight.

The drop test provides a straightforward and visual method to evaluate the toxicity of the cloned toxHPUFs towards the host bacteria, while the growth curves reveal the time-specific impacts of toxHPUFs.

To confirm whether the HPUF subsets obtained from preliminary screening (*see* Subheadings 3.5 and 3.6) exhibit toxicity towards the native host of your phage, select a suitable shuttle vector which works both in *E. coli* and the native host bacterium for your phage of study. Follow the abovementioned confirmation steps to test the toxicity in the native host. Remember literature searching on proper transformation of such shuttle vector to your bacterium is crucial to customize these steps.

Morphological impacts of toxHPUFs on the host bacteria can be examined, e.g., using time-lapse microscopy as described elsewhere [19]. The microscopical results provides clues on how this certain toxHPUF reprograms the bacterial cells (Fig. 2).

3.9 Structural and Functional Analyses of toxHPUFs

α-Helical structures are important in protein–protein interactions [20]. Structural prediction can further narrow down the number of toxHPUFs by eliminating the peptides lacking α-helical structures.

1. Begin the structural prediction of the list of toxHPUFs identified from previous wet-lab screening using AlphaFold2 [21] (*see* **Note 11**). Eliminate the ones that contain only β-sheets.
2. Use protein fold recognition server Phyre2 to find whether your toxHPUFs of interest contains functional domains matching to known proteins (*see* **Note 12**).
3. A similar search can be done on protein function prediction server HHpred [22] (*see* **Note 13**). Remember to set cutoffs of identity (e.g., 30%) and confidence levels over, e.g., 90% for Phyre2, and apply similar parameters for HHpred.
4. Compare the sequence identities of the proteins with known protein structures obtained from Phyre2 and HHpred.
5. Use the molecular visualization system PyMOL to superimpose functional and structural protein database files. The "super" function in PyMOL is particularly useful for this task (*see* **Note 9**).

6. The RMSD (root mean square deviation) scores are automatically calculated when using the "super" function to align your proteins in PyMOL. These scores measure structural alignment and overlap of the predicted protein structures. An RMSD score below 2 Å is regarded as a fairly good alignment [23].

By following these steps, one can get a clue on how your toxHPUFs of interest or at least partly could function on bacterial cells. Pay extra attention to the toxHPUFs with predicted functions related to alternating cell division, DNA replication, protein translation, and RNA transcription in the host bacteria.

3.10 Comparative Genomics and Functional Annotation

Following the abovementioned procedures, you should have now possessed a selection of toxHPUFs that demonstrate toxicity towards at least *E. coli*. If your targeted phage infects other bacterial species beyond *E. coli* and the identified toxHPUFs exhibit toxicity towards these bacteria as well, it would be beneficial to conduct comparative genomics between *E. coli* and your specific bacterial strain. This comparative analysis could aid in pinpointing common essential proteins, which could serve as potential targets for the toxHPUFs. Additionally, genome-wise functional annotation could help in providing a more detailed view on the functions of the potential targets.

Similarly, as described in Subheading 3.7, perform the genomic analysis on a supercomputer, such as Puhti.

1. Retrieve the protein fasta files of the *E. coli* screening strain (for instance, DH10B) and your bacterium from database like NCBI genome. If your bacterial genome was recently sequenced in your lab, perform genome annotation using a method described in previous work, such as [24]. Upload these files to your working directory on Puhti if necessary.
2. Activate relative biotool packages $ module load biokit.
3. Initiate all-against-all blastp using the fasta files of your strain and the *E. coli* screening strain. For high-quality results, set a small E-value like 10^{-50}: $ blastp -subject <file1.faa> -query <file2.faa> -outfmt 6 -evalue 1e-50 -out results.tsv.
4. Utilize the eggNOG-mapper web server to functionally annotate the genome of your bacterium (*see* **Note 10**).
5. Suppose your toxHPUF of interest has shown effects on cell division in microscopy, like Gp232 (Fig. 2). In that case, select the proteins that fall under the Clusters of Orthologous Groups (COGs) category "D" *Cell cycle control, cell division, chromosome partitioning* from the result file out.emapper.annotations. Ensure that you have done sufficient literature review on the necessity of the proteins in your bacterium.

6. From the blastp result file results.tsv, cross-check to see if any cell-division-related proteins from your eggNOG-mapper results display high similarity to those in *E. coli*. Concentrate on the E-values, percentage of identity, and bit score. A lower E-value signifies a better match, while a higher bit score suggests better sequence similarity. For instance, set the percentage of identity to 90%, and E-value to 0.0 to narrow down the target candidates.

Following the abovementioned in silico procedure, you can considerably reduce the number of bacterial proteins (which are potential targets of your toxHPUF) to approximately 20. These can then be further studied in protein–protein docking prior to more intensive wet-lab examinations.

3.11 Protein–Protein Interaction Analysis

There is a high likelihood that some of the potential target proteins lack existing X-ray crystal structures in the PDB, or the automated predictions in the AlphaFold database fail to offer satisfactory 3D models. If you encounter such scenarios, adhere to Subheading 3.9, **step 1**, and obtain the structures of your target proteins using AlphaFold2 using curated amino acid sequences.

1. Perform protein–protein interaction prediction between your toxHPUF and each of the target protein in HDock (*see* **Note 11**). Upload your pdb file of one target protein to the "Input Receptor Molecule" and your pdb file of toxHPUF to the "Input Ligand Molecule."
2. Alternatively, run the hetero-oligomer model in AlphaFold2 splicing your toxHPUF and target protein sequences. Use a colon sign to specify inter-protein chain breaks for modelling complexes.
3. Visualize the resulting interaction models in PyMOL using the following steps.
4. Determine the interface residues using the InterfaceResidues.py programme (*see* **Note 21**). Run the command interfaceResidue <your_heterodimer_name>, chain B, chain C.
5. If the two chains (toxHPUF and target) appear to be in close proximity, visualize the hydrogen bonding: Action → find → polar contacts → to other atoms in the object, if any are present (an example is shown in Fig. 3a).
6. Show the interface as sticks, while the rest is displayed as cartoon. Apply different color and shape schemes for each chain in order to get a clearer visualization.
7. Any clear separation between the two chains will exclude the target from further analyses (an example in Fig. 3b). Shape complementary and hydrogen bonding are crucial factors in protein–protein interface [25]. Therefore, a stable heterodimer

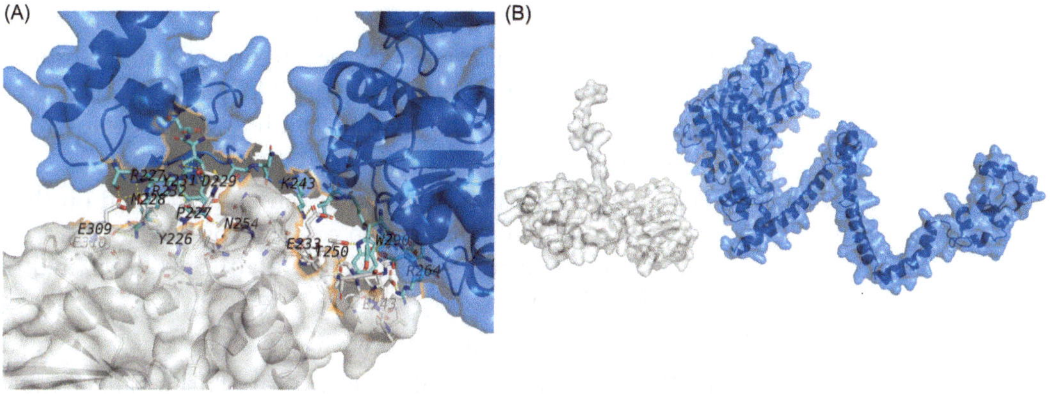

Fig. 3 PyMOL visualization of hetero-oligomers between Gp232 of *Yersinia* phage fR1-RT and potential target proteins (**a**) MreB and (**b**) MinC from *Y. enterocolitica* Ye03-R1. Gp232 in blue; the targets in grey; interface residues in sticks; residues responsible for hydrogen bonding are labelled

between toxHPUF and the potential target, secured by a reasonable number of hydrogen bonds, indicates the likelihood of such interaction occurring in reality.

Remember, while the multidisciplinary screening protocol presented in this chapter provides a general guideline, the specific conditions and steps to follow will depend on the exact requirements of the phage proteins that should be optimized accordingly.

Upon identifying a select group of potential targets and a toxHPUF of significant interest, the next step is to confirm their interactions in your lab. Future studies could employ methodologies such as bacterial two-hybrid systems [26] and pull-down assays [27], as described in other comprehensive sources. This will help validate and solidify the relationships between these elements. It is important to note that the understanding and confirmation of these interactions are crucial to furthering the development of new therapeutic strategies or diagnostic tools based on these interactions.

4 Notes

1. GenBank, the genetic sequence database at the National Institutes of Health (NIH), is accessible at https://www.ncbi.nlm.nih.gov/genbank/.
2. For preliminary screening in *E. coli*, pUC19-based plasmids like pU11L4 [8–10] or pCU1LK [11] with *lac* promoter can be used for cloning the HPUF fragments.
3. Other common lab strains of *E. coli* can also be used for the cloning, such as strain DH5α JM109, MC1061, or TOP10. Take the natural resistance of your cloning strain into consideration.

4. Puhti is accessible at https://www.csc.fi/web/guest/puhti.
5. For confirmation, it is good to use a tightly controlled expression system, such as pBAD30 or pBAD33 where the gene expression is under the control of the P_{BAD} promoter, or pRAB11 plasmid where the *tet* promoter is in control.
6. The predictions using AlphaFold2 can be performed on their interactive python notebook on colab: https://colab.research.google.com/github/sokrypton/ColabFold/blob/main/AlphaFold2.ipynb.
7. Phyre2 is accessible at http://www.sbg.bio.ic.ac.uk/~phyre2.
8. HHPred server is available at https://toolkit.tuebingen.mpg.de/tools/hhpred.
9. PyMOL is available at https://pymol.org/2/.
10. eggNOG-mapper web server is at http://eggnog-mapper.embl.de/.
11. HDock is accessible at http://hdock.phys.hust.edu.cn/.
12. NCBI Genome is accessible at https://www.ncbi.nlm.nih.gov/genome/.
13. Other high-resolution mass spectrometers can also carry out the MS analysis. Please enquire your local proteomic facility.
14. Ensure that the proteins regarded as phage particle-associated proteins are re-annotated to, e.g., "hypothetical structural proteins" in your phage genome for future reference.
15. Lower transformation efficiency (at lowest 10^8 CFU/μg vector) of electrocompetent *E. coli* cells may also be used for transferring the ligation mixture.
16. If your HPUFs of interest cannot be processed in one experiment run, normalize the CFUs of the HPUFs and toxic control gene to the CFU of the nontoxic control gene in an individual experiment and expressed as relative CFUs.
17. A pool can contain up to 20 HPUF genes. The exact pool size depends on your screening volume. For example, in the screening of 96 HPUFs, a pool of 16 HPUFs can provide an evenly distributed gene of interest in each pool.
18. For pUC19-based plasmid vector, 100 μg/mL of ampicillin is usually enough to maintain the plasmid. It is recommended to test antibiotic strength before using a new vector, for example, by drop test on plates with various antibiotic concentrations.
19. Alternatively, 300-bp pair-end sequencing can also be applied.
20. The bash file can be edited, e.g., using the nano command, $ nano file_name.sh.
21. Save the InterfaceResidue.py script from here: https://pymolwiki.org/index.php/InterfaceResidues.

Acknowledgement

The authors would like to thank professors Abram Aertsen and Rob Lavigne from KU Leuven, Belgium, for the time-lapse microscopy as shown in Fig. 2. This research was funded by the Academy of Finland (grant number 288701) and by Jane and Aatos Erkko Foundation to M.S. (Decision 2016). X.W. received a personal grant from the Jane and Aatos Erkko Foundation (grant number 200050).

References

1. Theuretzbacher U, Outterson K, Engel A, Karlen A (2020) The global preclinical antibacterial pipeline. Nat Rev Microbiol 18(5): 275–285. https://doi.org/10.1038/s41579-019-0288-0
2. Coates AR, Halls G, Hu Y (2011) Novel classes of antibiotics or more of the same? Br J Pharmacol 163(1):184–194. https://doi.org/10.1111/j.1476-5381.2011.01250.x
3. Mulani MS, Kamble EE, Kumkar SN, Tawre MS, Pardesi KR (2019) Emerging strategies to combat ESKAPE pathogens in the era of antimicrobial resistance: a review. Front Microbiol 10:539. https://doi.org/10.3389/fmicb.2019.00539
4. Wan X, Hendrix H, Skurnik M, Lavigne R (2021) Phage-based target discovery and its exploitation towards novel antibacterial molecules. Curr Opin Biotechnol 68:1–7. https://doi.org/10.1016/j.copbio.2020.08.015
5. Roach DR, Donovan DM (2015) Antimicrobial bacteriophage-derived proteins and therapeutic applications. Bacteriophage 5(3): e1062590. https://doi.org/10.1080/21597081.2015.1062590
6. Liu J, Dehbi M, Moeck G, Arhin F, Bauda P, Bergeron D et al (2004) Antimicrobial drug discovery through bacteriophage genomics. Nat Biotechnol 22(2):185–191. https://doi.org/10.1038/nbt932
7. Singh S, Godavarthi S, Kumar A, Sen R (2019) A mycobacteriophage genomics approach to identify novel mycobacteriophage proteins with mycobactericidal properties. Microbiology (Reading) 165(7):722–736. https://doi.org/10.1099/mic.0.000810
8. Mohanraj U, Wan X, Spruit CM, Skurnik M, Pajunen MI (2019) A toxicity screening approach to identify bacteriophage-encoded anti-microbial proteins. Viruses 11(11). https://doi.org/10.3390/v11111057
9. Spruit CM, Wicklund A, Wan X, Skurnik M, Pajunen MI (2020) Discovery of three toxic proteins of Klebsiella phage fHe-Kpn01. Viruses 12(5). https://doi.org/10.3390/v12050544
10. Kasurinen J, Spruit CM, Wicklund A, Pajunen MI, Skurnik M (2021) Screening of bacteriophage encoded toxic proteins with a next generation sequencing-based assay. Viruses 13(5). https://doi.org/10.3390/v13050750
11. Nyhamar E, Webber P, Liong O, Yilmaz Ö, Pajunen M, Skurnik M and Wan X (2023) Discovery of bactericidal proteins from *Staphylococcus* phage Stab21 using a high-throughput screening method. Antibiotics 12(7):1213
12. Zabarovsky ER, Winberg G (1990) High efficiency electroporation of ligated DNA into bacteria. Nucleic Acids Res 18(19):5912
13. Sambrook J, Russell D (2001) Molecular cloning: a laboratory manual, 3rd edn. Cold Springs Harbor Laboratory Press, Cold Springs Harbor
14. Coil D, Jospin G, Darling AE (2015) A5-miseq: an updated pipeline to assemble microbial genomes from Illumina MiSeq data. Bioinformatics 31(4):587–589. https://doi.org/10.1093/bioinformatics/btu661
15. Chuang LY, Cheng YH, Yang CH (2013) Specific primer design for the polymerase chain reaction. Biotechnol Lett 35(10):1541–1549
16. Woodall CA (2003) Electroporation of E. coli. In: E coli Plasmid vectors: methods and applications, p 55–59
17. Sanson B, Uzan M (1993) Dual role of the sequence-specific bacteriophage T4 endoribonuclease RegB. mRNA inactivation and mRNA destabilization. J Mol Biol 233(3):429–446
18. Leon-Velarde CG, Happonen L, Pajunen M, Leskinen K, Kropinski AM, Mattinen L et al (2016) Yersinia enterocolitica-specific infection by bacteriophages TG1 and varphiR1-RT is dependent on temperature-regulated expression of the phage host receptor OmpF. Appl Environ Microbiol 82(17):5340–5353

19. Young JW, Locke JC, Altinok A, Rosenfeld N, Bacarian T, Swain PS et al (2011) Measuring single-cell gene expression dynamics in bacteria using fluorescence time-lapse microscopy. Nat Protoc 7(1):80–88
20. Azzarito V, Long K, Murphy NS, Wilson AJ (2013) Inhibition of alpha-helix-mediated protein-protein interactions using designed molecules. Nat Chem 5(3):161–173. https://doi.org/10.1038/nchem.1568
21. Jumper J, Evans R, Pritzel A, Green T, Figurnov M, Ronneberger O et al (2021) Highly accurate protein structure prediction with AlphaFold. Nature 596(7873):583–589
22. Zimmermann L, Stephens A, Nam SZ, Rau D, Kubler J, Lozajic M et al (2018) A completely Reimplemented MPI bioinformatics toolkit with a new HHpred server at its core. J Mol Biol 430(15):2237–2243
23. Castro-Alvarez A, Costa AM, Vilarrasa J (2017) The performance of several docking programs at reproducing protein–macrolide-like crystal structures. Molecules 22(1):136
24. Wan X, Takala TM, Huynh VA, Ahonen SL, Paulin L, Bjorkroth J et al (2023) Comparative genomics of 40 Weissella paramesenteroides strains. Front Microbiol 14:1128028
25. Kuroda D, Gray JJ (2016) Shape complementarity and hydrogen bond preferences in protein-protein interfaces: implications for antibody modeling and protein-protein docking. Bioinformatics 32(16):2451–2456
26. Karimova G, Gauliard E, Davi M, Ouellette SP, Ladant D (2017) Protein-protein interaction: bacterial two-hybrid. Methods Mol Biol 1615: 159–176
27. Louche A, Salcedo SP, Bigot S (2017) Protein–protein interactions: pull-down assays. In: Bacterial Protein secretion systems: methods and protocols, p 247–255

Chapter 16

CRISPRpi: Inducing and Curing Prophage Using the CRISPR Interference

Jeffrey K. Cornuault

Abstract

We present here a CRISPR-interference-based protocol to trigger prophage induction, even for non-inducible prophages. This method can also be used to cure the prophage from the bacterial host. The method is based on silencing of the phage's repressor transcription, thanks to CRISPR interference. Plasmid electroporation is used to bring the CRISPRi system into the bacteria, specifically on a plasmid carrying spacers targeting the prophage repressor. This method enables prophage induction and curation in a week or two with a high efficiency.

Key words Prophage, Induction, CRISPRi, Curation, Indicator strain, Temperate, Lysogeny, Phage

1 Introduction

Lysogeny is widely prevalent among bacteria, evident by the fact that around 50% or more of bacterial species are lysogens [1, 2]. Lysogeny is usually detrimental to bacterial fitness due to spontaneous prophage induction killing cells at each generation [3]. Phages counterbalance that fitness cost by bringing "moron" genes with them that improve bacterial fitness in different ways, such as pathogenicity [4], stress resistance [5, 6], phage resistance [7], or even improving their colonization abilities [8]. Prophage studies are driven by the need to study prophage biology and their impact on host fitness. Such studies require deletion of the prophage from the bacterial genome. Deletion can be achieved through many different protocols [9], but some of them, such as mitomycin C or UV light exposure, are highly mutagenic and restoring the ancestral DNA sequence is not guaranteed. Furthermore, efficiencies of these prophage-curing protocols are hard to assess because of limited data. Phage amplification on an indicator strain is a critical step to study phage biology, yet the isolation of indicator strains can be a hard task, especially with the tendency of

Huan Peng et al. (eds.), *Phage Engineering and Analysis: Methods and Protocols*,
Methods in Molecular Biology, vol. 2793, https://doi.org/10.1007/978-1-0716-3798-2_16,

phage to be specialists. Such strains can, in theory, be created by just deleting prophages from their hosts.

Prophages remain in a lysogenic state thanks to the constant transcription of the phage repressor that shut down the expression of every gene involved in the lytic cycle. As long as the phage maintains the presence of repressor proteins in the cytoplasm, it remains in the lysogenic phase.

CRISPRpi (CRISPRi prophage induction) allows us to trigger a "natural" induction of prophage by getting rid of the phage repressor responsible for the maintenance of the lysogenic state. Prophages harbor a stereotypical genetic organization for their lysogenic module: the phage repressor and activator on opposite sides of transcriptional activity (Fig. 1) alongside the promoter sequence of these two genes located in the intergenic region. Some protein binding sites (called operator regions) can be localized in the same area in order to allow more regulation, such as phage Lambda with cI and the DNA loop tightening the lysogenic state [10].

CRISPR interference (CRISPRi) uses a mutant version of Cas9, called dCas9, that lost its ability to cut DNA [11]. Nonetheless, the dCas9 complex is still able to recognize its target and stay bound to the targeted DNA sequence, acting like an artificial transcriptional repressor. When induced, the prophage repressor is usually degraded or its expression repressed [12]. With CRISPRpi, we force the repression of the prophage repressor to allow for prophage activator expression, triggering the lytic cycle. By doing so, we avoid needing to research the induction signal, which can be laborious or even impossible, as some phages are well known for being non-inducible (P2 of *E. coli*, for instance [13]). Design of a good spacer is primordial to reach the maximal efficiency for silencing of the phage repressor. Annealing of two oligonucleotides containing the spacer, a direct repeat, and BsaI restriction site creates an insert ready to be ligated into the CRISPRi plasmid. CRISPRpi was used successfully on Lambda and P2 in *E. coli*. For each phage, the pCRISPathBrick plasmid was modified to incorporate the appropriate spacer in the plasmids. Plasmids were then transformed by electroporation into lysogenic *E. coli* [14] (electroporation yield is usually much higher than heat shock, even more so for large plasmids, allowing the recovery of more transformants [15, 16]). After transformation, prophage induction is assessed by a plaque assay. Surviving transformed bacteria are screened via PCR to detect clones where the prophage is absent. Overall, this protocol allows induction of prophages, even non-inducible ones, and allows us to recover a significantly high titer for phages, even if the final titer is lower compared to mitomycin C induction [9]. Furthermore, it produces a high-frequency prophage-free bacterium, allowing for easy isolation. Nonetheless, some limitations exist, such as bacteria for which no transformation

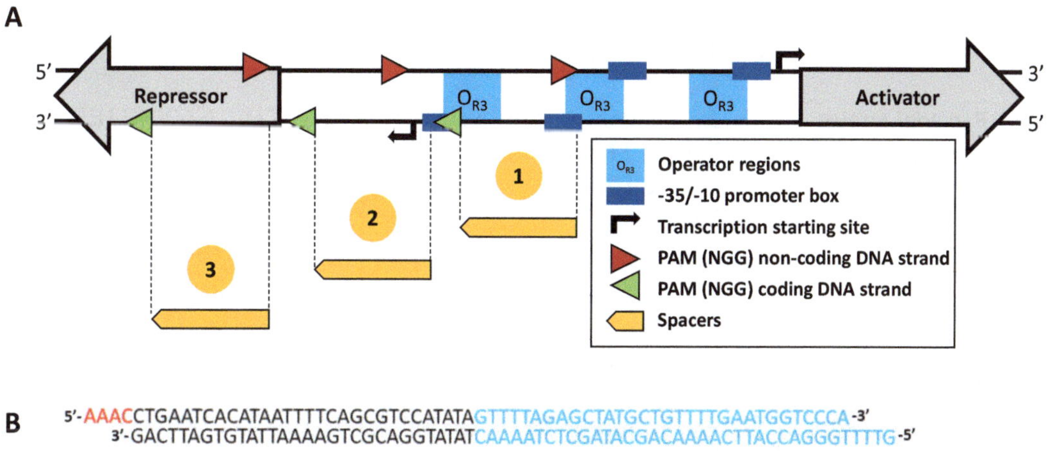

Fig. 1 Main features for the generation of spacers. (**a**) General organization of the lysogenic module and key elements that determine spacer design. (**b**) Annealing of the two oligonucleotides form an insert compatible for BsaI ligation in a BsaI linearized plasmid

protocol exists or a limited knowledge on the prophage repressor and its promoter sequences.

2 Materials

Prepare all solutions and media with deionized water. Autoclave all growth media.

2.1 Media, Antibiotics, and Reagents

1. Tryptic soy broth (TSB): casein peptone (pancreatic) 17 g/L, soya peptone (papain digest) 3 g/L, NaCl 5 g/L, $K2^{2+}HPO4^{2-}$ 2,5 g/L, glucose 2,5 g/L, pH adjusted to 7.3.
2. Lysogeny broth (LB): 10 g/L bactotryptone,5 g/L yeast extract, 10 g/L NaCl, pH adjusted to 7.0.
3. TSB and LB agar: Add 1% agar to liquid recipe.
4. Top TSB/LB agar: Add 0.75% agar to liquid recipe.
5. Chloramphenicol to a final concentration of 25 μg/mL for selective media.
6. Glycerol 100% (autoclaved).
7. Cold deionized water, DNA-free (H_2O_{dd}) (autoclaved).
8. Cold 10% glycerol in H_2O_{dd}.
9. 1 M NaCl (DNA-free, filter sterilized).
10. Nuclease-free water.
11. 3 M NaAc, pH 5.2.
12. 20% maltose solution (filter sterilized).

13. 100% ethanol.
14. 10 mM Tris–HCl pH 8.5.

2.2 Bacterial Strains and Plasmids

1. *E. coli strains*: NEB5α (from NEB), K-12 C600 (HER1025, Félix d'Hérelle Reference Center for Bacterial Viruses), K12S (HER 1037, Félix d'Hérelle Reference Center for Bacterial Viruses).
2. pCRISPathBrick [14], P15a ori, low copy number, chloramphenicol resistance (Plasmid pCRISPathBrick was gifted by Mattheos Koffas [Addgene plasmid #65006; http://n2t.net/addgene:65006; RRID: Addgene_65,006]).

2.3 Commercial Kit, Enzymes, and Molecular Biology

1. Any miniprep kit for plasmid extraction.
2. Primers in molecular biology water, concentration of 100 μM.
3. T4 polynucleotide kinase (PNK).
4. T4 DNA ligase.
5. 5× T4 ligase buffer, aliquoted.
6. BsaI restriction enzyme (or its HFI/HFII upgraded versions).
7. 10× Cutsmart buffer.
8. Antarctic phosphatase (AP).
9. 10X Antarctic phosphatase buffer.
10. DNA ladder.
11. Taq polymerase and corresponding reagents for PCR.
12. Primers for spacer cloning (see below), standard quality, resuspended in DNAse-free pure water.
13. Sequencing primers "Spacer ctrl F/R."
14. Primers specific for the targeted phage (F & R).
15. 2% agarose gels.
16. DNA quantification device.

2.4 Equipment and Materials

1. 37 °C shaker incubator.
2. 37 °C incubator.
3. A PCR thermocycler (Thermocycler Eppendorf Mastercycler_Nexus X2).
4. Ice-cold sterile centrifuge tube.
5. PCR tubes.
6. Electroporator.
7. Speed vacuum concentrator.
8. Bunsen burner.
9. 0.2-cm electroporation cuvette (cold, 4 °C).

10. Nanodrop or any DNA concentration measurement equipment.
11. Insulating ice bucket.
12. Refrigerated centrifuge able to handle large volume of liquid culture.
13. A tabletop centrifuge.
14. Glass test tubes.
15. Spectrophotometer.
16. Cuvettes for measuring OD.
17. If doable, a refrigerated water bath.
18. Freezer −80 °C.
19. Freezer −20 °C.
20. Fridge 4 °C.
21. Any gel electrophoresis equipment.

2.5 Glassware and Tubes

1. 15/50-mL sterile tubes.
2. 1.5/2-mL sterile centrifuge tubes.
3. Test tubes for 10 mL culture.
4. Erlenmeyer flask 500 mL.
5. MF-Millipore™ Membrane Filter, 0.025 μm, 25-mm diameter.
6. Syringe filter (0.22 μm).

2.6 Spacer Design (Fig. 1)

Design of the spacers is a critical step for this protocol. Several criteria have to be taken into account to design the spacers. First, look for PAM sequences that have to be located at the 3′ end of the protospacer sequence (5'-NGG-3′ for SpCas9 [*see* **Note 1**]) in the lysogeny module. Then, check their location with respect to the phage repressor and its promoter sequence. If the promoter is not known, some tools can be used to predict the promoter sequence (https://molbiol-tools.ca/Promoters.htm). Otherwise, manual observation could give some strong clues about the presence of a promoter sequence (look for the −35 [canonical TTGACA]/−10 boxes [canonical TATAAT]) [17] or relatives sequences. CRISPRi inhibits transcription either by blocking binding of RNA polymerase to the regulatory sequence (−35/−10) or by blocking access to the sense DNA strand, causing RNA polymerase to fall from the sense strand when undergoing transcription [18]. The latter is less efficient at inhibiting transcription than the prior. In the case of some phages (Fig. 1), almost identical direct repeat sequences, called operator regions (usually 3), can overlap the promoter sequence region. It is important to identify such sequences for the next steps [19].

Table 1
Oligonucleotide creation procedure for spacer cloning. In dark is the sequence of the protospacer and matching spacer. Other elements are then added such as PAM sequence (green), BsaI restriction site (red), BsaI and direct repeat sequence (purple), and direct repeat sequence (blue)

Step 1: Protospacer sequence and PAM	5'-CTGAATCACATAATTTTCAGCGTCCATATA-NGG-3'
Step 2: Sequence to blast against bacterial genome	5'-TCCATATA-NGG-3'
Step 3: Spacer raw sequence	5'-CTGAATCACATAATTTTCAGCGTCCATATA-3'
Step 4: Oligonucleotide forward sequence	5'-AAACCTGAATCACATAATTTTCAGCGTCCATATA GTTTTAGAGCTATGCTGTTTTGAATGGTCCCA-3'
Step 5: Oligonucleotide Reverse sequence	5'-GTTTTGGGACCATTCAAAACAGCATAGCTCTAAAAC TATATGGACGCTGAAAATTATGTGATTCAG-3'

Once all potential spacers are visualized and the repressor's promoter sequence is identified, here is a ranking for the best protospacer locations. Protospacer sequences located on the coding strand should always be favored over those on the noncoding strand (lower efficiency of the latter) [18] (Fig. 1a.), here are the best protospacer by priority order:

1. Protospacer covering promoter sequence (and operator sequences if existing) (Fig. 1a, spacer 1)
2. Protospacer between promoter and repressor gene (Fig. 1a, spacer 2)
3. Protospacer located in the 5′ end of the repressor ORF (Fig. 1a, spacer 3)

For the plasmid pCRISPathBrick, the size of the spacers is 30 nt; however, this value may vary depending on the plasmid used. In order to avoid any off-target effects, potential spacers should be blasted against the genome of the host bacterium to make sure that no bacterial gene expression will be modified, or to at least be aware of any potential side effects (Table 1). The 3′ end of spacers tends to be the most critical part for DNA recognition, so we recommend BLASTing the 8 last nucleotides of the spacer against the bacterial genome. It should be pointed that a recent publication showed that off-target binding can occur with only the last 4 or 5 nucleotides matching with the off-target sequence [20]. Once the candidate sequences are selected, follow the procedure shown in Fig. 1b to finish the design of the oligonucleotides to be used for spacer cloning via restriction ligation (*see* **Note 2**). Oligonucleotides are designed to be able to anneal together and form a sequence: spacer-direct repeat between two BsaI cohesive ends (Table 1 and Fig. 1b). The ligation then allows cloning of the spacer into the plasmid. When ordering the oligos, no specific quality is necessary, just order regular-quality oligonucleotides.

3 Methods

3.1 Plasmid Extraction of Native pCRISPathBrick (Day 1)

1-. Grow *E. coli* Dh5α strain carrying pCRISPathBrick overnight at 37 °C, 200 rpm in LB + chloramphenicol (*see* **Note 3**).

2-. Perform a plasmid extraction according to manufacturer's instruction (mini-kit or midi-kit).

3-. After the last washing step, elute the plasmid with warm deionized water (~55–60 °C) (*see* **Note 4**).

4-. If plasmid concentration is less than 100 ng/μL, increase DNA concentration by using a speed vacuum device.

5-. Keep DNA at 4 °C for short-term conservation, −20 °C for the long term. Avoid multiple freeze/thaw cycles that will break the plasmid.

3.2 Oligonucleotide Annealing (Day 1)

1. Dilute the two oligonucleotides forming your spacer to 50 μM in molecular biology water, one per tube.
2. For 5′-phosphorylation of the oligos, in one single tube, mix 2 μL of each diluted oligo with 10 μL of 5× T4 ligase buffer, 1 μL of T4 polynucleotide kinase (PNK), and 34 μL of distilled water in a PCR tube to a final volume of 50 μL (*see* **Note 5**).
3. Perform the phosphorylation for 30 min at 37 °C and heat inactivate the PNK for 20 min at 65 °C.
4. Add 2.5 μL of 1 M NaCl to the phosphorylated oligos.
5. Place the tube in a PCR thermal cycler and heat the sample for 5 min at 95 °C before cooling down to 25 °C at a rate of 0.1 °C per sec (*see* **Note 6**).
6. Quantify and keep at −20 °C for up to 2 months. This will serve as an insert in the pCRISPathBrick plasmid.

3.3 Digestion of pCRISPathBrick with BsaI (Day 1)

1. Digest 1 μg of purified pCRISPathBrick with BsaI by following the manufacturer's instructions with an overnight incubation at 37 °C and a final volume of 50 μL (*see* **Notes** 7 and **8**).
2. Dephosphorylate the 5′ ends of the digested vector with the Antarctic phosphatase (*see* **Note 9**) using the following mix: 50 μL of the previous digestion reaction, 6 μL of AP reaction buffer (10×), 2 μL of Antarctic phosphatase (5 units), complete with H_2O to reach a final volume of 60 μL.
3. Incubate at 37 °C for 30 min.
4. Stop reaction by heat inactivation at 80 °C for 2 min.

3.4 Purify the Digested and Dephosphorylated Vector by Precipitation with Salts and Ethanol (Day 1) (See Note 10)

1. Add 1:10 volume of NaAc 3 M pH 5.2 to vector DNA.
2. Add two volumes of 100% ethanol.
3. Mix thoroughly and incubate on ice for 15 min.
4. Centrifuge in a benchtop microcentrifuge at 16,000 × *g* for 15 min.
5. Carefully remove the supernatant and wash the pellet with 70% ethanol.
6. Air dry the pellet for 5–10 min.
7. Resuspend the pellet in 30 μl 10 mM Tris–HCl pH 8.5 (*see* **Note 11**).
8. Run the purified vector on a 0.8% agarose gel with the annealed oligos (from point 3.2.6) to ensure purity. Quantification can be made in a semiquantitative way with the DNA ladder, by comparing the intensity of your vector band with the ones on the DNA ladders, from the same size, or using a DNA quantification device (Nanodrop or Qubit).
9. Store the digested pCRISPathBrick at −20 °C until needed.

3.5 Ligation of the Targeting Plasmid (Day 1)

1. Set up ligation reactions with the T4 DNA ligase following the manufacturer's instructions. We typically use the following mix: 2 μL of 10× T4 DNA ligase buffer, 50 ng of pCRISPathBrick, enough annealed oligonucleotides to achieve a molar ratio of approximately 3–5:1 (*see* **Note 12**), 1 μL of T4 DNA ligase and complete to 20 μL with nuclease free water.
2. The molar ratio between an insert and plasmid is critical for an efficient ligation reaction. Molarity concentration of plasmid and insert have to be calculated using the following formula ([DNA concentration {ng/μL}*1000]/[length of DNA*650]) = DNA concentration pmol/μL. The molar ratio between the oligos and the plasmid is then calculated as insert pmol/plasmid pmol in the reaction.
3. Preparation of the ligation mix has to be handled on ice during the whole process. Do not forget to prepare another ligation reaction without the insert (negative ligation control) to evaluate vector recircularization.
4. Perform the ligations overnight at 16 C (in a PCR machine or a water bath) (*see* **Note 13**).
5. Heat-inactivate the ligase at 65 °C for 10 min, and then put on ice.

3.6 Dialysis of Ligation Product (Day 2)

1. Fill a petri dish with deionized water. Using clean tweezers, put a filtering membrane (MF-Millipore™ Membrane Filter, 0.025 μm, 25-mm diameter) on the water. Pipette the whole

ligation reaction on top of the membrane, close the petri dish, and wait 15 min (*see* **Note 14**).

2. Fifteen minutes later, recover the ligation reaction and keep on ice.

3.7 Preparation of Electrocompetent Cells (Day 2)

1. Grow 10 mL of *E. coli* Dh5α in TSB, 37 °C, 200 rpm, overnight.
2. The next day, inoculate 150 mL of TSB in a glass Erlenmeyer with 1.5 mL of the overnight culture. Incubate at 37 °C, 200 rpm, for 2–3 h.
3. When OD600nm is around 0.5–1, put cells on ice for at least 30 min (cells can be kept for a longer time on ice if lunch break is needed). For here, all elements that will be in contact with the bacteria have to be ice-cold.
4. In a pre-chilled centrifuge and tubes, spin down the cells at 10,000 g, 12 min, 4 °C. Empty the supernatant and resuspend the cells in 75 mL of ice-cold H_2O_{dd}.
5. Repeat the centrifugation and resuspend cells in 2 mL of ice-cold 10% glycerol solution.
6. Spin down the cells with a tabletop centrifuge at 12,000 g for 2 min, then throw away supernatant, and resuspend cells in 300 μL of ice-cold 10% glycerol solution. Keep on ice.
7. Aliquot cells in cold tube, 50 μL resuspended cells per tube (*see* **Note 15**). Keep on ice.

3.8 Electroporation of Dh5α (Day 2) (See Note 16)

1. For each tube, add 5 μL of ligation reaction and pipette up and down a few times to mix well. Avoid warming the tube with the palm of your hand. Keep the remaining ligation reaction at – 20 °C.
2. Transfer cells to a cold electroporation cuvette (0.2-cm gap).
3. Carefully clean the outside of the cuvette to remove any trace of water and put the cuvette in the electroporator.
4. Electroporate the cells (2.5 kV, 200 Ω, 25 μF) and immediately resuspend them in 950 μL of prewarmed TSB (37 °C).
5. Incubate for 1 h at 37 °C, 200 rpm, and plate 100 μL of 10^0 and 10^{-1} dilutions on TSB agar + chloramphenicol. Keep the leftover transformation reaction at 4 °C.
6. Incubate overnight at 37 °C and observe colonies the day after. If the ligation worked well, you should observe at least 10 times more colonies for the complete ligation reaction vs. the control with no inserts. Equal numbers could indicate a difficult cloning reaction where more clones should be tested (*see* **Note 17**).

3.9 Colony PCR and Sanger Sequencing (Day 3)

1. Carefully pick colonies with tips and resuspend them in 20uL of H_2O_{dd} (*see* **Note 18**). If the ligation yields numerous colonies, 10 colonies are enough for screening. Otherwise, try 20 colonies.
2. Prepare a PCR master mix (Taq polymerase) with sequencing primers "Spacer ctrl F/R" that is enough for 1 reaction/colony plus two extras, one for the negative control (water only) and one for the positive control (nonmodified pCRISPathBrick). Add 1 μL of each resuspended colony as the DNA template, one PCR tube per colony. Start PCR with 5 min at 95 °C to lyse cells and free their DNA for the PCR.
3. Perform the PCR (annealing T° = 48 °C for Taq polymerase, elongation time 30 s).
4. Analyze the PCR product on a 2% agarose gel: expected size of the native sequence is 139 bp, 200 bp if one spacer is integrated. Higher band sizes suggest multiple integration of the same spacer.
5. Send correctly sized PCR products for Sanger sequencing using spacer ctrl F/R as sequencing primers.
6. For the colonies with a good insert size, inoculate 10 mL of LB + chloramphenicol (25 μg/mL) with the 19 μL of colony resuspension and incubate overnight at 37 °C, 200 rpm.
7. Make a glycerol stock of clones (1 mL of overnight culture +300 μL of 100% glycerol, *see* **Note 19**) and store them at − 80 °C. Once the Sanger sequencing results return, throw away any stocks of clones with the incorrect sequence.

3.10 Ready-to-Use CRISPRi Plasmid Extraction (Day 5)

1. Grow one good clone per plasmid you want to test on lysogens and perform a plasmid extraction as previously described (point 3.1).

3.11 Lysogenic Strain Electroporation (Day 5)

1. Prepare electrocompetent cells of HER 1025, exactly as described before, and transform them with the CRISPRi plasmids targeting different areas of the prophage. Transform competent cells with 100 ng of plasmid (maximum volume 5 μL).
2. Do not forget to make controls for this experiment, such as the native plasmid with no spacers, no plasmids, or the plasmid with a spacer targeting a prophage's gene not involved in lytic cycle or prophage induction.
3. After a 1h incubation at 37 °C, inoculate 10 mL of TSB + chloramphenicol with 50 μL of transformation reaction (1 tube per transformation) and incubate overnight at 37 °C 200 rpm.

4. Plate the rest of the transformation reaction (100 µL) at dilutions ranging from 10^0 to 10^{-5} to measure the transformation efficiency (# transformants/µg of DNA).

3.12 Estimation of Prophage Induction (See Note 20) (Days 6–7)

1. After the overnight incubation, spin down the overnight culture (max speed, 10 min) and filter the supernatant with a syringe filter (0.22 µm). A lower OD600nm of the bacteria transformed with the CRISPRi plasmid targeting the phage repressor compared to the bacteria transformed with a CRISPRi plasmid with no spacers indicates efficient phage induction. This supernatant is your phage lysate. Phage lysate can be kept at 4 °C for several days if needed.
2. On the same day as the transformation (Subheading 3.11), start an overnight culture of *E. coli* HER1037 in LB +0.2% maltose (indicator strain), 37 °C, 200 rpm.
3. Make a dilution of the different phage lysates from Subheading 3.12, **step 1** (from 10^0 to 10^{-10}) in LB medium.
4. Melt LB top agar and let it cool down to 50–55 °C. Dispatch it in glass tubes, 3 mL per tube. Add 100 uL of overnight culture of the indicator strains and 30 µL of maltose 20% to each top agar tube.
5. For each tube add 100 µL of each supernatant dilution and plate it on LB agar plates.
6. Let the plates dry under flame for 10–15 min.
7. Incubate the plates overnight at 37 °C.
8. The following day, count the number of lysis plaques and determine the average phage concentration for each transformation reaction. A higher phage titer for bacteria transformed with CRISPRi targeting the repressor promoter, when compared to the control with no spacers, is expected.

3.13 Colony Screening by PCR to Identify Prophage-Free Colonies (Days 6–7)

1. After the overnight incubation, count the number of transformants for each plasmid. Calculate the transformation efficiency (# transformants/µg of DNA).
2. Isolate 20 clones of each transformation as indicated in Subheading 3.9, **step 1** and perform a PCR with primers targeting a specific gene of the prophage. Do not forget a negative and a positive control (indicator strain and lysogenic strain).
3. Run the PCR on a 1% agarose gel and visualize which clones are missing the phage sequence.
4. Start an overnight growth culture for clones missing the prophage and make glycerol stocks as indicated in 3.9.6/7.

4 Notes

1. PAM sequences are varied among different Cas9 proteins. If no spacer matching the criteria can be designed, you can check for dCas9 coming from bacteria other than *S. thermophilus*.
2. If you are considering cloning several spacers into the pCRIS-PathBrick plasmid, it might be worth it to order a Geneblock with the sequence of all your spacers and direct repeats. You can then clone the Geneblock into the plasmid using Gibson cloning. Nonetheless, the prevalence of direct repeats might increase the deletion of spacers via recombination, so this should be avoided.
3. As dCas9 plasmids can be fairly large, plasmid extractions can be low yield. Depending on the plasmid, incubating *E. coli* for 48 h at 30 °C sometimes gave us a higher yield. Other plasmids can be used instead of pCRISPathBrick for CRISPRi, some with inducible promoters, for instance. Readers should be really careful about these plasmids though. Since many inducible promoters are leaky and given that targeting a prophage repressor is highly toxic, a low background expression can lead to cell death or mutation accumulation. We tried to use pFD116 in its native form with a spacer against promoter of *cI* in λ prophage, and the leaky expression was enough to induce insane toxicity in lysogenic bacteria [21].
4. We prefer to use deionized water instead of elution buffer (Tris buffer) so that if a speed vacuum is used, there will be no salt accumulation in the sample. This allows easier digestion/ligation/electroporation of the plasmid DNA.
5. A 5× T4 ligase buffer is supplied with T4 DNA ligase. Divide the ligase buffer into single-use aliquots as ATP in the buffer can be degraded by repeated freeze-thaw cycles.
6. Oligonucleotides can also be cooled down by leaving them on the bench at room temperature for at least 5 h, but this procedure is less efficient for their annealing.
7. Since Nanodrop tends to overestimate DNA quantities, and that the vector purification step by ethanol has terrible yields, I recommend starting with 3 μg of plasmid instead of 1 μg and continuing the protocol as described.
8. If in a rush, the dephosphorylation of the linearized plasmid can be skipped, but you should be warned that not doing this step increases the amount of "background" colonies that will grow in the negative control of your ligation by a factor of 10. If your ligation has a low efficiency, it is going to be harder for you to find good clones where the plasmid integrates the new spacer.

9. BsaI (NEB) incubation time is 2 h 30 min, but if you work with the updated version, such as BsaI HF I or II (NEB), 35 min incubation is enough.
10. Overall, the yield of DNA precipitation by ethanol is terrible and a very low concentration of plasmid can be observed at the end of the process (from 1 to 25 ng/uL depending on the experiment). If linearization is achieved by a single BsaI restriction site (no removed insert), or by two BsaI restriction sites, creating a small-size insert, too small to be retained by DNA/PCR cleanup kits, then this whole part of the protocol can be replaced by the use of a DNA/PCR cleanup kit, enhancing the yield significantly. Doing a gel extraction of the linearized plasmid DNA can also be done, but the yield of the gel extraction kit remains low.
11. Soaking the DNA pellet in DNAse-free water overnight at 4 °C and then resuspending the pellet might improve the final DNA concentration.
12. Even though the manufacturer recommends a ratio around 1:5 maximum, we observed that using a ratio of 1:10 significantly increases the rate of positive clones, but also the rate of plasmids harboring two spacers instead of one. It also results in more colonies with good plasmids when the ligation step yields few colonies.
13. A large variety of ligation protocols exists. By personal experience, an overnight incubation at 16 °C is the safest and most efficient way to perform this cloning. If no refrigerated water bath is available, using a PCR machine for the incubation step is fine too.
14. Another way to perform the dialysis of ligation reaction is described in this study [22]. If the ligation reaction is not dialyzed, using a higher volume of ligation (more than 2 μL) might trigger an electric arc, seriously reducing your transformation efficiency.
15. Electrocompetent cells can be frozen at −80 °C for long-term storage, but the transformation efficiency will go down as the tubes get older. For this protocol, I have always used fresh competent cells.
16. Transformation of the ligation product can be achieved with chemically competent cells as well. Nonetheless, dCas9 plasmids tend to be fairly large and hard to transform. Furthermore, a low ligation efficiency means a low number of transformants. Electroporation is known to give a better transformation rate than heat-shocking, especially for large plasmids.

17. If the transformation yields no colonies, (i) spin down the leftover transformation reaction, resuspend it in 100 μL, and plate everything and (ii) do a PCR using the sequencing primers using 1 μL of ligation reaction as DNA template, with the native vector as a control in another tube. If the ligation was successful, you should observe a band at the expected size. If no band is there, that means the ligation failed, and no colonies should be expected.
18. If you are afraid of contamination of your clones by surrounding bacteria, streak each clone twice on LB agar +chloramphenicol and then use the streaked colonies for PCR screening.
19. Microwaving glycerol for 10–15 s warms it up enough to significantly reduce its viscosity and allows an easier pipetting. Make sure it is not too hot though, as it may kill your bacteria.
20. In some cases, no indicator strains exist to determine the phage titer. Alternative approaches such as qPCR quantification should be considered. A simple PCR with primers amplifying the attP site of the phage can also help to visualize prophage induction [23, 24].

Acknowledgments

J.K.C. is a recipient of a postdoctoral scholarship from Sentinel North. This research was supported by the Sentinel North program of the Université Laval and made possible, in part, by funding from the Canada First Research Excellence Fund. We thank Diana Balint for careful proofreading of the manuscript.

References

1. Touchon M, Bernheim A, Rocha EP (2016) Genetic and life-history traits associated with the distribution of prophages in bacteria. ISME J 10:2744–2754
2. Kim M-S, Bae J-W (2018) Lysogeny is prevalent and widely distributed in the murine gut microbiota. ISME J 12:1127–1141
3. De Paepe M, Tournier L, Moncaut E et al (2016) Carriage of λ latent virus is costly for its bacterial host due to frequent reactivation in Monoxenic mouse intestine. PLoS Genet 12: e1005861
4. Hayashi T, Makino K, Ohnishi M et al (2001) Complete genome sequence of Enterohemorrhagic Escherichia coli O157:H7 and genomic comparison with a laboratory strain K-12. DNA Res 8:11–22
5. Wang X, Kim Y, Ma Q et al (2010) Cryptic prophages help bacteria cope with adverse environments. Nat Commun 1:147
6. Aucouturier A, Chain F, Langella P et al (2018) Characterization of a Prophage-Free Derivative Strain of Lactococcus lactis ssp. In: lactis IL1403 reveals the importance of prophages for phenotypic plasticity of the host, p 9
7. Bondy-Denomy J, Qian J, Westra ER et al (2016) Prophages mediate defense against phage infection through diverse mechanisms. ISME J 10:2854–2866
8. Brown EM, Arellano-Santoyo H, Temple ER et al (2021) Gut microbiome ADP-ribosyltransferases are widespread phage-encoded fitness factors. Cell Host Microbe 29:1351–1365.e11

9. Cornuault JK, Moineau S (2021) Induction and elimination of prophages using CRISPR interference. CRISPR J 4:549–557
10. Oppenheim AB, Kobiler O, Stavans J et al (2005) Switches in bacteriophage lambda development. Annu Rev Genet 39:409–429
11. Qi LS, Larson MH, Gilbert LA et al (2013) Repurposing CRISPR as an RNA-guided platform for sequence-specific control of gene expression. Cell 152:1173–1183
12. Roberts JW, Roberts CW, Craig NL (1978) Escherichia coli recA gene product inactivates phage lambda repressor. Proc Natl Acad Sci USA 75:4714–4718
13. Bertani G (2004) Lysogeny at mid-twentieth century: P1, P2, and other experimental systems. J Bacteriol 186:595–600
14. Cress BF, Toparlak ÖD, Guleria S et al (2015) CRISPathBrick: modular combinatorial assembly of type II-A CRISPR arrays for dCas9-mediated multiplex transcriptional repression in E. coli. ACS Synth Biol 4:987–1000
15. Aune TEV, Aachmann FL (2010) Methodologies to increase the transformation efficiencies and the range of bacteria that can be transformed. Appl Microbiol Biotechnol 85:1301–1313
16. Yoshida N, Sato M (2009) Plasmid uptake by bacteria: a comparison of methods and efficiencies. Appl Microbiol Biotechnol 83:791–798
17. LaFleur TL, Hossain A, Salis HM (2022) Automated model-predictive design of synthetic promoters to control transcriptional profiles in bacteria. Nat Commun 13:5159
18. Bikard D, Jiang W, Samai P et al (2013) Programmable repression and activation of bacterial gene expression using an engineered CRISPR-Cas system. Nucleic Acids Res 41:7429–7437
19. Little JW, Michalowski CB (2010) Stability and instability in the lysogenic state of phage lambda. J Bacteriol 192:6064–6076
20. Rostain W, Grebert T, Vyhovskyi D et al (2023) Cas9 off-target binding to the promoter of bacterial genes leads to silencing and toxicity. Nucleic Acids Res 51:3485–3496
21. Depardieu F, Bikard D (2020) Gene silencing with CRISPRi in bacteria and optimization of dCas9 expression levels. Methods 172:61–75
22. Lemay M-L, Renaud AC, Rousseau GM et al (2018) Targeted genome editing of virulent phages using CRISPR-Cas9. Bio Protoc 8:e2674
23. Cornuault JK, Moncaut E, Loux V et al (2020) The enemy from within: a prophage of Roseburia intestinalis systematically turns lytic in the mouse gut, driving bacterial adaptation by CRISPR spacer acquisition. ISME J 14:771–787
24. Matos RC, Lapaque N, Rigottier-Gois L et al (2013) Enterococcus faecalis prophage dynamics and contributions to pathogenic traits. PLoS Genet 9:e1003539

Chapter 17

Isolation of Bacteriophages on *Actinobacteria* Hosts

Michelle Zorawik, Deborah Jacobs-Sera, Amanda C. Freise, SEA-PHAGES, and Krisanavane Reddi

Abstract

Bacteriophages are ubiquitous biological entities which can be found in a variety of habitats. Here, we describe protocols for the isolation of bacteriophages on a variety of *Actinobacterial* genera. Two approaches to phage isolation, direct isolation and enriched isolation, are described, which can be performed individually or in parallel. The protocols described can be adapted to isolate a wide array of bacteriophages.

Key words Bacteriophage, Plaques, Direct isolation, Enriched isolation

1 Introduction

Actinobacteria are free-living organisms, dwelling in a wide range of habitats including soil [1], freshwater [2, 3], the ocean, and extreme climate conditions such as the Antarctica [4], and desert [5]. In recent years due to the global rise in antibiotic resistance, the idea of phage therapy is gaining renewed momentum [6, 7]. In this chapter the isolation of bacteriophages from *Actinobacterial* hosts, including *Arthrobacter*, *Gordonia*, *Microbacterium*, *Mycobacterium*, and *Cutibacterium,* is described. The chosen hosts include opportunistic human pathogens (*Mycobacterium* [8], *Cutibacterium, and Microbacterium*) and bacteria with agricultural (*Arthrobacter*) and bioremedial (*Gordonia*) significance. Hence, isolation and study of these phages not only offers insights into viral and bacterial diversity and evolution, but also provides valuable tools for agricultural and clinical utilities which are urgently required, considering the pressing threat of antibiotic resistance [9, 10].

The methods involved in the isolation of bacteriophages were originated with the discoveries by Félix d'Hérelle and Frederick William Twort, circa 1917 [11]. Here, we describe protocols drawn from the SEA-PHAGES Discovery Guide (https://

Huan Peng et al. (eds.), *Phage Engineering and Analysis: Methods and Protocols*,
Methods in Molecular Biology, vol. 2793, https://doi.org/10.1007/978-1-0716-3708-2_17,

seaphagesphagediscoveryguide.helpdocsonline.com/home), the Hatfull and Modlin groups [6, 9, 12], and our own laboratories. These methods have been used extensively in integrated research-education programs by undergraduate students in the SEA-PHAGES program Jordan et al. [13] and Hanauer et al. [14] who have isolated over 23,000 individual phages including over 12,000 on a single strain of *Mycobacterium smegmatis* mc^{2}155 [9, 14]. In this chapter, two techniques are described, direct isolation and enriched isolation, each offering distinct advantages and considerations. During direct isolation, phages are extracted directly from the sample without preincubation with a bacterial culture [15]. This approach captures the diversity of phages in a sample without biasing isolation towards phages that may replicate most easily in the current conditions. During enriched isolation, phages are extracted after seeding the sample with host bacteria and incubating for an extended period of time, thereby increasing the yield of phages that replicate on that host. Both approaches may be performed in parallel to increase the chances of isolating novel phages [16]. The two methods in parallel make it likely that students will discover phages in a classroom setting. The many thousands of Actinobacteriophages isolated using these methods are catalogued in the PhagesDB database available at https://www.phagesDB.org [17].

2 Materials

Note: The materials, hosts, and the bacteriophages must be used and disposed as per your institution's chemical and biosafety guidelines.

2.1 Hosts

Actinobacterial host(s) of choice (*see* **Note 1** for a comprehensive directory of hosts). For the purpose of this chapter, bacteriophage isolation using *Arthrobacter, Gordonia, Microbacterium, Mycobacterium,* and *Cutibacterium* hosts is described.

1. ***Arthrobacter***: *A. globiformis* (NRRL B-2979 and NRRL B-2880), *A. sp.* (ATCC 21022), *A. sulfureus* (NRRL B-14730 and ATCC 19098), *A. atrocyaneus* (NRRL B-2883).
2. ***Gordonia***: *G. lacunae* (NRRL B-24551), *G. neofelifaecis* (NRRL B-59395), *G. rubripertincta* (NRRL B-16540), *G. terrae* 3612 (ATCC 25594 and CAG3), *G. westfalica* (NRRL B-24152).
3. ***Microbacterium***: *M. aerolatum* (NRRL B-24228), *M. foliorum* (NRRL B-24224), *M. liquefaciens* (LMG 16120), *M. paraoxydans* (NRRL B-14843 and NWU1), *M. terrae* (NRRL B-24214).

4. ***Mycobacterium***: *M. aichiense* (ATCC 27281), *M. aurum* (ATCC 23366), *M. phlei* (GT1S10, str.), *M. smegmatis* mc^2 155 (ATCC 700084) .
5. ***Cutibacterium (formerly Propionibacterium)***: *C. acnes* (ATCC 29399, ATCC 6919).

2.2 Culture Media and Materials

2.2.1 Materials

1. Sterile inoculation sticks.
2. Baffled 500-mL Erlenmeyer flask.
3. Liquid medium for the chosen host.
4. Shaker (set to 250 rpm).
5. Incubator set to optimum temperature of the host.
6. Frozen glycerol stocks of the chosen host.

2.2.2 Solutions and Buffers

1. Albumin dextrose (AD) supplement: Dissolve 4.25 g of sodium chloride and 25 g of albumin fraction V in a large beaker containing 400 mL of distilled water. Note: The albumin will settle at the top of the vessel; it is important not to stir vigorously as this could denature the albumin. If the solution appears frothy, then this is an indication that the albumin has denatured and should not be used. Once the components have dissolved, add 10 g of dextrose and allow this component to dissolve, bring the volume to 500 mL, and filter sterilize using a 0.22-μm polyethersulfone (PES) membrane sterilization unit. Store at 4 °C.
2. 1 M calcium chloride: To 90 mL of distilled water add 11.1 g of anhydrous calcium chloride, bring the volume to 100 mL, filter, sterilize, and store at 4 °C. When adding calcium chloride to various components, the final concentration will be 1 mM.
3. Carbenicillin (CB): Stock concentration is 50 mg/mL. The final concentration used in the media is 50 ug/mL. Dissolve 1.26 g of carbenicillin in 25 mL of distilled water, sterile filter, and store at 4 °C for 2 months.
4. Cycloheximide (CHX): Stock concentration is 10 mg/mL. The final concentration used in the media is 10 ug/mL. Note: CHX is a toxic agent. All manipulations must be performed in a fume hood using appropriate personal protective equipment. Dissolve 1 g of CHX in 100 mL of ethanol, filter sterilize using a nylon filter, aliquot into 1 mL aliquots, and store at −20 °C.
5. 40% glycerol: Dispense 40 mL of glycerol to a 100-mL cylinder, add distilled water to the 100 mL mark, cover with Parafilm, and invert the cylinder several times to mix the glycerol solution. Note: Since glycerol is viscous, it is best to use a measuring cylinder as the measuring tool.

6. 80% glycerol: Dispense 80 mL of glycerol to a 100-mL cylinder, add distilled water to the 100 mL mark, cover with Parafilm, and invert the cylinder several times to mix the glycerol solution. Note: Since glycerol is viscous, it is best to use a measuring cylinder as the measuring tool.
7. 1 M magnesium sulfate: Dissolve 246.47 g of magnesium sulfate in 700 mL of distilled water, bring the volume to 1 L, and filter sterilize.
8. 1 M Trisl pH 7.5: Dissolve 21.1 g of Trizma base in 975 mL of distilled water, pH to 7.5 with 12 M HCl, bring the volume to 1 L with distilled water, and store at 4 °C.
9. Phage buffer (PB): To 980 mL of distilled water, add 10 mL of 1 M Tris pH 7.5, 10 mL of 1 M magnesium sulfate, and 4 g of sodium chloride, filter sterilize, and store at 4 °C. Add 10 mL of 100 mM calcium chloride prior to use.
10. 20% Tween 80: Dispense 20 mL of Tween 80 in 60 mL of distilled water and mix, bring the volume to 100 mL, sterile filter, and store at 4 °C. Note: Since Tween 80 is viscous, it is best to use a measuring cylinder as the measuring tool.

2.2.3 Agar Plate Containing Growth Medium for the Chosen Host

1. Peptone yeast calcium agar (PYCa): In 990 mL of distilled water dissolve 15 g agar, 15 g peptone, and 1.0 g yeast extract, autoclave, cool, and aseptically add the following – 2.5 mL 40% dextrose, 4.5 mL 1 M calcium chloride. Under aseptic conditions, pour into sterile petri dishes, allow to set, and store in the refrigerator for 2 weeks. One mL of CHX can be added, to a final concentration of 10 ug/mL. Note: PYCa top agar is prepared in the same manner, adjust the top agar concentration as described in 2.2.4 and instead of pouring into sterile petri dishes, aliquot the medium into 50 mL/sterile tubes and store in the refrigerator. PYCa is used for *Arthrobacter*, *Gordonia*, and *Microbacterium* hosts.
2. Middlebrook 7H10 Agar (MHA): In 970 mL of distilled water add 12.5 mL glycerol, 19 g 7H10 Agar (Difco #262710), and 5 mL 40% dextrose. Autoclave and cool the solution to 55 °C, aseptically add 1 mL of CB, 1 mL CHX, and 1 mL 100 mM calcium chloride. Pour into sterile petri dishes, allow to set and store in the refrigerator for 2 weeks. MHA is used for *Mycobacterium* hosts. Two drops of anti-bubble can be added, though not necessary.
3. A Media: In 1000 mL of distilled water add 12 g of casitone, 12 g yeast extract, 4 g dextrose, 4 g potassium dihydrogen phosphate, 1 g magnesium sulfate heptahydrate, and 28 g agar. Autoclave, cool to 55 °C, pour into sterile petri dishes, allow to set, and store in the refrigerator for 2 weeks. A Media is used for *Cutibacterium* hosts.

2.2.4 Top Agar Suitable for Plating of the Chosen Host

To use stored top agar, melt the desired amount by repeatedly microwaving in 30-s intervals until boiling, but not overboiling. Once the top agar is completely melted, maintain it at 55 °C using a water bath or an incubator set at 55 °C. Agar concentrations can be varied between 0.35% and 0.7%. Less than that and the top agar may not solidify, greater than 0.7% and the phage may not be able to diffuse well for plaques to be seen.

1. Peptone yeast calcium top agar (PYCa TA): In 990 mL of distilled water dissolve 4.0 g agar, 15 g peptone, and 1.0 g yeast extract, autoclave, cool, and aseptically add the following– 2.5 mL 40% dextrose, 4.5 mL 1 M calcium chloride. Aliquot the medium to 50 mL/sterile tube and store in the refrigerator. Suitable for *Arthrobacter*, *Gordonia*, and *Microbacterium* hosts.
2. Middlebrook Top Agar (MBTA): Dissolve 4.7 g 7H9 broth media Difco 9 (Ref# 271310), 7 g agar, to a final volume of 1 L diH_2O. Heat to boiling to dissolve the agar. Dispense 5 mL aliquots in glass test tubes with tight-fitting caps and autoclave. Store MBTA in the refrigerator for 2 weeks. Suitable for *Mycobacterium* hosts. The concentration of this top agar is 0.7%. To use, add equal volumes of 7H9 broth + calcium (final concentration 1 mM $CaCl_2$) when plating.
3. A Media Top Agar: In 1000 mL of distilled water, dissolve 7 g agar, 12 g casitone, 12 g yeast extract, 4 g dextrose, 4 g potassium dihydrate phosphate, 1 g magnesium sulfate septahydrate, dispense 5 mL aliquots in glass test tubes with tight-fitting caps and autoclave. Store in the refrigerator for 2 weeks. Suitable for *Cutibacterium* hosts.

2.2.5 Liquid Medium Suitable for Cultivation of the Chosen Host

1. PYCa broth: In 990 mL distilled water dissolve 15.0 g peptone, 1.0 g yeast extract, and 2.5 mL 40% dextrose. Autoclave, cool, and aseptically add 4.5 mL 1 M calcium chloride, 1 mL CHX stock. PYCa broth is used for *Arthrobacter, Gordonia, and Microbacterium* hosts.
2. 7H9 broth: In 900 mL of distilled water add 4.7 g 7H9 broth base and 5 mL 40% glycerol, autoclave, and filter sterilize. For use in cultivating bacteria add the following: 90 mL 7H9 broth, 10 mL AD supplement, 100 μL CB, 100 μL CHX, and 1 mL 100 mM calcium chloride. If the medium is being used to recover glycerol stocks, add 250 μL 20% Tween 80. Note: Tween 80 must only be used in the initial culture of *Mycobacterium* hosts to prevent clumping of the organisms; subsequent cultures must not contain Tween 80 since this component will interfere with the plaque assay. 7H9 broth is used for *Mycobacterium* hosts.
3. Liquid A Media: In 1000 mL of distilled water add 12 g casitone, 12 g yeast extract, 4 g dextrose, 4 g potassium dihydrogen phosphate, and 1 g magnesium sulfate septahydrate. Autoclave. A Media is used for *Cutibacterium* hosts.

2.3 Sample Collection and Phage Isolation

1. Soil sample collection
   (i) Sterile spatula or similar digging tool.
   (ii) Sterile plastic container.
   (iii) Electronic device with GPS technology.
   (iv) Marker to label samples.
2. Skin microbiome sample collection
   (i) Sterile pore strip

2.3.1 Direct Isolation

1. Sample.
2. 15-mL sterile conical tube.
3. Liquid medium suitable for cultivation of the chosen host.
4. Syringe.
5. 0.22-μm sterile filter.
6. Shaker (set to 250 rpm).
7. Incubator, set to optimum temperature for chosen host.
8. If choosing *Cutibacterium* hosts:

(i) Sterile inoculation loop.
(ii) Sterile petri dish.

2.3.2 Enriched Isolation

Arthrobacter, Gordonia, and *Microbacterium* Hosts

1. Soil sample.
2. 50-mL sterile mini bioreactor tube.
3. PYCa liquid media.
4. Selected host culture.
5. Shaker (set to 250 rpm).
6. Incubator, set to 30 °C.

Mycobacterium Hosts

1. Soil sample.
2. Sterile 50-mL mini bioreactor tube.
3. Sterile distilled water.
4. 7H9 liquid media.
5. *Mycobacterium* host culture.
6. 100 mM calcium chloride solution.
7. Shaker (set to 250 rpm).
8. Incubator set to 37 °C.

Cutibacterium Hosts

1. Sterile inoculating loop.
2. Sterile microcentrifuge tubes.
3. Sterile pipette tips.
4. *Cutibacterium* host culture.

5. Liquid A Medium.
6. Airtight container.
7. AnaeroPack.
8. Incubator set to 37 °C.

2.4 Phage Detection

2.4.1 One-Plate Plaque Assay

1. Sterile culture tube.
2. Direct isolation phage filtrate (prepared in protocols 3.3.1 and 3.3.2).
3. Selected host culture.
4. Incubator.
5. 5-mL sterile serological pipette.
6. Top agar suitable for growth of the chosen host.
7. Sterile microcentrifuge tube(s).
8. Sterile phage buffer (PB).
9. P200 micropipette and sterile tips.

2.4.2 Spot Test

1. Agar plates containing appropriate growth medium for the chosen host.
2. Top agar suitable for growth of the chosen host.
3. Enriched isolation filtrate (prepared in protocol 3.3.3).
4. Sterile phage buffer (PB).
5. Lysate of phage known to infect the chosen host to serve as a positive control (if available).
6. P20 micropipette and sterile tips.
7. Incubator.

2.5 Phage Purification

1. Sterile microcentrifuge tubes.
2. Sterile phage buffer (PB).
3. Phage sample: using either a direct isolation lysate, an enriched isolation filtrate, or a phage lysate created during the previous purification round (prepared in protocol 3.5).
4. P20, P200, and P100 micropipettes and sterile tips.
5. Sterile culture tubes with caps.
6. Selected host culture.
7. Vortexer.
8. Agar plates containing appropriate growth medium for the chosen host.
9. Incubator.
10. 0.22-μm filter sterilization.

3 Methods

All procedures must be performed under aseptic techniques.

3.1 Host Culture Preparations

All experiments should be performed using fresh host cultures, under aseptic conditions.

3.1.1 Preparation of Host Glycerol Stocks

1. Pre-warm the appropriate agar plate and liquid media for 20 min at room temperature.
2. If the host strain arrives as a freeze-dried stock solution, add 1 mL of the appropriate liquid media to reconstitute the bacterium.
3. Label the plate and immediately streak the bacterium onto the agar plate.
4. Incubate at the appropriate temperature and oxygen requirements for 24–48 h.
   (a) Note: *Cutibacterium* must be incubated under anaerobic conditions see **step** 7.
5. Inspect the plate for single isolated colonies.
6. Aliquot 4 mL of the appropriate liquid media to a sterile culture tube under aseptic conditions and add a colony from the plate prepared in Subheading 3.1.1, **step 4** to the tube and incubate on a shaker at the appropriate temperature overnight.
7. For *Cutibacterium* use an airtight box and place the tube containing the *Cutibacterium* host culture inside the box along with an open AnaeroPack to create an anaerobic environment.
8. After the incubation period add 1 mL of sterile 80% glycerol, mix, and dispense into labeled cryovials.
9. Store the cryovials at −80 °C.

3.1.2 Recovery of Bacteria from Glycerol Stocks

1. Retrieve the desired bacterial host strain from the −80 °C freezer and immediately place it on ice.
2. Using a sterile inoculation stick, scrape the frozen surface of the glycerol stock and streak the sample onto a suitable prewarmed agar plate using a standard streak plate protocol such as the quadrant-streak method.
3. Immediately return the glycerol stock to the freezer.
4. Invert the plate and incubate it at the host's preferred growth temperature and oxygen requirements for 24–48 h, or until isolated colonies are observed.
5. Use the plate for immediate culture preparation, or store at 4 °C sealed with Parafilm until new cultures or streak plates are required.

3.1.3 Liquid Culture of All Hosts

1. Label a baffled culture flask and aliquot in 90 mL of the most suitable broth for the chosen host prewarmed to approximately 30 °C.
2. Using a sterile inoculation loop, pick an isolated colony from the streak plate and immerse it into the most suitable broth for the chosen host.
3. Incubate with shaking (250 rpm) at the appropriate temperature and oxygen requirements for the chosen host for 24–48 h (If working with *Cutibacterium*, do *not* shake, in order to avoid introducing air into the culture).
4. Evaluate the liquid culture for appropriate changes in turbidity and color, and then store at 4 °C until used for further experimentation. Note: *Mycobacterium* cultures typically take 1–4 days of incubation before turbidity can be visualized. Be careful to continue the incubation to avoid biofilm formation (and a clumpy suspension of cells).
5. For *Mycobacterium* use 7H9 broth plus Tween 80 as described in Subheading 2.2.5, **item 2** in a small volume (5 mL) to prepare a starter culture. Starter cultures can then be used to prepare the working cultures for plating (Subheading 3.1.3, **step 1**).

3.2 Collection of Putative Phage Samples

3.2.1 Arthrobacter, Gordonia, Microbacterium, and Mycobacterium Phage Collection from Soil

1. Collect an appropriate quantity of an environmental sample into a sterile container (*see* **Note 2**).
   (a) For solid samples, use a sterile spatula or similar digging device to collect the sample in a clean plastic bag, filling it at least halfway. Alternatively, invert the bag and use it as a glove to collect the sample without direct skin contact.
2. Label the container with a unique identifier and record relevant information, including the date and time of sampling, GPS coordinates (*see* **Note 3**), ambient temperature of the collection site, and any physical characteristics of the sample, such as the collection depth, approximate moisture level, and any other defining features in a notebook.
3. Store the sample at room temperature for immediate processing (*see* **Note 4**).

3.2.2 Cutibacterium Phage Collection from the Skin Microbiome

The entire protocol should be conducted under aseptic conditions using gloves to avoid contamination.

1. Wet the subject's nose with water, remove the plastic backing off the strip, and press the strip down on the subject's nose. Leave the strip in place for 15 min.
2. Aliquot 5 mL of liquid A Media into a sterile petri dish and place the strip inside.

3. Into a sterile microcentrifuge tube aliquot 1 mL of liquid A Media. Scrape the material from the strip with a sterile loop, ensuring that the material is visible on the loop, and then inoculate the tube of A Media with this material.
4. Filter sterilize the contents of the microcentrifuge tube into another sterile microcentrifuge tube.
5. Immediately proceed with Subheading 3.3.2 for direct isolation of *Cutibacterium* bacteriophages.

3.3 Phage Isolation

To increase the chances of isolating phages that can infect the desired host, follow both the direct isolation protocol and the enriched isolation protocol for each collected sample.

3.3.1 Direct Isolation for Arthrobacter, Gordonia, Microbacterium, and Mycobacterium Hosts

1. Fill a sterile, labeled 15-mL conical tube with the soil sample to the halfway mark.
2. Depending on the chosen host, add the appropriate liquid medium by first saturating the sample with the medium and then slowly continuing to add the broth, until the soil is submerged by 2–3 mL of liquid.
3. Cap the tube and gently invert it several times to suspend the sample.
4. Incubate the tube with shaking (250 rpm) for 1–2 h at the preferred growth temperature for the host.
5. Remove the tube from the incubator and allow it to rest until all solid matter has settled. Alternatively, centrifuge the tube at low speed (*see* **Note 5**).
6. Using a Luer-Lok syringe, carefully aspirate about 2 mL of the clear supernatant from the top of the tube.
7. Screw the syringe securely onto a sterile 0.22-μm filter and slowly depress the syringe plunger to filter the sample. Collect at least 1 mL of the resulting phage filtrate in a labeled sterile microcentrifuge tube (*see* **Note 6**).
8. Immediately use the phage filtrate to prepare a one-plate plaque assay, or store it at 4 °C for future processing (*see* **Note 7**).

3.3.2 Direct Isolation of Phages for Cutibacterium Hosts

1. Add 500 μL of *Cutibacterium* culture to two sterile microcentrifuge tubes. Do not vortex the *Cutibacterium* culture but simply swirl the culture to mix. These tubes will serve as the direct sample and the negative control.
2. To the tube labeled direct sample, add 300 μL of your microcentrifuge tube (from Subheading 3.2.2, **step 5**). To the tube labeled negative control add 300 μL of PB.

3. Incubate the tubes at room temperature for 30 min.
4. Dispense the tube labeled direct sample to a tube of molten A Media Top Agar (*see* **Note 8**) and overlay onto the surface of an A Media Agar plate. The plate must be gently swirled to spread the molten A Media Top Agar over the plate avoiding the creation of air bubbles.
5. Repeat this process for the negative control tube.
6. Allow the soft agar to solidify (*see* **Note 9**).
7. Once the plates have solidified, invert them and place them into an airtight sealable culture box with a fresh AnaeroPack. Incubate at 37 °C for 3 days.
8. After the incubation period inspect the plates for plaques.

3.3.3 Enriched Isolation

Enriched Isolation of Phages for *Arthrobacter, Gordonia, Microbacterium,* and *Mycobacterium* Hosts

1. Fill a sterile 50-mL sterile tube with the soil sample to approximately the 15-mL gradation mark.
2. Add liquid medium to the 35-mL mark by first saturating the sample with the medium and then slowly continuing to add the broth.
3. With the cap on tight, invert the tube to thoroughly mix all the contents, and shake at ~250 rpm for 1–2 h. This step serves to release the phages from the soil by interfering with the electrostatic charges between the phages and the soil particles.
4. After the incubation period allow the tube to rest until all solid matter has settled. Alternatively, centrifuge the tube at low speed (*see* **Note 5**).
5. Filter the supernatant through a 0.22-μm filter, capturing the filtrate into a sterile mini bioreactor or a sterile baffled Erlenmeyer flask.
6. Mix the chosen host culture and add 500 μL of the suspended cells to the mini bioreactor or baffled Erlenmeyer flask.
7. Incubate the tube with shaking (250 rpm) for 48–96 h at the optimal growth temperature for the host.
8. Remove the tube from the incubator and allow it to rest until all solid matter has settled. Alternatively, centrifuge the tube at low speed (*see* **Note 5**).
9. Using a Luer-Lok syringe, carefully aspirate about 2 mL of the clear supernatant from the top of the tube.
10. Screw the syringe securely onto a 0.22-μm filter and slowly depress the syringe plunger to filter the sample. Collect at least 1 mL of the resulting phage filtrate in a sterile labeled microcentrifuge tube (*see* **Note 6**).

11. Immediately use the phage filtrate to prepare a one-plate plaque assay/spot test, or store at 4 °C for future processing (*see* **Note 7**).

Cutibacterium Hosts

1. To a sterile 15-mL culture tube, add 3 mL of liquid A Media, 1 mL of *Cutibacterium* culture, and the remaining sample from Subheading 3.2.2, **step 5**.
2. Incubate the tube for 3 days at 37 °C in a sealed container containing an AeroPack. This is now the enrichment culture.
3. Dispense 1 mL of the enrichment culture into a sterile microcentrifuge tube and centrifuge the sample at maximum speed for 5 min to pellet the sample.
4. Filter the supernatant into a sterile microcentrifuge tube taking care not to disturb the pellet; place the tube on ice.
5. Add 100 μL of the prepared supernatant into a new microcentrifuge tube, this serves as the enriched sample and 100 μL of PB into a tube labeled as negative control.
6. Add 500 μL of *Cutibacterium* host to each tube and allow it to incubate for 30 min at room temperature.
7. Follow Subheading 3.3.2, **steps 4–8**.

3.4 Phage Detection

3.4.1 One-Plate Plaque Assay and Plaque Picking

To ascertain the presence of host-specific phages in the direct isolation samples, conduct one-plate plaque assays by using the phage filtrates obtained. If the assays yield positive results, proceed to prepare phage lysates for further purification by collecting phages from well-isolated plaques.

1. Add 500 μL of the direct phage filtrate to a labeled sterile culture tube.
2. Mix the host culture and add 250 μL of suspended cells to the same tube.
3. Cap the tube and gently mix the contents by flicking it or rotating it at an angle.
4. Incubate the sample for 15 min at room temperature. This is the phage and host adsorption step where the receptor binding proteins of the phage can attach to the receptors of the bacterial cell envelope and allow for the phage DNA to be injected into the bacterial cell. The adsorption time varies amongst phages and can be experimented with for each individual phage.
5. Using a 5-mL sterile serological pipette, transfer 5 mL of molten (at 55 °C) top agar into the adsorption tube and quickly aspirate the inoculated host–agar mixture back into the pipette (*see* **Note 8**).
6. Immediately pipette the mixture onto a suitable, labeled agar plate, and carefully but swiftly tilt the plate to spread the

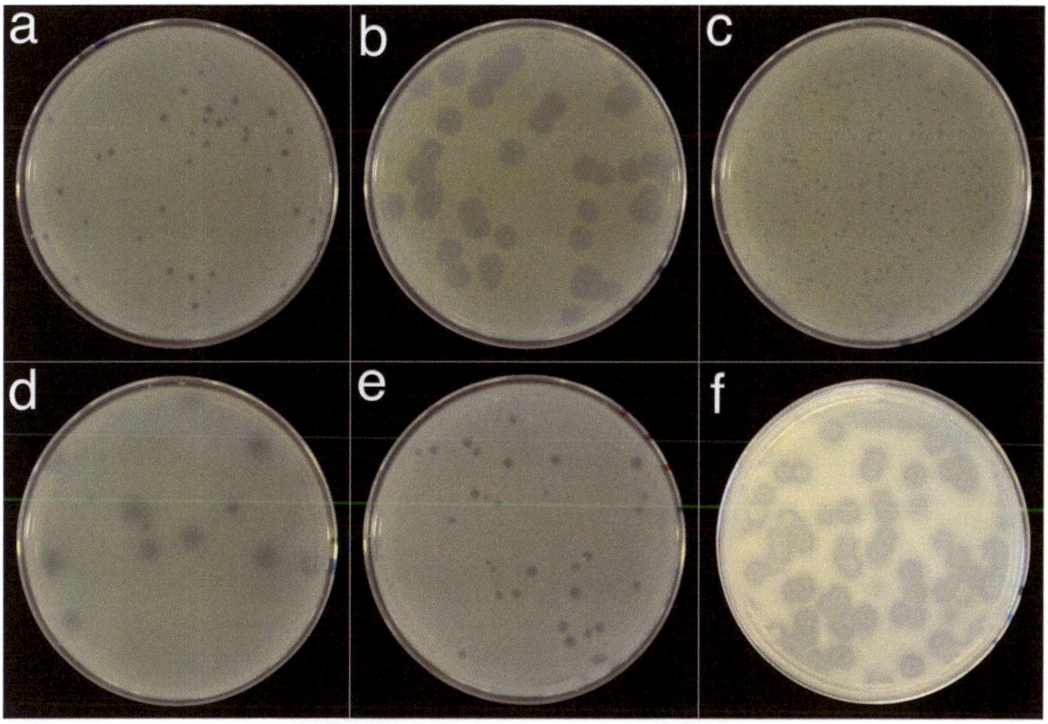

Fig. 1 Phages produce plaques with distinct morphologies. Standard plaque assays for multiple purified phage samples were conducted, employing the media types and incubation conditions tailored to the respective isolation host. Illustrated are (**a**) clear plaques with turbid margins; (**b**) large, turbid plaques; (**c**) clear pinpoint plaques; (**d**) large plaques with clear centers and turbid halos; (**e**) clear plaques with well-defined margins; and (**f**) large plaques with a turbid center and clearing around the turbidity

mixture evenly across the agar surface. Avoid introducing air bubbles.

7. Allow the plate to sit undisturbed until the top agar solidifies (*see* **Note 9**).
8. Invert the plate and incubate it for 24–48 h at the optimal growth temperature for the host.
9. Prepare a negative control by following **steps 2–9**, omitting the addition of phage filtrate (*see* **Note 10**).
10. Following incubation, examine the bacterial lawn on the plate for the presence of phages. Look for a plaque, a circular zone of cell death (clearing) that arises from one phage particle. Identify a well-isolated plaque and record information regarding its size and morphology (*see* **Note 11**) (Fig. 1). Sometimes the plaques will not be well isolated. Replate using a diluted sample (*see* **Note 12**).
11. Mark the location of the plaque by drawing a circle around it on the bottom of the plate and assign a unique phage identifier.
12. Label a sterile microcentrifuge tube with the phage identifier and aliquot 100 μL of PB (with added calcium) into the tube.

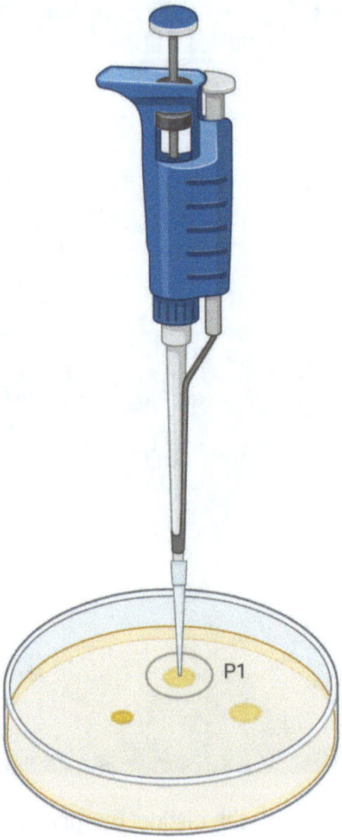

Fig. 2 Phage collection from plaques. Illustration of picking plaques, "created using BioRender.com." This technique is utilized to prepare phage lysates for subsequent purification. The process involves carefully collecting phages by gently picking the center of an isolated plaque using a sterile pipette tip and suspending the particles in PB. Avoid puncturing the bottom agar during this step. Subsequent phage lysates are created by picking phages from plaques that display the same morphology

13. Collect the phage from the plaque by gently picking its center at the agar surface with a sterile micropipette tip held at a ninety-degree angle, being careful not to puncture the bottom agar (Fig. 2). Care should be exercised not to pick up excess bacteria.
14. Place the pipette tip into the PB and transfer the phage by gently pipetting up and down.
15. Verify the presence of phages in the lysate immediately by conducting a spot test, or store it at 4 °C until ready to proceed.

3.4.2 Spot Test

To confirm the presence of phages in a sample, spot tests are conducted (Fig. 3). It is advisable to test all samples obtained from direct, enriched, and one-plate plaque assays with a spot test before moving on to the purification process, since sometimes air

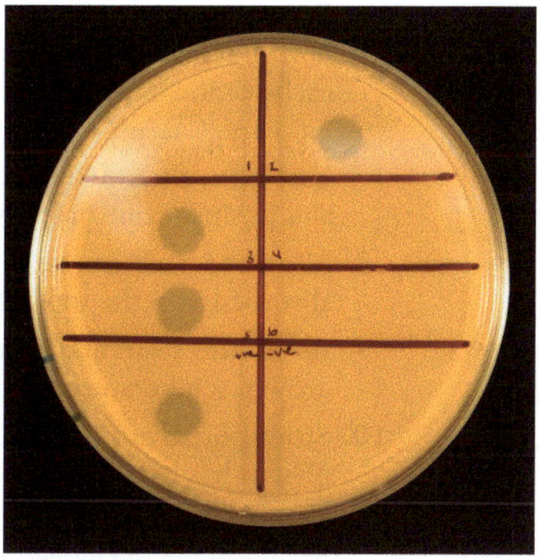

Fig. 3 Spot test verifies the presence of phages in samples 2, 3, and 5. To perform the spot test, *A. globiformis* was plated using PYCa top agar on a PYCa agar plate to form a lawn. Putative phage samples, along with a negative control (PB) and positive control (phage), were spotted onto the solidified lawn according to a grid pattern. The plate was incubated at 30 °C for 24 h. Clearances in the bacterial lawn were observed for samples 2, 3, and 5, as well as for the positive control, verifying the presence of phages in these samples. No clearing was observed for the negative control, indicating that the PB was not contaminated with phage

bubbles also look like putative plaques. Spot tests of any "putative" plaques will help to confirm the presence of phage in an efficient manner.

1. Obtain a pre-warmed agar plate suitable for the chosen host and label the bottom of the plate with a grid and spot designators. Be sure to include a space for positive and negative controls.
2. Transfer 500 μL of a well-mixed suspension of host cells into a culture tube.
3. Using a 5-mL serological pipette, transfer 5 mL of molten (at 55 °C) top agar into the culture tube and quickly aspirate the inoculated host–agar mixture back into the pipette (*see* **Note 8**).
4. Immediately pipette the mixture onto the agar plate. Tilt the plate carefully and swiftly to evenly spread the mixture across the agar surface. Avoid introducing air bubbles.
5. Allow the plate to sit undisturbed until the top agar solidifies (*see* **Note 9**).

6. Once the top agar has solidified, transfer 10 μL of phage lysate onto its designated section on the plate. Repeat this step for each section and sample tested (*see* **Note 13**) (Fig. 3).
7. Spot 10 μL of PB and 10 μL of phage lysate known to infect the host onto their designated spots on the plate in the negative and positive control spots, respectively.
8. Allow the plate to rest until all droplets have been completely absorbed into the top agar.
9. Invert the plate and incubate it for 24–48 h at the optimal growth temperature and oxygen requirements for the host.
10. Store potential phage samples at 4 °C until the presence of phages has been verified.
11. Following incubation, examine the bacterial lawn on the plate to determine the presence of phages. Look for a cleared spot in the designated space. If a sample shows a clearing, this is an indication that this may be a possible phage and the phage lysate can be used for purification. However, the controls should exhibit a clearing for the positive control but not for the negative control. Note that it is not advisable to pick from a spot test as the clearing is not a single plaque.

3.5 Phage Purification

To purify phages, perform a minimum of two to three iterations of plaque assays using phage lysates obtained from repeatedly harvesting plaques with consistent morphologies (Fig. 4). The purification process can be initiated from either a direct isolation phage lysate or an enriched isolation phage lysate, following the steps described below (*see* **Note 14**). This procedure must be performed under strict aseptic conditions.

3.5.1 Plaque Assay and Preparation of Lysate One

1. Prepare tenfold serial dilutions by arranging eight microcentrifuge tubes in a rack and labeling them from 10^{-1} to 10^{-8} and the lysate name (*see* **Note 15**).
2. Fill each microcentrifuge tube with 90 μL of PB.
3. Take 10 μL of the undiluted phage sample and transfer it to the 10^{-1} tube. Mix it with the PB by gently pipetting up and down.
4. Change the pipette tip and transfer 10 μL of the 10^{-1} solution to the 10^{-2} tube. Mix the contents by gently pipetting up and down multiple times.
5. Using a new, sterile pipette tip for each transfer, continue the dilution series as described until reaching the 10^{-8} tube.
6. To prepare the adsorption tubes, arrange ten sterile test tubes in a rack. Label nine of them as 10^{0}–10^{-8} and one as NC (negative control).

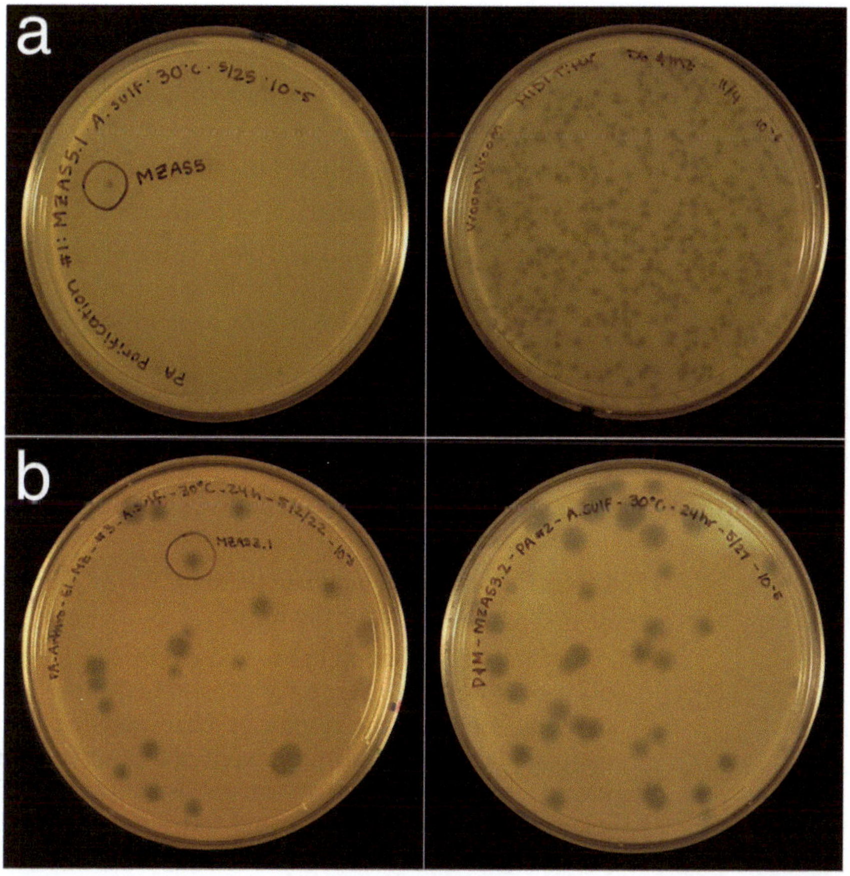

Fig. 4 Sequential plaque assays successfully purify *Arthrobacter* phages. Phages were purified through successive rounds of plaque assays. In the initial iteration, a soil sample was utilized; in subsequent rounds phage lysates obtained by picking plaques with identical morphology were used. All plates were prepared using PYCa media and incubated at 30 °C for 48 h. The left image in each panel presents the initially picked plaque, while the right image showcases the purified result. The presence of plaques with consistent morphologies resembling the originally chosen one signifies the successful purification of phages

7. Transfer 50 μL of the undiluted phage sample into the 10^0 tube; this tube serves as the 10^0 (or lysate zero.)
8. Using a new sterile pipette tip each time, aliquot 50 μL of each dilution into its corresponding adsorption tube.
9. Prepare the negative control by adding 50 μL of PB into the negative control tube.
10. Include a positive control by repeating **steps 1–7** using a phage lysate known to infect the chosen host as well as a negative control (*see* **Note 16**). Alternatively, prepare a spot titer test using the Spot Test Protocol 3.4.2 and using the same positive control phage.
11. Vortex the appropriate host culture and add 500 μL of the homogeneous suspension to each of the adsorption tubes,

including those for the negative control and the positive control.

12. Cap the tubes and gently mix their contents by flicking them or rotating them at an angle.
13. Incubate these adsorption tubes for 15 min.
14. Repeat the following steps for each adsorption tube:
    (a) Using a 5-mL serological pipette, transfer 5 mL of molten top agar (at 55 °C) into the adsorption tube. Quickly aspirate the inoculated host–agar mixture back into the pipette (*see* **Note 8**).
    (b) Immediately transfer the mixture onto the corresponding labeled agar plate. Carefully but swiftly tilt the plate to evenly spread the mixture across the agar surface.
    (c) Allow the plate to sit undisturbed until the top agar solidifies (*see* **Note 9**).
15. Invert the plates and incubate them for 24–48 h at the optimal growth temperature for the host.
16. After 24 h of incubation examine each plate for the number of plaques as well as to verify the consistency in the plaque morphology. Confirm the results by checking that the negative control plate has no plaques, while the positive control plate has visible plaques. If there are no visible plaques after 24 h, incubate the plates for a further 24 h and then examine the plates again for numbers of plaques as well as their morphology.
17. Identify a plate with well-separated plaques.
18. Depending on the phage sample used, select an isolated plaque that has the same morphology as the one previously picked from the one-plate plaque assay, and again record information about its size and morphology (*see* **Notes 17** and **18**).
19. Mark the location of the plaque by drawing a circle around it on the bottom of the plate and assign a unique phage identifier.
20. Label a sterile microcentrifuge tube with the phage identifier and add 100 μL of PB into the tube.
21. Collect the phage from the plaque by gently picking its center with a sterile micropipette tip held at a ninety-degree angle, ensuring not to puncture the bottom agar.
22. Place the pipette tip into the PB and transfer the phage by gently pipetting up and down. This preparation is now known as lysate one.

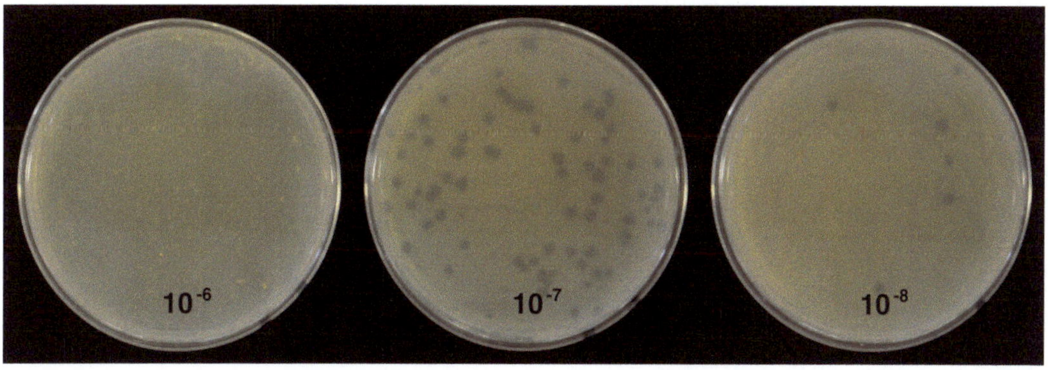

Fig. 5 Standard plaque assay produces a plate with a weblike pattern and shows tenfold reductions in phage concentrations. A standard plaque assay was conducted by serially diluting the phage tenfold and plating each dilution with *A. globiformis* using PYCa media. Plates were incubated at 30 °C for 24 h. The number of plaques decreased tenfold with each successive dilution (left to right), where the 10^{-6} dilution plate produced the characteristic webbed pattern

Phage Purification Using Lysate One

Since the environmental sample may have several different kinds of phage, phage purification allows one to purify the phage sample by performing several rounds of picking plaques, the end result being a "homogenous population" of phage.

1. Use the new lysate (lysate one) to perform a second round of purification plaque assays as described in Subheading 3.5.1, **steps 1–15**.
2. After incubation, assess the plates for consistent plaque morphologies resembling the previously selected one. If the plaques exhibit uniformity across the plates, it indicates successful phage purification (Fig. 5). Proceed to collect the final phage lysate (*see* **Note 17**). If there is only one plaque present on the plate, it is not recommended to pick this plaque, especially if it does not resemble the expected plaque morphology.
3. Identify the plate with confluent plaques, the webbed plate, and flood it with 8 mL of PB (*see* **Note 19**).
4. Allow the plate to sit undisturbed at room temperature for 2–4 h or refrigerate it overnight (12–14 h) at 4 °C.
5. After incubation, gently swirl the PB and remove the plate lid. Slightly tilt the plate to allow accumulation of the lysate on one side, and use a 10-mL serological pipette to aspirate the entire phage-infused buffer.
6. Carefully transfer the solution to the top of a 0.22-μm filter sterilization unit and vacuum-filter the PB to obtain the final phage lysate.

7. Determine the titer of the phage lysate using a standard plaque assay (*see* **Note 20**). A titer of 5×10^9 PFU/mL or greater is recommended for further analyses.
8. Immediately refrigerate the final phage lysate at 4 °C, where it can be stored for several months. If long-term storage is desired, DMSO can be added to samples and kept at −80 °C (*see* **Note 21**). Use the lysate to extract DNA for the further characterize the phage.

4 Notes

1. All listed hosts belong to the phylum *Actinobacteria* and have undergone evaluation for basic growth conditions and lawn formation. Previous phage isolation success can be accessed on PhagesDB (https://phagesdb.org/hosts/genera/3/) [17] and should be considered when choosing an isolation host.
2. To maximize the diversity of phages and increase the chances of finding those that target specific host bacteria, gather samples from a variety of environments and habitats where the desired bacteria thrive. From our experience, soil samples that were moist at the time of collection as well as near streams, lakes, or marshes had a greater success rate of isolating *Arthrobacter* phages. However, since the habitat of *Actinobacteria* is diverse and depends on various physicochemical factors such as pH and temperature, it is recommended that sampling should take place from vastly different habitats [18].
3. Record the GPS coordinates of the collection site using a smartphone or tablet equipped with location services. If electronic devices are unavailable during sample collection, this information can be recorded once a computer is accessible. The coordinates can be obtained by opening a web mapping platform such as Google Maps and locating the collection site on the map. Right-clicking on the collection site will bring up a menu where the latitude and longitude of the place can be recorded in decimal format.
4. It is advised to utilize the freshest possible sample when processing, despite the possibility of soil samples being collected several days prior. If collected ahead of time, the samples should be stored in a cool, dry place and processed in a timely manner to prevent them from drying out.
5. The settling of particulate matter may require up to 20 min, depending on the sample. To expedite this process, tubes can be centrifuged at 2000× *g* for 10 min. Although this method pellets most of the sample, certain components may remain

suspended. Avoid adding any floating particles to the syringe before filtering.

6. To prevent damage to the filter and avoid contaminating the phage filtrate, do not apply excessive pressure to the liquid if resistance is encountered. If the filter becomes clogged, replace it with a new filter.
7. Although isolation filtrates can be stored at 4 °C for up to 1 week, phage titers decline rapidly when stored at low concentrations. For optimal results, it is recommended to process samples immediately or within 24 h.
8. Melt the top agar by repeatedly microwaving it in 30-s intervals. After that, its temperature should be calibrated to 55 °C using a water bath or an incubator set to 55 °C. Before use, it is essential to check the temperature of the top agar to prevent any harm to the host and phage or premature solidification. To avoid early solidification of the agar, it is important that the samples are plated quickly and the top agar is not left in the pipette for more than a few seconds. If premature solidification occurs or the agar appears chunky, remelt the entire top agar before using it. When transferring the warm top agar, it is important to avoid introducing air bubbles as they may be mistaken for plaques later.
9. The time required for the top agar to solidify may vary depending on factors such as ambient temperature and humidity. To ensure proper solidification, gently tap the side of the petri dish or tilt it slightly while observing the agar for any movement. Exercise caution to avoid inverting the plate before the top agar has fully set to prevent it from running off the plate onto the lid. It is important to warm the agar plates to room temperature prior to use as this will reduce condensation as well as allow the top agar to solidify faster.
10. A negative control must be included when processing isolation samples. This control is prepared by plating only the host with top agar and provides valuable insight into two aspects: the ability to achieve a functional bacterial lawn and the expected appearance of the lawn when no phage is present. It is helpful to add a 10 uL spot of buffer once this negative control plate is solidified to help identify contamination sources (if contamination occurs).
11. Each putative plaque originates from a single phage in the original sample. As the progeny of the first lysed cell diffuses and kills more cells, the diffusion pattern is a sphere (or a circle) resulting in clearing called a plaque. To identify plaques effectively, remove the lid from a plate, let any condensation drip onto a paper towel, and hold the plate up to a light source. The areas where the bacterial lawn has been cleared will allow more

light to pass through, making them visible. When observing plaques, it is crucial to record important information such as size, turbidity (clear or turbid), margin type, and any unique morphological characteristics like halos or bullseyes (*see* Fig. 1 for examples), as this data will be used for purifying the phages. While a positive sample may contain multiple different phages based on the appearance of plaques with various morphologies, it is recommended to choose a single isolated plaque initially. If wishing to isolate multiple phages from the same sample, plaques with different morphologies should be chosen for subsequent picking. If no plaques are observed after the initial 24 h of incubation, it is advised to incubate the plates for an additional 24 h. If no plaques are found even after the extended incubation, it is likely that the soil sample does not contain phages capable of infecting the chosen host.

12. Keep in mind that the phage concentration might be high enough to clear the entire bacterial lawn (if this is the case the sample must be diluted and the experiment repeated), so it is important to evaluate negative control plates for the expected appearance of the lawn before drawing conclusions.
13. While it is possible to test multiple samples simultaneously, it is recommended to avoid placing more than eight spot tests on a single plate. When spotting phage samples, take care to position the pipette tip slightly above the agar to prevent puncturing it. To minimize splattering, slowly release the drop by gently pressing the micropipette plunger until it reaches the first stop. Keep in mind that labels on the bottom of the plate will appear reversed when flipped, so it is important to confirm the correct location beforehand.
14. To prevent duplication of phage isolates, consider that both direct and enriched isolations are derived from the same environmental sample and may contain identical phages in their respective filtrates. Therefore, it is advisable to purify only one phage per sample, using either the direct or the enriched isolation lysate. If the purification of multiple phages from a single environmental sample is desired, caution must be exercised in selecting plaques with distinct morphologies, irrespective of the initial isolation technique used to obtain them.
15. To ensure the purification and quantification of phages, it is necessary to generate plates with well-isolated and countable plaques. The number of tenfold dilutions prepared should therefore be adjusted based on the specific phage sample used and experiment performed. When dealing with samples that typically have low phage concentrations, like the filtrates obtained during the initial isolation step, diluting to 10^{-5} is generally enough to obtain isolated plaques that can be picked

for further purification. When working with high-concentration phage samples, such as the final phage lysate, obtaining plates with 20–200 plaques for accurate quantification usually requires diluting to 10^{-8}. However, if the titer exceeds 4×10^{11} PFU/mL, additional tenfold dilutions may be necessary.

16. For each purification plaque assay, it is important to include both positive and negative controls. These controls help establish the expected appearance of the bacterial lawn, validate the functionality of the assay, and aid in detecting any potential contamination. If the formation of the lawn is compromised or if the positive control phage fails to form plaques, it is necessary to prepare fresh materials and utilize a new phage lysate.
17. To purify phages effectively, it is crucial to create lysates from well isolated plaques. This is because phage particles can diffuse within the agar, including neighboring plaques. To minimize the effects of diffusion it is advisable to pick plaques early in the incubation timeline. In addition, selecting plates with a low number of plaques (but not a single plaque) that are adequately spaced reduces the risk of contamination from other phages in the chosen plaque and increases the likelihood of obtaining a single phage type, a single clonal population. Consistency in plaque morphologies must be maintained throughout the entire purification process.
18. When assessing the purity of a phage sample, it is considered most reliable to evaluate plaques based on morphological characteristics, such as margin types, turbidity, or other defining features like haloes. Plaque size should not be relied upon for evaluating purity, as it is known to fluctuate and sizes may appear smaller at higher concentrations. It should be noted that plaques produced by some phages can exhibit multiple morphologies. If multiple plaque morphologies persist, despite picking well-isolated plaques each time, this may just be the typical presentation for that phage. The best advice to ensure that the lysate generated is a single clonal population is to make sure you pick a well-isolated plaque early in its incubation period. Performing more than three rounds of purification is discouraged, as it can lead to the accumulation of mutations in the phage.
19. To create the final phage lysate, flood (add 5–8 mL of PB) the plate containing the highest concentration of phage with PB. It is preferable to use a webbed plate (Fig. 5) as it contains the maximum number of plaques and therefore the highest amount of phage particles. It may be difficult to determine the best webbed plate, so determining titers of any lysates from flooded plates will determine the highest yield.

20. After the completion of the incubation period, it is necessary to determine the concentration of the phage lysate, which is measured in plaque-forming units per milliliter (PFU/mL) This can be achieved by identifying a dilution that yielded a plaque count falling within the range of 20–200 and carefully counting the number of plaques or PFUs present on that plate. The acquired information should then be applied to the equation PFU/mL = (#PFU × dilution factor)/volume plated (in mL). For instance, if 35 plaques were counted on a plate that had been diluted to 10^{-7} and plated with 0.05 mL of the phage sample, the resulting titer would be 7×10^9 PFU/mL. To ensure optimal outcomes in future experiments and enable long-term storage, a high-titer lysate is required, which is defined as having a concentration of at least 5×10^9 PFU/mL. To prepare large volumes of high-titer phage lysate, the final phage lysate from Subheading "Phage Purification Using Lysate One", **step 8** must be diluted and plated to obtain multiple webbed plates, which can be achieved by noting the titer of the final phage lysate and diluting accordingly. Each of these plates will then be flooded with PB as described in Subheading "Phage Purification Using Lysate One", **steps 3–8**. This high-titer lysate can then be used to extract the phage DNA and the DNA subsequently sequenced.
21. While phage lysates can be kept at 4 °C for up to 4 months, their viability can diminish over time. For extended preservation, it is recommended to archive high-titer phage lysates with concentrations of at least 5×10^9 PFU/mL. To archive the phages effectively, label sample storage tubes with the name of the phage as well as the date and the titer. To a 15-mL conical tube add 4.0 mL of the high-titer phage lysate. Add 280 μL of DMSO to the tube. Be cautious to keep away from flames as DMSO and its vapors are flammable. Cap the tube and mix the contents thoroughly by vortexing. Transfer the lysate/DMSO mixture into the storage tubes. Avoid overfilling or underfilling them. Ensure proper closure of the sample storage tubes and immediately freeze them at −80 °C to maintain phage viability for decades.
22. The materials, hosts, and the bacteriophages must be used and disposed as per your institution's chemical and biosafety guidelines.

Acknowledgements

We would like to thank the numerous members of the bacteriophage community that have contributed to phage isolation methods for many decades, which are the basis of these protocols. Many of the protocols described here are based on those developed for the isolation of mycobacteriophages in the laboratories of Bill Jacobs and Graham Hatfull. Further development and classroom adaptations were done in the Phage Hunters Integrating Research and Education (PHIRE) program in the Hatfull lab and by program staff and the faculty community within the Science Education Alliance Phage Hunters Advancing Genomics and Evolutionary Science (SEA-PHAGES) program supported by the Howard Hughes Medical Institute. We especially thank Lu Barker, Kevin Bradley, Danielle Heller, Tuajuanda Jordan, Laura Marinelli, Marisa Pedulla, Welkin Pope, Marianne Poxleitner, and Vic Sivanathan for their contributions in the development and documentation of these methods. We thank Dominic Garza for the isolation of phages. We would also like to acknowledge Erin Sanders O'Leary and Jordan Moberg Parker for pioneering, researching, and teaching phage discovery at UCLA.

References

1. Lewin GR, Carlos C, Chevrette MG, Horn HA, McDonald BR, Stankey RJ, Fox BG, Currie CR (2016) Evolution and ecology of actinobacteria and their bioenergy applications. Annu Rev Microbiol. 70:235–254. https://doi.org/10.1146/annurev-micro-102215-095748. PMID: 27607553; PMCID: PMC5703056
2. Manivasagan P, Kang KH, Sivakumar K, Li-Chan EC, Oh HM, Kim SK (2014) Marine actinobacteria: an important source of bioactive natural products. Environ Toxicol Pharmacol 38(1):172–188. https://doi.org/10.1016/j.etap.2014.05.014. PMID: 24959957
3. Zothanpuia, Passari AK, Leo VV et al (2018) Bioprospection of actinobacteria derived from freshwater sediments for their potential to produce antimicrobial compounds. Microb Cell Fact 17:68. https://doi.org/10.1186/s12934-018-0912-0
4. Silva LJ, Crevelin EJ, Souza DT et al (2020) Actinobacteria from Antarctica as a source for anticancer discovery. Sci Rep 10:13870. https://doi.org/10.1038/s41598-020-69786-2
5. Xie F, Pathom-aree W (2021) Actinobacteria from desert: diversity and biotechnological applications. Front Microbiol Sec Syst Microbiol 12:765531. 1–27. https://doi.org/10.3389/fmicb.2021.765531
6. Hatfull GF (2022) Mycobacteriophages: from Petri dish to patient. PLoS Pathog 18(7): e1010602. https://doi.org/10.1371/journal.ppat.1010602
7. Hyman P (2019) Phages for phage therapy: isolation, characterization, and host range breadth. Pharmaceuticals 12(1):35. https://doi.org/10.3390/ph12010035
8. Ait Barka E, Vatsa P, Sanchez L, Gaveau-Vaillant N, Jacquard C, Klenk H-P et al (2016) Taxonomy, physiology, and natural products of actinobacteria. Microb Mol Biol Rev 80. https://doi.org/10.1128/mmbr.00019-15
9. Hatfull GF (2020) Actinobacteriophages: genomics, dynamics, and applications. Annu Rev Virol 7:37–61. https://doi.org/10.1146/annurev-virology-122019-070009
10. Durand GA, Raoult D, Dubourg G (2019) Antibiotic discovery: history, methods and perspectives. Int J Antimicrob Agents 53:371–382. https://doi.org/10.1016/j.ijantimicag.2018.11.010
11. Fruciano DE, Bourne S (2007) Phage as an antimicrobial agent: d'Herelle's heretical theories and their role in the decline of phage prophylaxis in the West. Can J Infect Dis Med Microbiol. 8(1):19–26. https://doi.org/10.1155/2007/976850. PMID: 18923687; PMCID: PMC2542891
12. Marinelli L, Fitz-Gibbon S, Hayes C, Bowman C et al (2012) *Propionibacterium acnes* bacteriophages display limited genetic diversity and broad killing activity against bacterial skin isolates. mBio 3(5). https://doi.org/10.1128/mbio.00279-12
13. Jordan JC, Burnett SH, Carson S, Caruso SM et al (2014) A broadly implementable research

course in phage discovery and genomics for first-year undergraduate students. mBio 5(1): e01051–e01013. https://doi.org/10.1128/mbio.01051-13

14. Hanauer DJ, Graham JM, SEA-PHAGES, Betancur L et al (2017) An inclusive Research Education Community (iREC): impact of the SEA-PHAGES program on research outcomes and student learning. Proc Nat Aca Sci 114(51):13531–13536. https://doi.org/10.1073/pnas.1718188115
15. Hatfull GF, Pedulla ML, Jacobs-Sera D, Cichon PM et al (2006) Exploring the mycobacteriophage metaproteome: phage genomics as an educational platform. PLOS Genetics 2(6):e92. https://doi.org/10.1371/journal.pgen.0020092
16. Pope WH, Bowman CA, Russell DA, Jacobs-Sera D et al (2015) Whole genome comparison of a large collection of mycobacteriophages reveals a continuum of phage genetic diversity. eLife 4:e06416. https://doi.org/10.7554/eLife.06416
17. Russell DA, Hatfull GF (2017) PhagesDB: the actinobacteriophage database. Wren J, editor. Bioinformatics 33:784–786. https://doi.org/10.1093/bioinformatics/btw711
18. Batinovic S, Wassef F, Knowler SA, Rice DTF, Stanton CR, Rose J, Tucci J, Nittami T, Vinh A, Drummond GR, Sobey CG, Chan HT, Seviour RJ, Petrovski S, Franks AE (2019) Bacteriophages in natural and artificial environments. Pathogens 8(3):100. https://doi.org/10.3390/pathogens8030100. PMID: 31336985; PMCID: PMC6789717

Index

Huan Peng, Jianfeng Liu and Irene A. Chen (eds.), *Phage Engineering and Analysis: Methods and Protocols*, Methods in Molecular Biology, vol. 2793, https://doi.org/10.1007/978-1-0716-3798-2,

O

P

Q

R

S

T

U

V

MIX
Papier aus verantwortungsvollen Quellen
Paper from responsible sources
FSC® C105338

If you have any concerns about our products,
you can contact us on
ProductSafety@springernature.com

In case Publisher is established outside the EU,
the EU authorized representative is:
Springer Nature Customer Service Center GmbH
Europaplatz 3, 69115 Heidelberg, Germany

Printed by Libri Plureos GmbH
in Hamburg, Germany